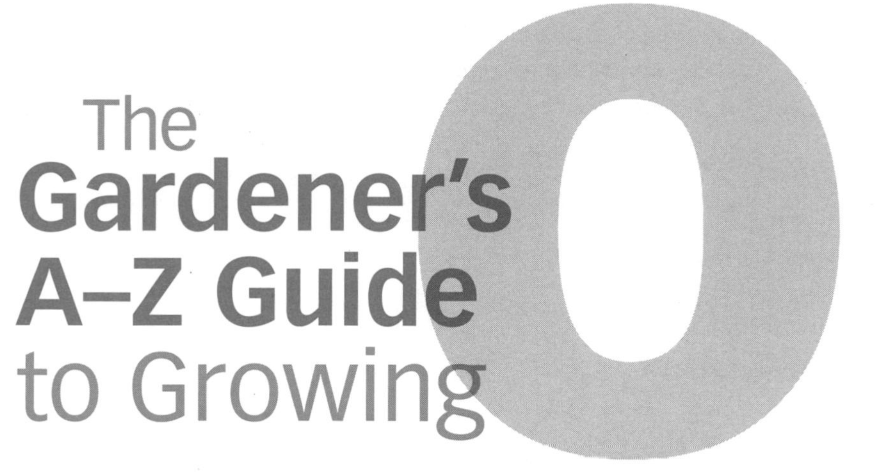

The Gardener's A–Z Guide to Growing Organic Food

Tanya L. K. Denckla

Woodcuts by Stephen Alcorn

Storey Publishing

The mission of Storey Publishing is to serve our customers by publishing practical information that encourages personal independence in harmony with the environment.

Edited by Gwen Steege and Marie A. Salter
Art direction by Cindy McFarland
Cover designed by Wendy Palitz
Interior designed by Wendy Palitz and Kent Lew
Text production by Susan Bernier, Kent Lew, Alexandra Maldonado, and Jennifer Jepson Smith
Woodcuts © Stephen Alcorn
Indexed by Susan Olason / Indexes & Knowledge Maps

The author gratefully acknowledges permission to reproduce the following copyrighted material:
Vegetable storage conditions (temperature, humidity, and length of storage) from James E. Knott, *Knott's Handbook for Vegetable Growers*, 4th Edition, ed. Donald N. Maynard and George J. Hochmuth. New York: Wiley, 1997. This material is used by permission of John Wiley & Sons, Inc.
Food storage conditions and times from Jeff Ball, *The Self-Sufficient Suburban Gardener.* Emmaus, Penn.: Rodale, 1983.
Excerpts from apple and plum rootstock charts and "Intercropping for Pest Reduction: Successful Scientific Trials," from Robert Kourik, *Designing and Maintaining Your Edible Landscape Naturally*. Santa Rosa, Calif.: Metamorphic Press, 1986.

Printed in the United States by Versa Press
10 9 8 7 6 5 4

Library of Congress Cataloging-in-Publication Data

Denckla, Tanya L. K., 1956–
The Gardener's A–Z guide to growing organic food / Tanya Denckla.
p. cm.
Rev. ed. of: The organic gardener's home reference. ©1994.
Includes bibliographical references and index.
ISBN-13: 978-1-58017-370-4 (alk. paper)
1. Vegetable gardening—Handbooks, manuals, etc. 2. Organic gardening—Handbooks, manuals, etc. 3. Fruit culture—Handbooks, manuals, etc. 4. Herb gardening—Handbooks, manuals, etc. 5. Nuts—Handbooks, manuals, etc. I. Denckla, Tanya L. K., 1956– Organic gardener's home reference. II. Title.
SB324.3.D47 2003
635'.0484—dc21 2003042728

In Memorium

To W. Donner Denckla, M.D.,

who conceptualized open-field hydroponics and provided many of the memorable sayings throughout this book, and who planted many seeds in many lives.

contents

acknowledgments

Special thanks to my husband **Cecil Cobb,** for his steadfast support of my working nights and weekends, and for his loving and gentle reminders to tend the garden of my life; **Marlene Denckla,** for her research work on updating the information on vegetables and herbs. I am also grateful for her continual technical support and enthusiasm through the first years of researching and writing the book. Thanks also to **Alia Anderson,** for her research work on updating the information on fruit and **Todd Davidson,** for his research work on updating the information on disease and insect pest controls.

For contributions to the previous versions of the book, many thanks to **Jennifer Taylor,** for updating the entire contents of the self-published volume, enabling it to metamorphose into the first edition; **Doug Britt,** president of Ag Life, for reviewing the insect and disease charts; **John Brittain,** president of Nolin River, for numerous helpful suggestions on nut varieties; **Rosalind Creasy,** author and edible landscaper, for a detailed review and helpful contributions to the entire manuscript, as well as for suggestions about regional issues; **Galen Dively, Ph.D.,** entomologist, University of Maryland, for a detailed review of the macropest charts, and many valuable contributions; **Frank Gouin, Ph.D.,** horticulturist, Chairman of the University of Maryland's Department of Horticulture, for reviewing the herb entries, and for kindly arranging review of the vegetable section; **Patrick J. Hartmann,** president of Hartmann's Plantation, Inc., for suggesting appropriate blueberry cultivars and for help with *Phytophthera cinnomomi*; **Richard A. Jaynes, Ph.D.,** geneticist, president of the Northern Nut Growers Association, for reviewing the nut entries and for thoughtful suggestions on nut varieties; **Clay Stark Logan,** president of Stark Bro's Nurseries and Orchards, Co., and **Joe Preczewski, Ph.D.,** director of Stark Bro's Field Research and Product Development, for reviewing the fruit entries and suggesting appropriate fruit varieties; **Alan MacNab, Ph.D.,** plant pathologist, Pennsylvania State University, for valuable contributions to vegetable disease remedies and especially for raising critical issues of presentation; **Charles McClurg, Ph.D.,** horticulturist, University of Maryland, for reviewing the vegetable entries; **John E. Miller,** president of Miller Nurseries, for reviewing the fruit entries and suggesting appropriate fruit varieties; **Tom Mills,** president of Indiana Walnut Products, for suggesting nut varieties; **Carl Totemeier, Ph.D.,** horticulturist, retired vice-president of the New York Botanical Gardens, for a very detailed review and contributions to the entire manuscript; and to **Jeff Ball,** author, communicator, president of New Response, Inc.; **Judy Gillan,** of the Organic Foods Production Association of North America (OFPANA); **Lewis Hill,** author, former owner of Hillcrest Nursery; **Richard Packauskas, Ph.D.,** entomologist, University of Connecticut; and **Robert D. Raabe, Ph.D.,** plant pathologist, University of California, Berkeley.

preface

I marvel at how tiny seeds of all kinds germinate—mere potential transformed into dynamic, never quite predictable life. Such has been the history of this compendium.

The seed for this book was planted in metropolitan Washington D.C., where I played at gardening in a tiny backyard. Starting with a few peach and pear trees, I slowly expanded the garden to include some perennial flowers, herbs, and annual vegetables. In those first few years my little retreat thrived and, to my untrained eye, seemed a bucolic environment for unlimited growth. Laissez-faire was my motto. Gardening seemed ever so easy.

It was about year three that the fruit trees signaled a transformation from simple fun to complex war. Peach tree borer, fireblight, Japanese beetles — as each new pest invaded and the trees suffered increasingly more serious damage, I experienced a parental panic attack. No mere fungus or insect would harm my plants. To my former husband Donner's horror, I pulled out the "big gun" insecticides, fungicides — anything with a scientific name on the label. And I sprayed everything, everywhere, sprayed mercilessly. I had entered full battle mode, re-enacting the ancient war or Man versus Nature. Control would be established.

Needless to say, I might have won the skirmish that summer, but the war was already lost. The garden ecosystem, frail and imbalanced from a lack of proper nurturing for sustained growth, was further weakened not strengthened, by my panic spraying. As a natural consequence, pests renewed their attack with a vengeance and increased strength the next year. Two seasons later, with regretful hindsight, to cleanse the garden of its continued abscess and to end the futile struggle, I cut the fruit trees to the ground.

Never again did I spray wholesale. From Donner, a physician who ardently believed in a systems approach to all living things, I gained understanding and respect for the delicate ecosystem in a garden. I learned the value of spot spraying. Reading more, I learned to avoid high nitrogen, phosphorus, and potassium fertilizers that can kill earthworms and other beneficial garden helpers. Spiders, once a sure way to make me shriek, became a welcome sight. At heart, I learned an attitude that, if nothing else, transformed angry garden warrior into neophyte garden steward.

Real strong on attitude. However, still real weak on knowledge. At this stage, some friends and I were ready to buy a country property where we hoped to build a model self-sufficient farm. To divide labor, we decided I would plan the food and flower gardens. With renewed gusto, I delved into accumulating heaps of garden magazines and books. Months of reading, however, resulted in frustrations and feeling so overwhelmed that I was almost ready to quit before a single seed had been planted. Gardening, surely, could not be so complicated. Why was a clear picture of, say, tomato, so hard to achieve? One book might be good on tomato pests but lacked information on planting times or varieties. Another might have a lot of information on tomato but scattered it through various charts and sections, causing page-flipping fingeritis and more frustration, not to mention the confusion brought on when two books gave conflicting information. I felt I would never be able to really plan a garden.

As a last ditch attempt to rationalize what I was reading, and to avoid twenty books strewn over the floor to find one answer, I started to centralize information on my home computer. Slowly, a database took shape. Someone noted it was becoming almost, well, a book. And much later, a full four years of research and compilation later, someone asked for a copy of the book when it was done.

Book? Another seed began to germinate. By now my small Washington garden had recuperated from the earlier pharmaceutical rampage. Though it was not a great showpiece, it was thriving from a minimalistic and organic approach. The farm garden had been put on hold as terra-forming and massive building commenced. A little head scratching led to some research into publishing, consultation with expert growers, and hours of discussion with farm cohorts — all led me to take a huge gulp of air and plunge into the ocean of self-publishing.

And what a wonderful swim it was! Reviews streaming in, letters from grateful gardeners, and orders more numerous than I could have imagined for such an ugly-duckling reference book. Most rewarding has been the thanks from readers. From small backyard gardeners to larger market gardeners, you have thanked me for creating a sourcebook that helps you do what you love to do, better and quicker. What a satisfying feeling, to know the book accomplishes what was intended: to make gardening a little easier and more fun.

Now a city woman turned southern country "girl," I enjoy gardening in Virginia's central Piedmont region in Fluvanna County, outside Charlottesville. In my four raised beds, I gain a little more expertise and experiment with a few more unusual edible plants every year. I annually add notes in the columns of this book, and every year it becomes a shade browner from garden soil. Since publishing the book, I have expanded my work with the environment in several directions. I helped co-found and led a community forestry and beautification organization, called Greener Harrisonburg, which planted more than 600 street trees in the space of a few years. I've also become an active mediator, facilitator, and trainer in environmental public policy and community issues and have helped found the Virginia Natural Resources Leadership Institute, a yearlong training program for people involved in all aspects of Virginia's environment, with the goal of learning about the toughest environmental problems in Virginia and ways to resolve these issues collaboratively. While my life has grown in new directions, I still regard this book as my foundation, my beginning point for learning about ways to work collaboratively with the complex natural systems of our environment.

In the early days of this book, I assumed and mistakenly stated that this is not a primary text. Since then, beginners have used this book with no other helping text to start their first gardens. It has also been used as a text by the Oglalla Lakota College for courses in organic gardening, and I am told it has been used by Cooperative Extension Services in different parts of the United States and even in other countries such as the Ukraine. So please use this as a resource for getting started as well as a resource to keep you going through the years, whether you are a beginner or a professional.

Most important, I hope this reference inspires and enables you to read less and plant more seeds.

Plant it, and you will grow.

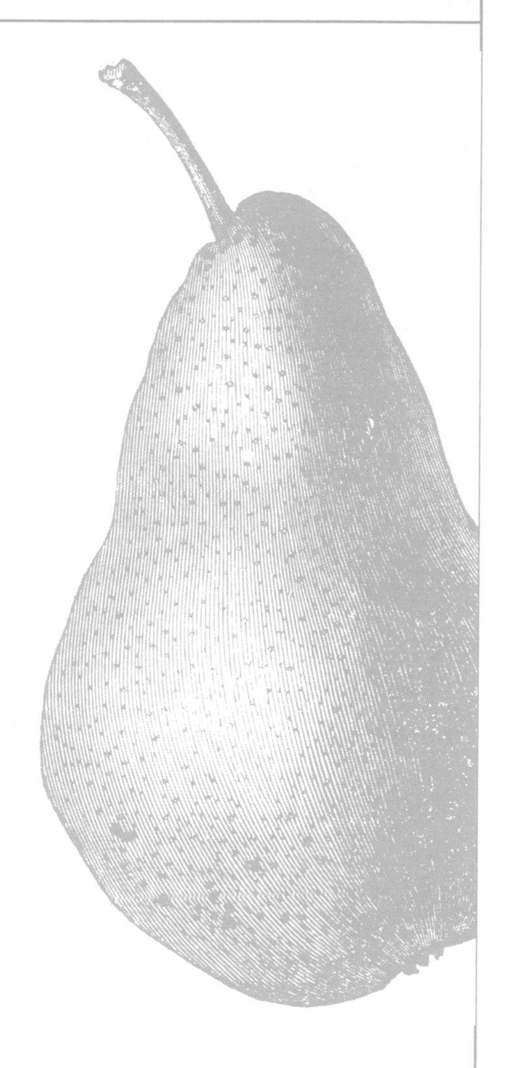

Chapter One

for starters

The industrial revolution brought mechanization and precise replication to the world of manufacture. The "green revolution" of the mid-twentieth century brought similar change to the world of agriculture. To tackle the daunting task of feeding the world's hungry, science and agriculture joined forces to create new crop hybrids, fertilizers, and pesticides that produced, first in the United States and then elsewhere, yields that were truly miraculous. This new approach to farming brought hope and inspiration to millions worldwide and fundamentally changed our perceptions of how a farm — and a garden — should work, produce, and be maintained.

As a result, throughout much of the world today, a farm is expected to operate like a well-oiled machine, generating with unrivaled efficiency vast quantities of a single crop, all produce achieving identically high standards of quality, appearance, marketability, and shippability. And as so often happens with aspirations so grand, individuals took inspiration from the goals of large-scale farms and adopted a similar approach, using commercial fertilizers, herbicides, and pesticides to manage their backyard gardens.

Since the early successes of the green revolution, however, the world and our understanding of it have fundamentally changed. The information age and globalization have given rise — and credibility — to the movement for sustainability, which seeks to work in ways that ensure resources are constantly replenished and renewed, not depleted and damaged. The sustainable model is all encompassing and can be applied to agriculture, business, and natural and human resources.

Closely tied to the sustainable model and rapidly superseding the machine metaphor is the notion of ecosystem — a dynamic, living system of forces that work in balance with each other. We now understand that just as rivers and forests have ecosystems, so do states, bioregions, communities, cities with urban forests, farms, and — yes — even our own backyard gardens.

So, clearly, an organic garden is not a machine. It is a living system of balanced forces between, for example, predator and prey, and these forces are always in flux. Soil composition, air quality, water, birds, insects, and weeds are just a few of the forces that influence the nature and health of a garden. Your role as garden steward is to work with these forces, encouraging the balance to shift in your favor, not to dominate them with the goal of achieving perfection.

Efforts to control or impose a balance with the hope of achieving picture-perfect produce can eventually backfire. Many growers have sought this perfection because consumers preferred and then demanded unblemished produce. A common result is that the grower's land may not have pests, but it also may not have earthworms in the soil, birds in the fields, or beneficial predatory insects. This may seem benign, but it isn't. Many of the practices used to obtain high yields of unblemished produce eventually promote the loss of three things: topsoil, water penetration, and biologically available nutrients. These losses, in turn, necessitate that the soil depend more and more on human-provided nutrients.

Our desire for perfect produce, among other factors, and the agricultural practices used to achieve it have led to unprecedented conditions that might be characterized as *open-field hydroponics. Hydroponics* is a method for growing crops in greenhouses in which plants are anchored in an inert material, such as gravel or sand, and are fed macronutrients and micronutrients mechanically. The term *open-field hydroponics,* therefore, suggests a similar situation in fields where the soil is increasingly inert and provides structural support but contributes little nutritionally to plant growth.

Sustainable Agriculture

Many small and large-scale growers, alarmed by soil depletion and related trends, have adopted sustainable agriculture as a goal, for both ecological and economic reasons. Definitions may vary, but *sustainable agriculture* generally refers to practices that are viable over long periods of time, both environmentally and economically. Sustainable agriculture, for example, strives to achieve soil that can produce crops reliably without nutrient depletion and with minimal amendments. In sustainable agriculture, the gardener makes a philosophical shift away from control to cooperation and evolves from master to steward.

On Being a Good Garden Steward

The role of garden steward is not difficult to achieve, particularly if embraced at the outset. One of its most important precepts is *feed the soil, not the plants.* The role of steward is perhaps most demanding during the planning phases, when crucial decisions must be made about where to put the garden, what varieties to plant, when and where to plant them, how to feed the soil, where to put the compost pile (or whether to have one), what kind of mulch to use, and, perhaps most significantly, how perfect you want your produce to appear.

Feed the soil, not the plants.

These decisions will help you resist the temptations of quick chemical fixes later. Some farmers have likened chemical sprays and fertilizers to cocaine — once started, an addictive cycle begins that is difficult to break. The quick "fix" can take years to "unfix." Research trials in Israel and elsewhere have indicated that farms transitioning away from chemical inputs initially obtain lower yields for a period of 3 to 5 years, after which yields increase to match and sometimes exceed those obtained with chemicals. This is a good reminder that the effects of chemical inputs are long-term, and that a long-term perspective is essential when it comes to garden health.

As a garden steward with a long-term perspective, you may decide to adopt your own standards, such as embracing a policy of limited spraying or no spraying. For example, you may decide to avoid even organic sprays that have been approved, such as copper-based fungicides, because they can sometimes kill earthworms, a gardener's best friend. At the same time, your goals may require that you use some sprays for limited and highly targeted purposes.

The key to a successful plan is to clearly define your goals and the reasons for them early on and then to stick to the plan. Without specific goals, you may succomb to convenience, zapping this or that pest as needed during the growing season.

The Self-Sustaining Garden

In addition to choosing to become garden steward, a backyard gardener might aspire to create a self-sustaining garden. A self-sustaining garden or farm differs from sustainable agriculture in one major respect: it supplies all of its own essential nutrients for balanced growth, from organic matter for compost to micronutrients for healthy plants, whereas the sustainable garden or farm may import many of these materials from off-site. A self-sustaining garden doesn't require importation of beneficial insects because they are already there. It doesn't require the application of imported lime to buffer the effects of acid soil, because earthworms and compost generated on-site are present to do the buffering. A self-sustaining garden is essentially the absolute opposite of open-field hydroponics. The former has been effective for thousands of years, whereas within just 40 years, by the end of the twentieth century, the latter had displayed severe limitations.

Which Goal Is for You: Sustainable or Self-Sustaining?

On a large scale, the self-sustaining farm and sustainable agriculture farm have one additional way in which they are distinct. Both seek to produce healthful food in a way that is beneficial to the farm's ecosystem, but the sustainable agriculture farm is a commercial enterprise in business to sell its products off-farm, while the produce from a self-sustaining farm is used to feed and sustain the farm's owner and workers. Ideally, a self-sustaining farm provides its owners and workers with all of their food and shelter needs. Those embracing the self-sustaining model often see it as the ultimate example of living lightly on and in balance with the land. Profit and marketability of the plant and animal products are not important considerations for the self-sustaining farmer and, in fact, may not even figure into the equation.

On a small scale, becoming a steward of a sustainable garden or a self-sustaining garden is eminently achievable. Most of us will probably need to begin with a sustainable garden, bringing in soil amendments, beneficial insects, and whatever else might be needed to improve the quality and sustainability of our soils and growing environment. But our vision need not end there. Why not aim to create a healthy garden ecosystem that eventually provides for itself, does not require chronic soil amendments or importation of beneficial insects, and that becomes self-sustaining? With our increasingly demanding lifestyles, in the long run this approach may ultimately be the most time and energy efficient and rewarding.

As steward of a self-sustaining garden, your first job is to recognize that the forces at work in your garden will never be in per-

Safeguard Earthworms

To protect earthworms, take the following steps:

- Till minimally, because tilling can disturb and kill earthworms and other soil microorganisms through mechanical abrasion, drying out, and disruption of their environment.
- When you do till, keep it shallow. An optimal tilling depth is 3 inches. To expose the eggs and cocoons of some insects to hungry birds, you may need to till up to 6 inches. Generally, unless you are creating a first-year bed and need to rototill up to 12 or 14 inches, you should never till deeper than 6 inches.
- Avoid heavy doses of chemical fertilizers because they can harm soil microorganisms and decrease earthworm activity. Excessive nitrogen fertilizer, regardless of its source, harms the soil, and the USDA has shown that it can reduce vitamin C content in some green vegetables. Use compost instead to provide a slow-release food to the soil and plants.
- Avoid uncomposted manure of any kind because it contains disease pathogens and seeds. If you must use manure, make sure it has been composted up to 160°F.
- Water regularly, and avoid excesses. Flooding or overwatering, and drought or drying out of the soil can kill soil microorganisms as well as earthworms.

fect balance. There will always be some plant damage. Because your garden is not a machine, and because your plants do not come from a factory, you will see variations in your garden conditions and plant productivity from year to year.

Your second job is patience. It usually takes at least several years to establish an ecosystem that operates in your favor — an ecosystem with earthworms, insect-eating birds, beneficial predatory insects, soil with organic matter sufficient to drain well and retain water to prevent runoff, and soil nutrient levels that support healthy plant growth.

The advantage of a self-sustaining garden is that it requires the least amount of money and time in the long-term. You may need to invest in a first colony of earthworms, build or buy birdhouses, buy compost and organic matter before your garden produces it for you, and perhaps even buy irrigation soaker hoses and row-cover material. But in several years, these investments should reward you with a healthy garden that doesn't require lots of imported materials or time-consuming pest controls.

As steward of a sustainable garden that is evolving toward a self-sustaining garden, your strategy is simple: focus on building healthy soil and ecosystems. Key ways of achieving this are explored in detail in chapter 5 (see pages 298–300.) Above all, do not panic when you see pests or diseases in your garden. The plant world is amazingly resilient, and studies have shown that plants can lose up to 20 percent of their foliage without a significant reduction in yields.

Allies in a Self-Sustaining Garden

A self-sustaining garden has many natural allies. Chief among them, in order of importance, are earthworms, compost, mulch, and flying creatures.

Earthworms

Earthworms are a gardener's best friend. Their tunneling and production of nitrogen-rich castings (excrement) accomplish many important soil-improving tasks, free of charge. Earthworms do all of the following for your soil:

- Aerate soil, improving the availability of oxygen to plant roots
- Improve water retention capacity, decreasing the need to water
- Keep the soil loose and friable, improving plant root capacity for growth
- Raise important minerals from the subsoil to the topsoil where plants can use them
- Counteract leaching out of nutrients by improvement of water retention
- Break up hardpan soils, which are inhospitable to plant growth
- Homogenize soil elements so they're more evenly available to plants
- Create fertile channels for plant roots
- Liberate essential nutrients into a form that is soluble and available to plants
- Neutralize soils that are too acid or too alkaline for healthy plant growth
- Balance out organic matter in the soil, so you needn't worry about exceeding the 5 to 8 percent optimal level
- Enhance the soil's environment for growing healthy self-sustaining plants

Compost

Compost is another major player in the self-sustaining garden. Essentially, *compost* is any organic material, including manure, that has decayed into a simpler form by the action of aerobic or anaerobic bacteria, depending on the composting method. *Humus* is any partially decomposed organic material, vegetable or animal, that is mixed into the soil to improve soil quality. *Mulch* is any material used to cover the soil; it can be nutrient-poor, such as newspaper, nutrient-neutral, such as plastic, or nutrient-rich, such as compost.

Compost can be used wherever humus or mulch is recommended. The process of composting reduces the original bulk of the organic material by one-fourth to one-tenth. So where a thick mulch is desired, you might prefer to use uncomposted material such as straw or chopped leaves. On the other hand, if you have access to large amounts of compost, it is a highly beneficial mulch because it feeds the soil. Compost is easy to make in any home garden or farm and does the following:

- Feeds the soil and its creatures gently, unlike chemical fertilizers that can kill earthworms and other beneficial organisms
- Lasts a long time, because it releases nutrients slowly in a readily available form, unlike chemical fertilizers that provide a quick boost and then peter out

Basic Composting Principles

Size of the pile. For rapid composting on a small scale, build a pile at least 3'×3'×3'.

Materials that go into it. Combine roughly equal parts of dry plant material and green plant material to achieve the desired carbon-to-nitrogen ratio of 30:1. *Dry plant material* can include autumn leaves, straw, dried and cut-up woody material, sawdust, and shredded white paper, newspaper, paper bags, and cardboard boxes and cartons. *Straw* consists of the hollow dried stems of grain-producing plants, whereas *hay* ("green" material) is the entire plant (stem and leaves) and therefore much higher in nitrogen. Some studies suggest that when added to compost birch and blackthorn (*Prunus spinosa*) leaves can help restore exhausted soil. *Green plant material* can include grass clippings, old flowers, weeds, fresh fruit and vegetable and kitchen wastes. Leafy materials that were cut green and allowed to dry are still considered green.

If possible, use a wide variety of materials in the compost, because this provides a better balance of pH, nutrients, and microbial organisms. Shred or cut material (including kitchen waste), when possible, before adding it to the pile; smaller particles decompose faster, but having some larger materials helps improve aeration.

For the most rapid results, do *not* add material to the heap once it has started composting.

What not to add to it. Do *not* add carnivorous animal manure, wood ashes, charcoal, animal meat, soils with a basic pH (such as those in California), or diseased plants. Also do *not* add bug-infested plants.

Caring for the compost pile. Cover the pile with a tarp, black plastic, or a lid of some kind; sunlight kills the bacteria that do the composting work, so a pile exposed to sunlight will not compost well in the outer layers.

Water the pile regularly to keep it evenly moist. It should have the consistency of a damp sponge: moist but not soggy, a moisture level of about 50 percent. A dry pile will not compost at all, whereas an over-watered or soggy pile will simply rot. One option is to uncover the pile during rain and re-cover it promptly to trap the moisture.

Turn the pile regularly, every 2 to 7 days, to aerate it and provide sufficient oxygen to the bacteria. Backyard piles often fail because oxygen cannot penetrate the pile. It is possible to overventilate, but this problem generally occurs only when the pile is ventilated through the sides and bottom and if it is turned on a daily basis. Some sources warn that turning piles reduces nitrogen levels in the final product, all the more reason to avoid overzealous turning.

You may want to add commercially available composting activator, a fine powder that contains helpful bacteria to speed along decomposition. The composting activator is not necessary but can be used to hasten results.

- Improves soil drainage by adding porous organic matter (humus)
- Improves water retention, again by the addition of organic matter (humus)
- Provides food that usually has a neutral pH (unlike some chemical fertilizers) and also buffers the soil against rapid pH changes
- Builds organic matter in the soil, which improves oxygen diffusion
- Feeds earthworms

Compost is a multipurpose tool. Some people apply large quantities to their garden, but a healthy minimum is to spread 1 inch of compost through your garden in spring each year before planting. It's helpful to apply another layer during the growing season. You can use compost as a fertilizer, before planting, at planting, after planting, and in fall after harvest; as a mulch, to help retain soil moisture, to keep summer soil cool, to keep winter soil warm, to smother weeds, and in some cases to discourage pests; and as a side-dressing during the growing season, as an extra food-boost to the plant.

You can buy commercially produced compost (many of the suppliers listed in appendix D offer it), but it's also simple to make. Compost is an ideal way to recycle many different household items such as kitchen waste, newspaper, typing paper (no gloss, no colored ink), paper napkins, unwaxed cardboard, sawdust, and other carbon-rich materials.

There are many ways to compost and results can vary, from production of 50 pounds in the backyard to thousands of tons on a commercial scale. It can be made in dug holes or in windrows, in silos or in barrels, in layers on top of the ground or in huge concrete vats. Because there are so many different methods and possible ingredients, a full discussion of composting is beyond the scope of this book. (See the Basic Composting Principles box for some basic guidelines.)

Understanding that bacteria are living organisms that chemically "chew up" the organic material in a compost pile will help you identify the optimal conditions for compost. Aerobic (oxygen-requiring) bacteria work faster, are usually less smelly, and generate higher temperatures in the compost pile, which can help kill certain disease pathogens. To favor these bacteria, you must turn the pile regularly, introduce oxygen into the pile (such as with perforated pipes), or make the pile small enough for oxygen to readily diffuse into the pile. In contrast, anaerobic bacteria do not require oxygen to work, take longer to accomplish the job, and do not generate high temperatures. In an anaerobic compost pile, earthworms encourage some aerobic activity and thereby expedite the composting process. Their castings also make it a richer source of nutrients. Aerobic methods can produce compost in as little as 14 days, but the pile needs to be built all at once. Anaerobic methods may take up to 1 year to produce compost, but the pile can be built slowly over the summer.

Both aerobic and anaerobic composting methods require moisture, because dry bacteria are "dead" bacteria. Both also need a large excess of carbon compared to nitrogen, usually a ratio of 30:1. In the absence of a large excess of carbonaceous material, bacteria digest their meal in a malodorous way. An odoriferous compost pile is usually working anaerobically and either needs more carbon material, needs to be turned, or both.

Certain materials are to be avoided. Under most circumstances, extremely high sources of nitrogen (for example, animal meat and carnivorous animal manure) give the pile indigestion, produce bacterial flatulence, and attract nuisance insects, but more important can contain disease pathogens. Diseased plant materials should never be added to a compost pile. Destroy diseased plant material by burying it in an area away from the garden or by burning it, if permitted. Although some diseases can be destroyed by the high temperatures generated in aerobic composting, backyard gardeners cannot rely on this result. The hottest temperatures occur at the center of the compost pile, so unless the pile is turned adequately and evenly mixed, certain parts of the pile may never reach the temperatures needed to destroy the disease pathogens. (For a fuller discussion of composting, consult sources in the bibliography, such as Stu Campbell, *Let It Rot!* [North Adams, Mass.: Storey, 1998.])

Along Came a Spider. . .

To attract spiders, apply a thin layer of mulch before hot summer weather sets in. In spring, spiders "balloon" or drift through the air on fine silk threads in search of hospitable summer homes. Mulch offers the cool, damp environment spiders need to remain hydrated, and garden pests provide their food. Research has shown that spiders regularly migrate hundreds of miles. One study found that 65 percent of the spider population in the study sample originated from far distances, rather than from neighboring gardens.

Mulch

Any material spread to cover the soil completely is considered *mulch.* Examples include compost, straw, chopped leaves, wood chips, cocoa bean shells, pebbles, plastic, and landscape fabric. Mulch plays many roles in a self-sustaining garden: it suppresses weed growth, keeps moisture in the soil by reducing evaporation, insulates the ground to keep it warm in winter and cool in summer, and can help prevent the emergence of certain harmful insects. Thick mulch around potatoes, for example, can help deter the Colorado potato beetle. In moister climates, some gardeners have reported using thick mulches of 4 to 8 inches or more during the growing season and never having to water or fertilize again. In hot, dry climates, a thick layer of mulch is not advisable because it absorbs rain and irrigation water, preventing water from reaching the soil. Thick mulch can also be counterproductive in areas that experience slug problems because slugs are attracted to cooler, moister environments.

Contrary to what some believe, grass clippings can be an effective mulch as long as they have not been sprayed with herbicides or pesticides and aren't loaded with weeds that are in the process of reseeding. In fact, in experiments at the University of Tennessee, grass clippings were shown to promote an abundance of pest-controlling spiders, another good friend of organic gardeners. Grass clippings can sometimes retard germination and emergence, as is the case with carrots, so delay application until after shoots have emerged through the soil.

Colored mulch, such as painted plastic, can serve other purposes as well, among them raising yields and deterring specific insects. In South Carolina, red mulch increased yields of cowpeas as much as 12 percent over the yields achieved with white or black mulch. Red is also believed to increase tomato yields, particularly early in the season, but has also been found to attract some pests. Orange mulch repels sweet potato whitefly but may attract other pests. Studies at the University of Florida have shown that of all available colors, red attracts the most white flies, blue attracts the most aphids and thrips, and white aluminum is the least attractive to all three pests.

Flying Creatures

Birds and bats that eat insects play an important role in a self-sustaining garden. They help keep the garden free of flying and crawling insects, such as ants, aphids, caterpillars, beetles and beetle larvae, crickets, flies, grasshoppers, leafhoppers, moths and moth eggs, sowbugs, wasps, and weevils.

Because of development and associated land-use changes, habitat for many bird species is shrinking. Pesticides also play a role and have been implicated in the diminishing habitat of the Eastern Bluebird. Providing habitat for birds significantly reduces your need for pest controls and can help restore much-needed sources of uncontaminated food for birds.

To attract more birds to your garden, incorporate in your landscape tall trees, smaller flowering trees, shrubs, berries, flowers, and native tall grasses that provide a range of seasonal nesting and feeding opportunities.

In the United States, some notable tall trees that provide habitat for a variety of birds include ash, balsam fir, birch, black gum, elm, hemlock, hornbeam, maple, mountain ash, oak, pine, spruce, and wild cherry. Smaller trees attractive to birds are those that flower and offer berries and fruits, such as crabapple, dogwood, hawthorn, holly, mesquite, mulberry, palmetto, persimmon, red cedar, serviceberry, and sumac. Shrubs and cactus attractive to birds for food and cover include algerita, autumn olive, all berry bushes (blackberry, blueberry, elderberry, hackberry, huckleberry, raspberry, snowberry), bitterbrush, buckthorn, filaree, grape, lote bush, juniper, madrone, manzanita, mountain laurel, multiflora rose, pinyon, prickly pear, spicebush, viburnums, Virginia creeper, and wisteria. Most wildflowers attract birds, as do bee balm, lupines, and sunflowers. A small area of tall ornamental or native grasses can provide yet another important type of bird habitat.

Bats are far from the blood-sucking horrors legend would suggest. They have a voracious appetite for the insects they eat at night and are thought to be helpful in keeping down mosquito populations. To attract more bats to your property, erect bat houses; these simple wooden structures are widely available at garden and home centers, or you can build your own.

Bees, wasps, and other beneficial insects are vital garden friends. Bees are nature's best pollinators, making possible the fruits and vegetables we all enjoy. Wasps, like other beneficial insects, not only prey on various destructive insects but also parasitize eggs, larvae, and adult insects. To attract wasps and other beneficial insects, plant companion herbs (especially *Umbelliferae* herbs), flowers, and clovers at the edge of your garden. (For more on beneficial insects, see page 316.)

Planning and Maintaining a Healthy Garden

To be self-sustaining, your garden should be able to defend itself from severe damage from most pests most of the time. Such natural defense is promoted by four major factors: sun, water, soil, and air circulation. All of these factors have a role; each is necessary but none is sufficient alone.

The first thing to consider is location. Choose a sunny spot, preferably one with morning and afternoon sun. In humid, moist, or rainy regions where fungus can be expected, morning sun is especially important to dry the dew as quickly as possible. Try to locate the garden where it will get good air circulation. Avoid low areas susceptible to pockets of fog and high locations exposed to harsh winds.

Nontoxic Preservative for Untreated Wood

If you want to preserve untreated wood for raised beds, decks, picnic tables, or swing sets, the U.S. Department of Agriculture (USDA) suggests the following nontoxic preservative. Studies show it to be as effective as the highly toxic preservative pentachlorophenol, and it lasts for up to 20 years.

- 1 ounce paraffin wax
- Enough solvent (mineral spirits, paint thinner, or turpentine at room temperature) to make a total volume of 1 full gallon
- 3 cups exterior varnish or 1½ cups boiled linseed oil

In a double boiler, melt the paraffin. (*Never* heat paraffin over a direct flame!) Away from the heat, vigorously stir the solvent, then slowly pour in the melted paraffin. Add the varnish or linseed oil and continue to stir thoroughly.

Apply by dipping the untreated lumber into the mixture for 3 minutes or by applying a heavy coat. The wood can be painted when thoroughly dry.

If the soil is clay or sandy, mix in peat humus to improve drainage and water retention. Whatever the condition of the soil, consider building raised beds. Although they require a modest initial investment of time and money, raised beds will pay you back in many different ways over the years. They are an excellent first step toward attaining a self-sustaining garden. Consider some of the advantages of raised beds:

• They minimize soil compaction, because you never walk on the growing medium.

• They offer oxygen more readily to plant roots, because there is little soil compaction.

• They drain better, because there is little soil compaction.

• They retain water better, because there is little soil compaction.

• They are easier to plant, weed, and maintain, because there is little or no soil compaction.

• They have greater yields because of better penetration of air, water, and sunlight; they also achieve higher germination rates for early plantings.

• They allow earlier and later planting, because they warm up earlier in spring and hold heat longer in fall.

• They allow greater root development, because of low soil compaction, good drainage, and oxygen diffusion.

• They save space. Plants can be spaced closer because you need walking space only between the beds, not between each row.

There are different types of raised beds and different ways to prepare them, some of which do not involve double digging, an arduous method used in biodynamic/French intensive gardening that creates the ultimate raised bed. (For more information on double-digging, see John Jeavons, *How to Grow More Vegetables Than You Ever Thought Possible* [Berkeley, Calif.: Ten Speed Press, 1991].) Raised beds can have rounded tops or flat tops, curved sides or straight sides. (A good discussion of raised beds and how to build them can be found in Marjorie Hunt and Brenda Bortz, *High Yield Gardening* [Emmaus, Penn.: Rodale, 1986].) Our preferred method is to build a contained raised bed, 12 to 16 inches high, which allows you to sit on the edge while planting or weeding. An ideal size for a raised bed is 4 feet by 12 feet. This size allows you to easily reach into the center of the bed without having to stretch your arms too far or step on the soil.

You can build contained raised beds with a variety of nontoxic substances (e.g., stone, brick, concrete blocks) or with untreated wood that is naturally rot-resistant (e.g., cedar, cypress, and locust). In a food garden, *never* use lumber treated with chromated copper arsenate (CCA), creosote, or pentachlorophenol. Since 1984, the EPA has directed that treated wood products not be used where they might come into contact with food, animal feed, or drinking water, either directly or indirectly, because of the possibility and increasing evidence that toxic substances will leach into the soil and enter the food chain. (For a nontoxic alternative, see box on page 11.)

Studies on treated wood have found that the chemicals chromium, copper, and arsenic leach out in significant amounts into soil, food crops, and beehives and onto hands in "wipe" tests on playground equipment. If treated lumber has been used for your raised beds, create a plastic or other barrier between the lumber and soil. You may also want to test your soil to determine the amount of leachate already present.

A quick and easy way to build noncontained raised beds without having to double dig is to use a tractor outfitted with a four- or five-bottom plow and rotary hoe. Plow a row in one direction, then turn around and plow the same row in the other direction. This flips the soil twice and creates a raised bed. Using the rotary hoe, straddle the bed with the tractor and run the hoe over the bed to break up the surface and create a smooth bed. That's all there is to it.

Bed planning and maintenance are essential components of a pest-control program. See pages 298–300 for a detailed discussion of essential steps to take during the planning stage to aid in pest prevention and in finding remedies.

The Case for Organic Gardening

Myths about organic gardening have been gradually dispelled during the past decade, so it's probably safe to say that the only reasons for *not* growing a backyard garden organically are habit, ignorance, and an unrealistic desire for the picture perfect. Let's look at some common misconceptions.

"It's expensive." Contrary to popular belief, growing organic is cost-effective and *not* more costly than conventional methods; money that might be spent on commercial fertilizers, herbicides, and pesticides can be easily redirected to organic compost, soil amendments, and traps and remedies. Growing with fertilizers and pesticides requires regular yearly inputs, keeping associated expenses high, whereas a focus on building soil health over time can reduce the need for soil amendments and pest remedies, minimizing expenses.

"It's difficult." Growing organic is *not* harder, it simply requires attention to different things in the garden. On a small scale, effort once put into spraying the garden with pesticides and herbicides is now put into soil improvements and weeding and harvesting. Organic growing is also easier in many respects, because as the garden becomes a more vibrant and balanced ecosystem, the problems and pests become less invasive and less damaging and require less attention.

"It takes more time." An organic garden requires no more time to maintain than does a conventional garden. A conventional gardener may spend time shopping for, mixing, and spraying herbicides to eliminate weeds from hardpan soil; an organic gardener may spend the same amount of time simply pulling the weeds. Because soil improvement is one goal of organic gardening, over time the soil becomes loose and friable, and weeding, in turn, becomes easy and pleasant.

In the end, small and large growers contend that, while managing organic gardens and farms may require more intensive hand work than conventional methods, the overall time needed for garden management is roughly the same. It's simply a matter of where one chooses to spend one's time: spraying herbicides or mechanical weeding, spraying fertilizers or spreading compost, spraying pesticides or encouraging beneficial insects.

"The results are ugly." Organic growers have learned the importance of devising ways to grow, harvest, and ship produce to minimize blemishes and spoilage. They know which cultivars work best for markets and roadside stands. Gradually, with the tremendous expansion of markets for fresh organic produce as well as canned and frozen organic foods, the expectations of consumers have changed. Organic produce is no longer the province of just small food co-operatives and farmers markets; specialty and many mainstream food chains also carry organic produce. People are now accustomed to seeing garden produce of all kinds and all shapes — all organic, all beautiful. The fruits of your home organic garden can also be beautiful, perhaps even *more* beautiful than those from conventional gardens, because organic gardening opens a whole new world of possibilities: heirlooms and open-pollinated varieties rarely seen in the food stores, such as lavender potatoes, purple peppers, huge yellow strawberry-shaped tomatoes, and more.

A Practical Shift

While organic agriculture may have a long way to go before it becomes fully mainstream, the impact of organic farming is extending to conventional agriculture. Slowly but surely, those who don't farm or garden organically are changing their methods to reflect lessons learned from organic growing.

For years, studies have shown that what we do to the earth affects our living systems. What farmers put on their fields affects the quality of drinking water in their wells; what they spray affects the health of their children, their workers, and themselves. Their entire savings is tied up in the land, and they rely on its continued productivity. It is not surprising, then, that conventional farming practices are moving toward organic methods: rotation of crops to interrupt insect cycles, broader use of integrated pest management, reduction of pesticide and herbicide use, and more scientific management of soils.

No longer is organic farming so easily dismissed as a "kooky" or ridiculous idea. Organic fields speak for themselves. Prince Charles of Wales, a visionary in sustainable agriculture and sustainable development, transitioned his homeland farms to organic. When neighboring farmers witnessed the successful yields of his fields, he was asked to help other farms make the transition from conventional to organic methods.

The discussion has shifted from the philosophical — *whether* organic can and should be done, the grand debate of the late twentieth century — to the practical, *how* it might be done cost effectively, particularly on large-scale farms dependent on economies of scale and on machinery that might not be conducive to organic methods.

Food Quality

Organic agriculture has had a profound impact on our food consciousness as a society. But is organic food really safer? A study led by Consumers Union found pesticide residues in 23 percent of organic fruits and vegetables. While this compared favorably with residues found in 73 percent of conventional produce, it begs the question, Why does organic produce have pesticide residues if none is used in its production? The statistics can be misleading. Only trace concentrations were found in organic produce, and most samples tested well below the level permitted by the U.S. Environmental Protection Agency, which is only 5 percent of that allowed in conventional produce. Still, the fact that residues show up at all, given the strict controls on organic production, is disturbing proof of the lasting environmental impact of our actions.

Pesticides remain in the soil for years; substances banned long ago such as DDT (dichlorodiphenlytrichloroethane) and chlordane accounted for 40 percent of the residues identified in organic produce in the Consumers Union study. Researchers speculate that the other 60 percent of residues might have resulted from pesticide "drift" (pesticides carried from neighboring farms by irrigation water) or contamination during storage and shipping.

It remains to be seen how much time must pass before the food supplies of the United States and the world are free of pesticides and herbicides. Still, by increasing the awareness of the need to improve the safety of our food supply by using fewer pesticides and herbicides, organic production is having a positive impact.

The good news is that backyard gardeners can enjoy all the benefits of organic gardening and eating organic food today, without waiting for commercial producers to make the transition in the years or decades ahead. Organic methods free us from worries about children and pets being exposed to poisons in the garden and remove concern about unintended negative impacts on native wildlife. You can garden knowing that a strawberry plucked fresh from the plant can be safely popped in your mouth, warm from the sun, unwashed and enjoyed. This can be a small taste of heaven on earth, and it's yours if you choose.

Coming of Age or Getting Old?

Some believe that the federally funded National Organic Program, which sets standards for the growing, handling, and labeling of organic food, lends credibility to organic agriculture. Others consider the federal standards a perversion of the principles of organic agriculture, because, they say, many of the standards are vague and weaker than needed, particularly in the area of animal management.

Such farmers argue that the term *organic* is no longer meaningful because it has been co-opted by some who want to make the label commercially available. Likewise, the vague language of the standards benefits those seeking loopholes. The current standards, for example, require "access" to such things as open air, sunlight, shade, pasture, and exercise areas, but do not stipulate that animals be raised on the pastureland.

Whatever your opinion, it's clear that sustainable practices for building healthy soil and growing healthy food are beneficial. The key is to remain focused. If your goal is to grow food that is safe and healthy for your family, in a way that is good for the planet, then the sustainable methods discussed in this book are for you. The practices recommended here are true to the philosophical roots of organic gardening and are good for soil and water quality, allowing you to grow safe, healthy organic food.

How to Use the Plant Charts

You can use this book in many different ways. Information is organized and cross-referenced for easy retrieval. Plants are divided into three categories: vegetables, fruits and nuts, and herbs. You will find an overview of each plant under its own entry, and an overview of insects and diseases in the charts beginning on page 24. For major topics, particularly those concerning fruit and nut trees, we cover only the basics. For more complete discussions of these topics, consult the books listed in the bibliography. Following are explanations of the categories used in the plant entries; when entries are specific to a plant type, they are so noted.

Plant Names

The genus and species, family name (Latin and English), and the common name are listed for each plant. Latin names have been included to help you problem solve. Plants in the same family don't always resemble each other. When planning crop rotations and preparing for insect and disease problems, therefore, it can be useful to be aware of any family ties.

Site (Fruits & Nuts Only)

Avoid planting fruit and nut trees in low pockets or in areas where fog and frost may collect. Most fruit trees require full sun. Morning sun is especially important to dry dew as rapidly as possible, which helps to prevent fungus problems. Some fruit trees that are susceptible to late-frost damage are better sited on a north-sloping hill, or 12- to 15-feet away from a north wall, in order to delay budding as long as possible. To avoid nematode and verticillium wilt problems, you might choose a site where grass or cover crops have been grown for the previous 2 years.

Temperature

For Germination: The optimum soil temperature for seed germination, in degrees Fahrenheit. Seeds can germinate at temperatures outside the noted optimum range, but generally plan on longer germination at lower temperatures and shorter germination times at higher temperatures.

For Growth: The optimum air temperature for plant growth, in degrees Fahrenheit.

Soil & Water Needs

pH. On this scale of soil alkalinity/acidity, a 7.0 is neutral. Most plants will grow well in a pH range from 6.0 to 7.0. Home test kits are widely available; see appendix D for a list of suppliers.

Fertilizer. Organic gardeners typically fertilize twice in one season, often just with compost: once before planting and again in the middle of the growth cycle. (See pages 307–310 for different types of organic fertilizers.) For fruit trees, special attention is paid to the quality and extent of new growth, because this is one of the easiest ways, all other factors (e.g., light, soil, moisture, temperature) being equal, to determine whether a tree is obtaining proper nourishment. If tree growth is not within the appropriate range, it may need more nitrogen, assuming all other conditions are favorable. If tree growth is greater than the appropriate range, it might be receiving too much nitrogen.

Side-dressing. Any addition to the soil (such as fertilizer, compost, or soil amendments) after the plant is already set in the soil is considered *side-dressing*. To add, create a narrow furrow 1- to 3-inches deep at the plant's drip line, or 6 inches from the plant base, whichever is greater; sprinkle the amendment into the furrow and cover with soil.

Water. The individual plant's watering needs (*heavy* = 1 gallon per square foot, or 1.4"–1.6" per week; *medium* = ¾ gallon per square foot, or 1"–1.2" per week; and *low* = approximately ½ gallon per square foot, or 0.8" per week.

Measurements

Planting Depth. Best planting depth for germination and root lodging.

Root Depth. When possible, a range of average to maximum-recorded root depth has been given. If a range isn't given, the average root depth has been provided. The better worked your soil is, the deeper the roots extend, but ideally you shouldn't till more than 3 to 6 inches once a bed is established.

Height. The maximum height the plant is likely to grow to under optimal growing conditions. Different varieties are likely to have different heights, hence a range is given for planning purposes. Place taller plants toward the middle of a bed and shorter plants on the outer edges.

Breadth. The maximum breadth that the plant is likely to grow under optimal growing conditions. Different varieties may have different breadths, hence a range is given for planning purposes. Plan parallel rows to allow sufficient space for the maximum breadth of both plants.

Space between Plants. In beds: This is the closest spacing recommended between plants under optimal conditions in raised beds and any other garden in which plants are spaced close together. (*Note:* If your climate is damp and humid, you should probably use broader spacing to allow for more air circulation and sun penetration, which prevent fungal and other diseases.)

In rows: Spacing recommended between plants in conventional rows, as opposed to raised beds.

Space between Rows. Spacing needed between conventional rows.

Growing & Bearing (Fruits & Nuts Only)

Bearing Age. The average number of years required for a fruit or nut tree to bear fruit from the time of planting in your garden. Bearing age may vary among cultivars of the same fruit or nut; be aware of this at the time of purchase.

Chilling Requirement. The number of hours a fruit or nut tree needs below a certain temperature, usually 45°F, before it will blossom. Trees with lower chilling requirements bloom earlier and, therefore, are more susceptible to late-spring frosts.

Pollination. Many fruit and nut trees require pollination, some by the same variety or cultivar of tree and some by a different cultivar in the same family. For cross-pollination to occur, the trees must be flowering at the same time, so check a catalog or nursery to be sure that you have correctly matched trees that can pollinate each other. Some fruits and nuts may be *self-fruitful* and therefore do not require pollination by another tree.

Propagation. Whether perennial or annual, herbs can often be propagated at home. Some herbs, for example, go "to seed" at the end of its growing season and can be propagated by saving the dried seed for replanting next year. Other perennial herbs are propagated by dividing the roots or taking cuttings for rooting in water or a moist growing medium. Some perennials tend to peter out after several years and need to be divided and replanted to retain their vigor.

Shaping (Fruits & Nuts Only)

Training: It is possible to train certain berries and fruit trees to grow in a two-dimensional form by pruning and training the plant along a multi-strand wire structure. For grapes, raspberries, and blackberries, this may be done to increase air circulation to reduce fungal diseases, improve ease of harvest, and improve ability to manage and prune the fruiting canes. For fruit trees, this is usually done for aesthetic purposes, but it can also ensure good air circulation.

Pruning: Major pruning is almost always done when a tree is dormant. Prune new transplants only *after* the first year of growth. Pruning at the time of transplanting to bring a tree's top growth into balance with root growth was once recommended, but research now shows that such pruning can stunt a tree's growth for years. Summer pruning is usually limited to balanced thinning to ensure high-quality fruit.

A major goal of pruning is to remove diseased, damaged, dead, and disfigured material. Judicious pruning can also increase light penetration and air circulation for better-quality fruit and better disease control. Compost pruned material, unless it is diseased, in which case it should be destroyed.

The information provided on pruning is primarily intended as a reminder for the experienced gardener. Both a science and an art, pruning is above all a visual experience and, in my opinion, cannot be taught without diagrams and hands-on experience. (For books with more extensive discussions of pruning, consult the bibliography. *Pruning Made Easy* by Lewis Hill [North Adams, Mass.: Storey, 1997] is an excellent choice.)

First Seed-Starting Date (Vegetables and Herbs Only)

The seed-starting dates in most catalogs and books are based on a national average. Some may even break down starting dates by USDA hardiness zone. Such dates are not necessarily the best for your backyard, however, because your microclimate may differ significantly from these average temperatures. The formulas included enable you to calculate the seed-starting dates appropriate for your particular garden, no matter your location and no matter how idiosyncratic your microclimate. They also give you the flexibility to plant varieties of the same vegetable that have different maturation times. These formulas should increase your confidence that seed started will grow safely to maturity. After you've done several trials, feel free to substitute into the equation your own numbers for each variable.

Germinate. The average number of days for a seed to *germinate,* which is when green shoots emerge through the soil.

Transplant. The average number of days *after*

germination that the plant needs to grow inside until it is transplanted outside. (*Note:* This number includes the 4 to 7 days that may be needed for hardening off.)

LFD. "Last frost date" in spring, an average date available from your local Extension agent. If possible, keep records to determine the average LFD for your own area.

Days Before/After LFD. The average number of days, before (or after) the last frost date, that a plant can tolerate living in outdoor soil. Use this number to determine the approximate dates you should transplant seedlings outside. This number also gives you a more precise measurement of the plant's frost hardiness. Because different sources suggest different setting-out dates, I've given the broadest possible range.

Maturity. The average number of days required for a plant to reach horticultural maturity for harvest. For plants that are transplanted, the plant will need the number of days noted under "Maturity" to complete maturation for harvest. If you don't want to start your seeds inside and prefer to direct-sow outside, calculate the total days to horticultural maturity by *adding* together the Germinate, Transplant, and Maturity.

SD Factor. "Short day" factor for late summer or fall plantings. Horticultural maturity times noted in seed catalogs assume long days and warm temperatures. For late-summer or fall plantings of many vegetable species, you need to adjust horticultural maturity times by 2 weeks to accommodate shorter and cooler days. Some species, like the radish, require short days to form and don't need such adjustments. When possible, I have noted which species require a SD Factor adjustment.

Frost Tender. An additional 2-week adjustment for frost-tender vegetables, which need to mature at least 2 weeks before frost in order to produce a full harvest.

FFD. "First frost date" in autumn, an average date available from your local Extension agent. If possible, keep records to determine the average FFD for your own area.

Insect and Disease Pests

No matter how many insects and diseases are listed in each plant entry, you are unlikely to encounter more than a few in your garden at one time. The diseases listed in each entry are limited to those that are transmitted by an insect vector, are highly infectious, or for which there are remedies over and above the standard preventive measures described on pages 307–315. Insect, animal, and disease pests are described in more detail in chapter 5.

Important Note on Allies, Companions, and Incompatibles

For planning purposes, all allies, companions, and incompatibles in this text are cross-referenced. If a source says that squash hinders potato growth, for example, that does not necessarily mean that potatoes hinder squash growth. Still, if one plant is alleged to harm the growth of another, they should be considered incompatible. So, in this text, squash is listed as an incompatible in the potato entry and vice versa. The same cross-referencing is done for companions and allies.

Allies

In the context of this book, *allies* are plants purported to actively repel insects or to enhance the growth or flavor of the target plant. Chapter 6 offers important caveats on allies and a listing of the reputed function of each, which may be helpful in determining if it will be useful in your garden.

Some Evidence. The effects of these allies have been tested in field trials. If the source and location of the field test are unknown, the claim is listed as *anecdotal* because its relative merits are unknown. This is an important distinction, because a plant that may function as an ally when tested in the Tropics may not offer the same benefits in northern Maine. The names of research sources are listed when possible.

Anecdotal. Allies classified as *anecdotal* have not undergone scientific testing or field trials and often hark back to tradition, folklore, and anecdotes. Presumably, for most of these anecdotal sources, positive correlations have been made, which have prompted the perpetuation of the information. Keep in mind that observations made at an unknown place and time are far from hard evidence. Use of allies categorized as anecdotal would constitute experimentation.

Companions

Companions are alleged to share space and growing habits well but do not necessarily play an active role in each other's insect protection or growth. Use your garden as a test bed. Some companions may work well, and others not at all. One might even prove a true ally by offering insect protection.

Incompatibles

Incompatibles are alleged to play an actively negative role in each other's growth by diminishing vigor or flavor, increasing the risk of insect or disease invasion, or decreasing yields. Although these claims may not be conclusive, it may be safer to avoid planting alleged incompatibles near one another.

Harvest

Outlines when and how to harvest and includes special tips on how to recognize that a vegetable is ready for harvest. Special preparations or directions for curing the vegetable before eating or storing are also provided.

If for health purposes you are concerned about nitrate levels in your vegetables, not to worry. Studies show that the way you harvest can help to reduce nitrate levels. The morning of harvest, use a long pitchfork or another implement to break up the soil around the vegetables, severing the roots; this stops nitrogen uptake. Then simply harvest crops late the same day, when nitrate levels are at their lowest.

Storage Requirements

Here you'll find tips on the best ways to store the edible harvest.

Fresh: How long fresh harvest will keep, at optimum storage temperature and humidity.

Preserved: The number of months the harvest will keep by specific preservation methods.

Note: Portions of the storage information are reprinted from Donald N. Maynard and George J. Hochmuth, eds., *Knott's*

Handbook for Vegetable Growers, Fourth Edition (New York: Wiley, 1997), and from Jeff Ball, *The Self-Sufficient Suburban Gardener* (Emmaus, Penn.: Rodale, 1983).

Selected Varieties

Recommended varieties have been selected to offer a broad range of growing qualities, including the following: disease resistance; ease of growing in organic gardens; flavor; storage qualities; special height, size, or habit; high yields; self-pollination (fruits); open-pollinated types for seed savers (vegetables); ease of harvest, including factors such as size, bruising, or peeling; heirloom varieties grown years ago but not often found at local nurseries; unusual color or ornamental qualities; and hybrids with good yield potential.

Most entries are ordered alphabetically. Where warranted, fruits have been clustered in categories by time to fruit maturation (early, midseason, late).

For vegetables, *days to maturation* (e.g., 51 days) can refer to either the time from seeding to harvest or the time from transplanting to harvest, depending on which catalog is used as a source. Maturation time varies greatly by region, climate, soil conditions, annual rainfall, and a host of other variables. Be sure to use *days to maturation* only as a relative number with which to compare different varieties.

The success of a variety in a specific garden cannot be predicted or guaranteed. In addition, varietal characteristics may differ, particularly among heirloom and open-pollinated types, depending on which company you order from. Obviously, the ultimate test of a variety is to actually try growing it in your garden.

Note: The selections of seeds available through catalogs may change yearly. Varieties suggested here may not always be available through the sources noted. Do a search on the Internet for other potential sources of seed.

Rootstocks (Fruits & Nuts Only)

Most fruit and some nut trees sold today are grafted onto a rootstock. Grafting seeks to combine the desired fruit variety with a rootstock that controls the ultimate size of the tree. A more rapid form of propagation than seed starting, grafting also is beneficial because it hastens the onset of bearing. Rootstocks are numerous and not easy to sort through because most nursery catalogs and books mention certain characteristics and omit others, making it difficult to get an overview. Choice of rootstock should depend on such things as desired tree size, the natural growing height of the tree variety, soil drainage, soil fertility, and the specific insect and disease problems in your area. When buying a tree, find out what rootstock it is attached to, if any, and the growing characteristics of the rootstock, as this will affect things such as tree size, how early it bears, and disease resistance. To determine which rootstock is best for your garden, consult reliable nurseries, many of which are listed in appendix C.

Chapter Two

vegetables

artichoke

Cynara scolymus
Compositae or Asteraceae (Sunflower Family)

ARTICHOKE is a cool-season crop, tender to frost and light freezes. Plan an average of three plants per person. In warm climates with mild winters, artichoke is grown as a perennial; in cold climates as an annual. Choose a sunny, sheltered location. Add plenty of compost or rotted manure to the soil before planting and again when the plants are 6 to 8 inches high. In cold climates, plant artichokes in large containers to keep the roots alive through winter. In warm climates, cut the plants to the ground in the fall. In cooler areas, to prepare for winter either cut the plants to the ground and bring containers indoors, or cut them to 15 to 20 inches above the ground, bend over the stalks, mulch heavily with leaves, and cover with a waterproof tarp or basket. Some recommend removing side shoots during the growing season. This increases the size of the central head but reduces overall yield. One gardener harvested almost thirty heads from just the side shoots of one plant. If not harvested for your table, the bud will blossom into beautiful purple-blue flowers suitable for arrangements.

Temperature

For germination: 60°F–70°F
For growth: 60°F–65°F

Soil & Water Needs

pH: 6.5

Fertilizer: Heavy feeder; lots of well-rotted manure or compost

Side-dressing: Every 3–4 weeks

Mulch: Apply over winter, and when plants are 6"–8" high.

Water: Heavy

Measurements

Planting Depth: Set buds just above the soil surface.
Root Depth: More than 4'

Height: 3'–6'
Breadth: 3'–6'

Space between Plants:
In beds: 2'–3'
In rows: 4'–6'
Space between Rows: 6'–8'

Propagation

Seed or suckers. To propagate by sucker, use a trowel to slice off the parent plant 10" tall suckers, each with a section of root. In warm climates, plant the suckers in a 4" hole. In cold climates, plant the suckers in a pot to overwinter indoors.

Pests

Aphid, plume moth, slug

Diseases

Curly dwarf (virus), southern blight, verticillium wilt

Allies

None

Companions

Brassicas

Incompatibles

None

Harvest

When heads are still closed, about the size of an orange, and while the stem 2" below the bud is still supple, cut off 1"–2" of stem with the head. Heads that have already opened are tough. Always harvest the central bud first. After harvest, cut the stems to the ground, or cut to 12" above the ground to encourage side shoots. Side shoots produce buds smaller than the first central bud.

First Seed-Starting Date

Start 6 weeks before the LFD in 4" deep pots in an area where the temperature is about 65°F. When germinated, put in full sun where the temperature is around 50°F for 9–10 days. Transplant when four true leaves have appeared.

Storage Requirements

Store in a paper bag in refrigerator to increase humidity and avoid drying out.

Fresh

Temperature	Humidity	Storage Life
32°F	95%–100%	2–3 weeks

Preserved

Method	Taste	Shelf Life
Canned Hearts	good	12+ months
Frozen Hearts	good	4+ months
Dried	N/A	N/A

Selected Varieties

Grand Beurre: Produces early enough to bear as an annual in cold climates
Not available in the United States

Green Globe: Excellent flavor; does well in long growing seasons, mild winters, and damp climates
Widely available

Imperial Star: Developed to grow from seed as an annual; best choice for short season areas; yields three times more artichokes than older varieties
Widely available

Purple of Romagna: New purple variety popular with chefs; this heirloom needs mild climate and is more tender than green varieties
Bakers Creek

Violetto Purple Variety: Pretty violet color
Write to Cook's and Shepherd's; seeds may be occasionally available, although not necessarily listed in catalogs

Greenhouse

None

asparagus

Asparagus officinalis
Liliaceae (Lily Family)

ASPARAGUS is a perennial, early spring crop. Plan an average of ten plants per person. Best planting time is early spring. Plant in a sunny spot protected from the wind. Because asparagus roots often extend both downward and outward 5 to 6 feet, plant in deeply rototilled soil that has incorporated green manure and compost. Traditionally, roots are planted in furrows 8 to 10 inches deep and 10 inches wide. Spread the roots, cover the crowns with 2 to 3 inches of sifted compost humus, and water well. As the plant grows through summer, add more soil, but do not cover the tip. If you prefer to plant individually, dig holes 8 to 10 inches deep and 5 inches wide, then proceed with the same method for furrow planting. Every spring, asparagus rows should be "ridged" by drawing up several inches of topsoil or, better, newly applied compost. This counters the tendency of the crown to get too close to the surface. After harvest, sow a cover crop of cowpeas or other legume between the asparagus rows, which discourages weeds and adds to the organic matter when dug under. University of Minnesota trials have shown that fall plantings of 9- to 11-week-old seedlings equal or exceed the growth of spring transplants. Carl Cantaluppi of the University of Illinois confirmed that you can increase yields by up to 40 percent by planting crowns at a depth of 5 to 6 inches, rather than at 10 to 12 inches. He also claims that asparagus is not a heavy nitrogen feeder because the ferns return most of the nitrogen to the soil. Decide for yourself.

Temperature

For germination: 60°F–85°F
For growth: 60°F–70°F

Soil & Water Needs

pH: 6.0–8.0

Fertilizer: Heavy feeder; apply compost to first-year beds in autumn, and again after harvest to established beds in spring. Apply fish emulsion twice yearly. Beds may need P and K before planting, and N after planting.

Mulch: Use straw or light material during winter and remove it in spring. Use compost during the growing season.

Water: Heavy

Measurements

Planting Depth: 8"–10" (see page 27)
Root Depth: More than 4'

Height: 3'–8' (fern growth, depending on soil and climate)
Breadth: 2'–4' (fern growth, depending on soil and climate)

Space between Plants:
In beds: 12"
In rows: 15"–18"
Space between Rows: 3'–4'

Average Bearing Age

3 years from seeds, 2 years from crowns

Pests

Aphid, asparagus beetle (early May), cucumber beetle, garden centipede, gopher, Japanese beetle, mite, slug, snail, spotted asparagus beetle

Diseases

Asparagus rust, fusarium wilt

Allies

Uncertain: Basil, goldenrod, nasturtium, parsley, pot marigold, tomato

Incompatibles

Onion family, weeds (during first 6 weeks of asparagus growth)

Harvest

When spears are ⅜" thick and 6"–8" high, cut spears ½" below the soil surface to lessen the chance of disease and pest infestation. Heads should be tight and spears brittle. Stop harvesting when stalks are less than ⅜" thick. When grown from roots, do not harvest the first year. Let the plants go to foliage, and when they brown in the fall, cut them to ground level. The second year, harvest spears for about 4 weeks. In following years, the harvest will continue for 8–10 weeks.

Storage Requirements

Wrap spears in moist towels or stand upright in a glass of water, then refrigerate in plastic bags. Blanch asparagus before freezing it.

Fresh

Temperature	Humidity	Storage Life
32°F–35°F	95%–100%	2–3 weeks

Preserved

Method	Taste	Shelf Life
Canned	good	12+ months
Frozen	excellent	12 months
Dried	fair	12+ months

asparagus

Selected Varieties

Connover's Colossal: Open-pollinated; good for general use and growing from seeds
Bountiful Gardens

Jersey Giant: One of the new all-male varieties (meaning it produces more harvest); vigorous, with very high yields; resists rust; tolerates fusarium crown and root rot; adapted to a wide variety of climates from New England to Washington State and south to the Carolinas
Burpee, May, Nourse, Park

Jersey Knight: Yields twice as many spears as Washington; all-male variety is resistant to all major asparagus diseases
Cook's, Park

Larac: New hybrid white asparagus; high yields of thick spears; adaptable to many different types of soil
Park

Mary Washington: Heirloom; good flavor; rust resistant
Widely available

Purple Passion: Stringless spears are sweet and crisp; color fades in cooking
Shepards, Johnny's

Rutgers SYN-4-56: Hybrid; offspring of Jersey Giant; high yields; resists rust; tolerates fusarium; some female plants but most are male
Nourse

UC 157: A new hybrid developed by the University of California for the West Coast and the South; good flavor and quality
Burpee, Shepards

Waltham: Hybrid; uniform spear size; heavy producer; rust resistant
Widely available

Greenhouse

None

dried beans

Phaseolus vulgaris
Leguminosae or Fabaceae (Pea Family)

BEANS are a warm-season crop, tender to light frosts and freezes. Plan an average of ten to twenty plants per person. Cold, wet weather fosters disease. To prevent disease, do not sow or transplant too early, touch the plants when wet, or touch healthy plants after working with diseased ones. Most dried beans, whether bush or semivining, require long growing seasons. To direct-sow them, layer grass mulch 4- to 6-inches deep on the bed in fall. This will decompose to about 2 inches by spring, keep the soil warm 6 inches deep, and won't pull nitrogen out of the soil, allowing you to plant earlier in the spring. (See Snap Beans, page 40, for comments on presoaking and inoculation.) Dried beans are very high in protein. Like other legumes, soybeans and cowpeas are excellent green manure crops that enrich soil with organic matter and nitrogen.

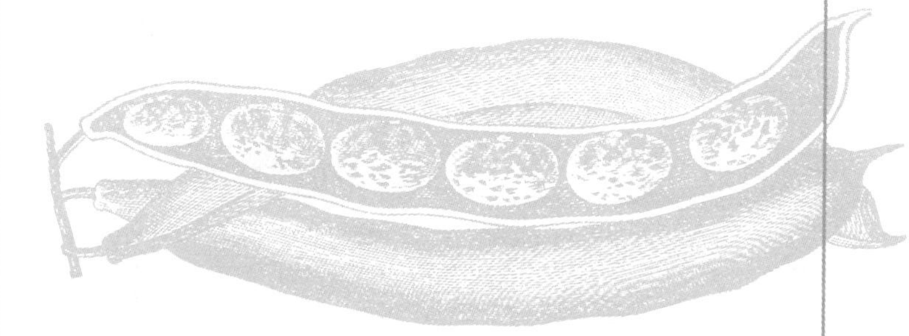

Temperature

For germination: 60°F–85°F
For growth: 60°F–75°F

Soil & Water Needs

pH: 6.2–7.5

Fertilizer: Light feeder. Because bean plants fix N when inoculated properly, they should require low N. After the plant flowers, apply fertilizer low in N; medium P and K. Avoid low K at all times.

Water: Average and constant

Measurements

Planting Depth: 1"
Root Depth: 36"–48"

Height: 10"–24"
Breadth: 4"–8"

Space between Plants: 2"–6"
Space between Rows: 12"–30"; 8" on center in raised beds

Pests

Aphid, bean leaf beetle, beet and potato leafhopper, cabbage looper, corn earworm, cucumber beetle, cutworm, flea beetle, garden webworm, Japanese beetle, leaf-footed bug, leaf miner, Mexican bean beetle, mite, root-knot nematode, seedcorn maggot, slug, tarnished plant bug, thrips, webworm, weevil, whitefly, wireworm

Diseases

Anthracnose, bacterial blight and wilt, bean mosaic, common mosaic, curly top, damping off, powdery mildew, rust, southern blight, white mold, yellow mosaic. (*Note:* If legal, burn diseased plants.)

Allies

Some evidence: Goosegrass, red sprangletop, sorghum mulch (for cowpeas)
Uncertain: Catnip, celery, corn, goldenrod, marigold, nasturtium, oregano, potato, rosemary, savory

First Seed-Starting Date

Transplant or direct sow when soil temperature is 60°F.

Germinate	+	Transplant	–	Days After LFD	=	Count Back from LFD
4 to 10 days	+	21 to 28 days	–	0 to 10 days	=	25 to 28 days

Last Seed-Starting Date

Germinate	+	Transplant	+	Maturity	+	SD Factor	+	Frost Tender	=	Count Back from FFD
4 to 10 days	+	0 (direct)	+	98 to 125 days	+	14 days	+	14 days	=	130 to 163 days

Companions

Beet, cabbage, carrot, celery, corn, cucumber, eggplant, peas, potato, radish, strawberry

Incompatibles

Fennel, garlic, gladiolus, onion family

Harvest

Wait until the plant's leaves have fallen in autumn to pick dry pods or to pull the entire plant. Harvest before the first frost. Soybeans and limas, however, should be picked when any split pods are spotted because beans often drop from the shells as they dry. Cure for several weeks in a well-ventilated area, piling them on screens or slatted shelves. Beans are dry and ready to thresh when they don't dent when bitten.

Following are four methods of threshing.

- Thrash the plant back and forth inside a clean trashcan.
- Place the plant in a large burlap bag with a hole in the corner and flail.
- Put plants in a cone-shaped bag, tie the bottom, and walk or jump on the bag.
- Put the beans into a bag with a hole in the bottom; tie the bottom closed. Hang bag from a tree and beat well, then untie the hole. With the help of a good wind, the chaff will blow away and the beans will fall into a container placed below.

dried beans

Storage Requirements

Remove all bad beans. Place on shallow trays and heat at 170°F–180°F for 10–15 minutes. Cool. Store in a cool, dry area in tight jars. To avoid weevil damage, see page 414.

Selected Varieties

All varieties listed are open-pollinated.

Aprovecho Select Fava: Bush; large seeds; sweet flavor; good fresh or dry; hardy to below 20°F; matures 2–3 weeks before pole beans; good in maritime Northwest and East Coast
Abundant Life

Black Coco: Bush; milder black bean; buttery flavor; good fresh or dried
Territorial

Black Valentine: 50 days; bush, tall spreading plant; heirloom; meaty black bean; enjoyable flavor
Widely available

Borlotto: 68 days; beautiful bush plant with speckled rosy red and cream pods; delicious fresh or dried; an Italian heirloom rarely seen in markets
Shepherd's, Cook's

Butterbeans or Green Vegetable Soybean (Glycine max): 88 days; very digestible; good fresh or frozen; stocky and highly branched plants
Widely available

Buttergreen: 45 days; bush; very short season; good dried or young as snap beans; resists bean mosaic; mellow flavor; introduced in 1989
Burpee

Chickpea or Garbanzo: 100 days; bush; usually only one seed per pod
Burpee

Cowpea (Blackeyed or Southern pea) — Calico Crowder: 79 days; very flavorful; tan with maroon splashes; good for southern and warm coastal areas
Southern Exposure

Desi Bush: Small seeds native to India; better for home garden than large white commercial type; for interior Northwest and short-season dry areas; not good for coastal fog areas
Abundant Life

Dolores de Hidalgo: Prolific in low desert winters; produces small beans; suitable for high and low desert
Native Seed

Fava (Tarahumara Habas): Frost hardy; suitable for high and low desert
Native Seed

French Horticultural: 66 days; bush produces some runners; 6"–8" pods; good fresh, dried, or frozen
Widely available

Jacob's Cattle or Trout: 85–95 days; bush 24" tall; short-season baked bean type; a beautiful white with splashes of maroon; very tasty and meaty; good fresh or dry; an heirloom favorite in Vermont and Maine
Widely available

Kabuli Black: 95 days; black-seeded (Ethiopian origin); small seeds; better for home garden than large white commercial type
Garden City

Midnight Black Turtle Soup: Best variety of this Latin favorite; tall bush keeps pods off the ground; developed by Cornell Univeristy
Johnny's, Garden City

Pima Bajo: Excellent green or dried; suitable for low desert planting
Native Seed

Pinto: 85 days; bush 14"; short half-runner plant; beans are tan with brown speckles; susceptible to mosaic damage; medium-to-high yields
Abundant Life, Burpee, Field's, Johnny's, Park

Queen Ann: 68 days; compact 26" plants with no runners; no significant insects or diseases; high and reliable yields; good fresh, fried, frozen, or canned
Southern Exposure

Red Kidney: 95 days; bush 16"–22"; some mosaic and Japanese beetle damage; bush 16"; high yields
Widely available

San Juan Pinto: Suitable for high desert planting
Native Seed

Santa Maria Pinquito: 120 days; bush; heirloom variety of Pink Bean, used primarily in soups, stews, and Santa Maria barbecue; keep shape when cooked

Soldier: Heirloom; kidney-shaped white bean with yellow-brown eye; very rich and meaty flavor
Widely available

'Tender Cream': Possesses both excellent eating qualities and resistance to cowpea curculio, root-knot nematodes, cowpea mosaic virus, and several common diseases

---

Vermont Cranberry: Bush; beautiful with swirls of red on a cream background
Cook's, Johnny's

Notes

"Plants of the Southwest" and Native Seeds offer unusual heirloom varieties native to the Southwest.

lima beans

Phaseolus limensis
Leguminosae or Fabaceae (Pea Family)

LIMA BEANS are a warm-season crop, very tender to frost and light freezes. Plan an average of ten to twenty plants per person. For every 2 pounds of filled pods, you should get 1 pound of shelled beans. Limas are more sensitive to cold soil and calcium deficiency than are snap beans. Limas don't like transplanting, so it's often recommended to sow them directly in the beds; however, seeds will not germinate if the soil isn't warm enough. (See Snap Beans, page 40, for comments about presoaking, inoculation, and cold, wet weather.) For direct sowing, plant five to six seeds in a hill and thin to 3 to 4 inches. Bush beans usually mature more quickly than pole beans and are determinate, with one clean harvest. Pole limas generally have better flavor and are indeterminate, with a continuous harvest, but they require a trellis and some extra effort. Bush limas don't do well in wet weather; they develop an unpleasant earthy taste if pods touch the ground. Corn plants can provide a substitute support for nonrampant pole beans; plant the beans between 6- to 8-inch-tall corn plants.

Temperature

For germination: 65°F–85°F
For growth: 60°F–70°F

Soil & Water Needs

pH: 6.0–7.0

Fertilizer: Light feeder. Beans fix N when inoculated properly, so most need low N; medium P and K.

Side-dressing: 4 weeks after planting, apply a balanced or low N fertilizer, or compost.

Water: Average; constant

Measurements

Planting Depth: 1½" to 2"
Root Depth: 36"–48"

Height:
Pole: 8'–15'
Bush: 10"–18"
Breadth:
Pole: 6"–8"
Bush: 4"–8"

Space between Plants:
In beds:
Pole: 6"
Bush: 4"–6"
In rows:
Pole: 10"–18"
Bush: 6"–8"
Space between Rows:
36"–48"

Support Structures

Use a 6' post, A-frame, tepee (3 poles tied at the top), or trellis for pole beans

Pests

Aphid, bean leaf beetle, beet and potato leafhopper, cabbage looper, corn earworm, cucumber beetle, cutworm, flea beetle, garden webworm, Japanese beetle, leaf-footed bug, leaf miner, Mexican bean beetle, mite, root-knot nematodes, seedcorn maggot, slug, tarnished plant bug, thrips, webworm, weevil, whitefly, wireworm

Diseases

Anthracnose, bacterial blight and wilt, bean mosaic, common mosaic, curly top, damping off, powdery mildew, rust, southern blight, white mold, yellow mosaic. (If legal, burn diseased plants.)

First Seed-Starting Date

Germinate	+	Transplant	–	Days After LFD	=	Count Back from LFD
7 to 18 days	+	21 to 35 days	–	14 to 28 days	=	14 to 25 days

Last Seed-Starting Date

Germinate	+	Transplant	+	Maturity	+	SD Factor	+	Frost Tender	=	Count Back from FFD
7–18 days	+	0 (direct)	+	60 to 80 days	+	14 days	+	14 days	=	95 to 126 days

Allies

Some evidence: Goosegrass, red sprangle-top
Uncertain: Catnip, celery, corn, marigold, nasturtium, oregano, potato, rosemary, savory

Companions

Carrot, corn, cucumber, eggplant, lettuce, peas, radish
Bush only: Beet, all brassicas, strawberry

Incompatibles

Fennel, garlic, onion family
Pole only: Beet, all brassicas, kohlrabi, sunflower

Harvest

For the best fresh flavor, pick beans when young. To encourage the plant to set more beans, pick when beans are bulging through pods. For dried beans, wait until pods turn brown or leaves drop in fall. Pick pods and cure for several weeks in a well-ventilated area, piling them on screens or slatted shelves. Beans are dry and ready to thresh when they don't dent when bitten. See Dried Beans, page 33, for threshing methods.

Storage Requirements

Blanch before freezing. Store dried beans in jars in a cool, dry place.

Fresh

Temperature	Humidity	Storage Life
37°F–41°F	95%	5–7 days

Preserved

Method	Taste	Shelf Life
Canned	fair	12+ months
Frozen	excellent	12 months
Dried	excellent	12+ months

Selected Varieties

All varieties listed are open-pollinated.

POLE

Christmas or Speckled Calico, Giant Florida: 85 days; rich "butter bean," nutty flavor; large beautiful red-speckled beans
Widely available

Hopi: White, gray, yellow and red types; native to the Southwest; suitable for high and low desert
Native Seed

Hyacinth: Introduced in 1989; baby white limas; some drought tolerance
Park

King of the Garden: 100 days; early, 5" pods; heirloom; high yields; good fresh or dried
Heirloom Seeds, Park, Southern Exposure

Prizetaker: 90 days; very large bean of good quality
Burpee

BUSH

Baby Fordhook: 70 days; 14" bush; thick seeded; small lima; high quality
Burpee

Fordhook 242: 75 days; AAS Winner; heat resistant; upright; early butterbean; good fresh, canned, or frozen
Widely available

Henderson Bush: 65 days; baby white (when dried); flat pods; withstands hot weather; good canned or frozen; good in all climates
Widely available

Greenhouse

Fordhook 242, Henderson Bush, King of the Garden

snap beans

Phaseolus vulgaris
Leguminosae or Fabaceae (Pea Family)

BEANS are a warm-season crop, tender to light frosts and freezes. Plan an average of ten to twenty plants per person. Bush beans are usually determinate with one clean harvest, so plant every 10 days for continuous harvest. Pole beans are usually indeterminate with a continuous harvest for 6 to 8 weeks if kept picked, so only one planting is necessary. Bean roots don't tolerate disturbance so handle seedlings minimally. Plant outside at the same depth they grew in the pot.

Some gardeners recommend presoaking seeds before planting, but research indicates that presoaked seeds absorb water too quickly, causing the outer coats to spill out essential nutrients, which encourages damping-off seed rot. Yields can increase 50 to 100 percent by inoculating with *Rhizobium* bacteria. To inoculate, simply roll seeds in the powder. Cold, wet weather fosters disease, so do not sow or transplant too early, touch plants when wet, or touch healthy plants after working with diseased ones. Pinch off the growing tips of pole beans when plants reach the top of their support system.

Temperature

For germination: 60°F–85°F
For growth: 60°F–70°F

Soil & Water Needs

pH: 6.2–7.5

Fertilizer: Because bean plants fix N when inoculated properly, they should require low N; after they flower, apply light N; avoid low K.

Water: Low until the plant flowers, then average

Measurements

Planting Depth:
Spring: 1"
Fall: 2"
Root Depth: 36"–48"

Height:
Pole: 8'–15'
Bush: 10"–24"
Breadth:
Pole: 6"–8"
Bush: 4"–8"

Space between Plants:
In beds:
Pole: 6"
Bush: 2"–4"
In rows:
Pole: 12"
Bush: 4"–6"
Space between Rows:
(pole or bush) 18"–36"

Support Structures

Use 6' posts, A-frame, tepee (three poles tied together at the top), or trellis to support pole beans.

For an alternate support, plant nonrampant pole beans between corn that isn't too densely planted, when the corn is 6 to 8 inches tall.

Pests

Same as for Lima Beans (see page 37), with the addition of the European corn borer

Diseases

Anthracnose, bacterial blight and wilt, bean and common mosaic, curly top, damping off, powdery mildew, rust, southern blight, white mold, yellow mosaic. (*Note:* If legal, burn diseased plants.)

Allies

Some evidence: Goosegrass, red sprangletop
Uncertain: Catnip, celery, corn, marigold, nasturtium, oregano, potato, rosemary, savory

First Seed-Starting Date

Germinate	+	Transplant	–	Days After LFD	=	Count Back from LFD
4 to 10 days	+	21 to 28 days	–	7 to 14 days	=	18 to 24 days

Last Seed-Starting Date

Germinate	+	Transplant	+	Maturity	+	SD Factor	+	Frost Tender	=	Count Back from FFD
4 to 10 days	+	0 (direct)	+	48 to 95 days	+	14 days	+	14 days	=	80 to 133 days

snap beans

Companions

Pole and bush: Carrot, chard, corn (corn rows can be wind breakers for dwarf beans), cucumber, eggplant, peas, radish, strawberry
Bush only: Beet, all brassicas

Incompatibles

Pole and bush: Basil, fennel, garlic, gladiolus, onion family
Pole only: Beet, all brassicas, sunflower

Harvest

For best flavor, pick early in the morning, after leaves are dry. Harvest before seeds bulge, when beans snap off the plant and snap in half cleanly. Continual harvest is essential for prolonged bean production.
Bush, snap: Pick when ¼"–⅜" diameter.
Filet: Pick daily and, for peak flavor, when no larger than ⅛" diameter, regardless of length.

Storage Requirements

Blanch before freezing.

Fresh

Temperature	Humidity	Storage Life
40°F–45°F	95%	7–10 days

Preserved

Method	Taste	Shelf Life
Canned	fair	12+ months
Frozen	excellent	12 months
Dried	excellent	24 months

Selected Varieties

All varieties listed are open-pollinated.

BUSH SNAP

Blue Lake 274: 55 days; 16" semierect plant; 6" pods; white seeds; bears uniformly maturing crop; long season; high yields; good fresh or frozen; resists bean mosaic
Widely available

Pencil Pod Black Wax: 52 days; 15" plant; heirloom; stringless, tough yellow skins with black bean; fullest rich black bean flavor when mature; early and extended producer; good fresh or canned
Widely available

Romano or Roma: 50–60 days; 2' plant; long, flat, stringless Italian bean; mosaic resistant; good snap or dried, canned, or frozen; also available as pole
Widely available

Royal Burgundy: 51 days; unusual purple flowers and pods; beans turn green when cooked; tolerates cold soil
Widely available

Tendercrop: 53 days; reliable in poor weather; high yields; dark green, slender bean; thick clusters off the ground; resists mosaic, powdery mildew, and pod mottle virus; good for canning
Widely available

Topcrop: 49 days; AAS Winner; stringless, slender, meaty bean; mosaic resistant; hardy
Widely available

POLE SNAP

Blue Lake: 62 days; vine to 8'; 6" pods; springless; excellent flavor; white seeds also good as shell beans; very sweet and tender
Widely available

Kentucky Wonder: 65 days; 9" stringless pod; heirloom; very popular; good flavor; resists rust; also available as bush
Widely available

POLE RUNNER

Perennial in warm climates; dig up bulbous root and replant in spring.

Scarlet Runner (*P. coccinus*): 70 days for fresh beans; 115 days for shell beans; vine to 10'; pretty ornamental red flowers are edible; 8" pods; good soup bean; tolerates cool weather; pods won't set in hot weather
Widely available

Scarlet Runner (*Tarahumare tecomari*): Red flowering variety that produces purple, lavender, black, and mottled beans; suitable for high desert (over 4000')
Native Seed

Selected Varieties

FILET (pencil-thin)

Aramis: 65 days; bush; good taste raw and cooked; disease resistant; very uniform high yields; stringless
Garden City, William Dam

Camile: Most productive; disease resistant; good sweet flavor; best lightly steamed
Cook's, Horticultural Products

Emerite Filet Pole Bean: Outproduces bush beans; not grown commercially
Kitchen Garden, Shepherd's

Finaud: Small 6" bush; a true filet that doesn't need to be picked every day
Cook's

Fin de Bagnols: Very slender pods; early; uniform; tender; sweet, delicate flavor; grow like bush beans
Cook's

Morgane: High yields; 5–6 harvests from single planting; delicious flavor and disease resistant
Cook's

FLAGEOLET

These rich, meaty shelling beans are often eaten fresh like limas.

Chevrier: A delicacy when eaten fresh, flageolet are also grown as shell beans for drying; pick when seeds are ½" in the shell
Cook's

Flaro: New flageolet bush shell bean
Nichols, Vermont Bean

Greenhouse

Tendercrop, Topcrop

eating beet

Beta vulgaris
Chenopodiaceae (Goosefoot Family)

BEETS are an annual cool-season crop, half-hardy to frost and light freezes. Plan an average of ten to twenty plants per person. Most beet cultivars are open-pollinated and multigerm, where one seed yields a clump of four to five plants that need to be thinned. These multigerm seeds, also known as "seed balls," germinate better if soaked an hour before planting. There are three mains types of eating beets: Long (Cylindra), Medium (Semiglobe), and Short (Globe). Cylindra types mature slowly and, because they grow as long as 8 inches, require deep soil. They can also be a good organic matter crop. For all types, look for cultivars resistant to bolting and downy mildew. Yellow and white beets are sweeter than red varieties. Newer hybrids are usually sweeter than older varieties and offer more green leaves. Most beets contain 5 to 8 percent sugar, whereas newer hybrids such as Big Red run 12 to 14 percent. Hybrids tend to mature 7 to 14 days earlier, are more upright, and tend to have higher yields. Like kale and some other vegetables, in hot weather most beets get tough, woody, and develop an "off" flavor. An exception, according to some, is Detroit Crimson Globe. If your summers are hot, generally choose a variety that matures in 45 to 60 days. In greenhouses, beets are often grown for their greens only.

Temperature

For germination: 50°F–85°F
For growth: 60°F–65°F

Soil & Water Needs

pH: 5.8–7.0 (5.3 for scab)

Fertilizer: Heavy feeder. Needs high P; avoid high N; good tops may mean the roots are poorly developed and the plant is getting too much N.

Side-dressing: Every 2 weeks provide a light and balanced feeding; when tops are 4"–5" use low N.

Water: Average and evenly moist

Measurements

Planting Depth: ¼"
Root Depth: 24"–10'

Height: 12"
Breadth: 4"–8"

Space between Plants:
In beds: 3"–4"
In rows: 6"

Space between Rows: 18"–24"

Pests

Mostly pest free; occasional beet leafhopper, carrot weevil, earwig (seedlings), garden webworm, leaf miner, mite, spinach flea beetle, whitefly, wireworm

Diseases

Mostly disease free; occasional cercospora, downy mildew, leaf spot, rust, scab

First Seed-Starting Date

Germinate	+	Transplant	+	Days Before LFD	=	Count Back from LFD
5 to 10 days	+	0 (direct)	+	9 to 18 days	=	14 to 28 days

Last Seed-Starting Date

Germinate	+	Transplant	+	Maturity	+	SD Factor	+	Frost Tender	=	Count Back from FFD
5 to 10 days	+	0 (sow direct)	+	55 to 80 days	+	14 days	+	0 days	=	74 to 104 days

Allies

Uncertain: Garlic, onion family

Companions

All brassicas, bush beans, head lettuce

Incompatibles

Field mustard, all pole beans

Harvest

In late June, or before the hot season enters its prime, scrape some soil away from the beets to check their size. Pull or dig when the beets are 1"–2" across. They can become tough and woody-flavored when allowed to grow much larger, depending on the variety.

eating beet

Storage Requirements

Remove all top greens, leaving about 1" of stem with the beet. Do not wash. Pack beets in straw or moist sand. Beets can also be left in ground and dug up from under the snow.

Fresh

Temperature	Humidity	Storage Life
32°F	98%–100%	4–7 months topped; 10–14 days bunched

Preserved

Method	Taste	Shelf Life
Canned	good	12+ months
Frozen	fair	8 months
Dried	fair	12+ months

Selected Varieties

Albino White: 50 days; unusual white beet won't "bleed"; mild; sweet
Widely available

Big Top: Vigorous greens can be harvested early
Widely available

Chioggia: 50 days; open-pollinated; gourmet red-and-white-striped beet; try the Improved variety
Widely available

Crosby's Egyptian: 60 days; open-pollinated; heirloom; flattened shape; sweet and rich
Heirloom Seeds, Southern Exposure

Cylindra: 60 days; up to 8" (like carrot); needs deep soil
Widely available

Detroit Dark Red: 60 days; most widely-sold beet; reliable; high-quality
Widely available

Golden: 50–55 days; beautiful yellow color; won't bleed into other foods; holds its flavor well into the growing season, and when allowed to grow large, both root and greens are tasty; a poor germinator that requires more thick sowing
Widely available

Kleine Bol: 50 days; a true baby beet that grows fast
Shepherd's

Lutz Green Leaf: 70 days; open-pollinated; good yields; improves in storage and stays tender; high in vitamins A and C
Widely available

Pronto Heirloom Dutch: Best baby beet; sweet roots can be left in ground without becoming tough or woody
Cook's, Kitchen Garden

Red Ace: 52 days; hybrid; midseason; milder than "Detroit Dark Red"; sweet and rich; stays tender when older
Widely available

Greenhouse

Greenhouse beets are usually grown for their greens. Varieties include: **Detroit Dark Red, Green Top Bunching, Lutz Green Leaf**

Temperature

For germination: 50°F–85°F
For growth: 60°F–65°F

Soil & Water Needs

pH: 6.0–7.5 (7.2 deters clubroot)

Fertilizer: Heavy feeder. Before planting, add compost to the soil. If clubroot is a problem, raise the pH by adding lime or taking other measures (see Acid Soil on page 307).

Side-dressing: When buds begin to form, side-dress the plant with compost.

Water: Medium and evenly moist

Measurements

Planting Depth: ¼"
Root Depth: 18"–36"

Height: 18"–4'
Breadth: 15"–24"

Space between Plants:
In beds: 15"
In rows: 18"–24"
Space between Rows:
24"–36"

Pests

Aphid, cabbage butterfly, cabbage looper, cabbage maggot, cutworm, diamondback moth, flea beetle, harlequin bug, imported cabbageworm, mite, root fly, slug, weevil, whitefly

Diseases

Alternaria leaf spot, black leg, black rot, clubroot, damping off, downy mildew, leaf spot, rhizoctonia, yellows

Allies

Some evidence: Candytuft, shepherd's purse, wormseed mustard
Uncertain: Catnip, celery, chamomile, dill, garlic, mint, nasturtium, onion family, radish, rosemary, sage, savory, tansy, thyme, tomato, wormwood

Companions

Artichoke, beet, bush beans, chard, cucumber, lettuce, peas, potato, spinach

First Seed-Starting Date

Germinate	+	Transplant	+	Days Before LFD	=	Count Back from LFD
3 to 10 days	+	42 days	+	14 days	=	59 to 66 days

Last Seed-Starting Date

Germinate	+	Transplant	+	Maturity	+	SD Factor	+	Frost Tender	=	Count Back from FFD
3 to 10 days	+	21 days	+	55 to 74 days	+	14 days	+	0 days	=	93 to 119 days

Incompatibles

Pole lima and snap beans, strawberry, tomato (latter may also be an ally)

Harvest

Harvest when the heads are dark green, or dusky violet for purple varieties. If heads turn yellow, you've waited too long. For most varieties, small compact heads offer the best flavor. The exception is Romanesco, whose head can grow up to 1' in breadth and whose natural color is chartreuse. Harvest the central head first. Some varieties will produce side shoots that develop small head clusters; these plants will provide for 1 to 2 months, or until frost. Cut the stalk so that several inches remain on the plant.

Selected Varieties

DeCicco: 65 days; open-pollinated; old Italian variety; small heads; long harvest with side shoots
Widely available

Green Goliath: 55 days; open-pollinated; high yields; early and extended harvest; many side shoots; good for freezing
Widely available

Green Valiant: 66 days; hybrid; large heads; good for intensive planting; good for fall planting; frost resistant; dense heads and heavy stalks; develops side shoots
Johnny's, Territorial

Italian Green Sprouting (aka Calabrese): 65–80 days; open-pollinated; forms many side shoots after central harvest; prolific
Widely available

Mercedes: 58 days; hybrid; reliable for fall planting and harvest before winter
Shepherd's

Packman: Can be used as a spring or fall crop and produces vigorous uniform large heads and side spears; consistent performer
Harris Seed

Storage Requirements

Fall crops are better than summer crops for freezing.

Fresh

Temperature	Humidity	Storage Life
32°F	95%–100%	10–14 days
32°F–40°F	80%	1 month

Preserved

Method	Taste	Shelf Life
Canned	fair	12+ months
Frozen	good	12 months
Dried	fair	12+ months

Selected Varieties *(cont'd)*

Premium Crop: 60–80 days; hybrid; large heads; good for summer and fall crops; uniform; compact; holds firmness well
Widely available

Purple Sprouting: 220 days (from seed); open-pollinated; one of oldest heirlooms (pre-1835); profusion of tender, very sweet shoots; hardy to –10°F; in some areas must overwinter before it flowers in spring
Widely available

Rapini or **Broccoli Raab** or **Ruvo Kale** ***(B. rapa ruvo)***: 60 days; not a true broccoli; Italian specialty item; doesn't develop real heads; grown for leaves, ribs, and flower buds. Sow very early spring with radishes, or in fall. Sauté with garlic and olive oil, or use like kale
Nichols

Romanesco: Unusual yellow color; open-pollinated; spiraling large head; needs a long season, so start seeds early; also needs extra feeding; different strains have different maturation times for short or long seasons: make sure to order a strain that will do well in your area
Bountiful Gardens (for Northern climates), Burpee (75 days), Gurney's (105 days), Johnny's (75 days for cultivar "Minaret"), Shepherd's (100 days)

Silvia: Developed in 1989; more resistant to powdery mildew than most
Letherman's

Southern Comet: AAS Winner; large central head followed by many side shoots; for spring or fall; ready about 2 months after transplanting; good for high-density planting
Cook's, Park

Spartan Early: 47–76 days; open-pollinated; extra early; good-sized, solid center heads; doesn't go to seed; good flavor
Widely available

Waltham 29: 74 days; open-pollinated; for early fall planting; will mature in cold weather; long harvest; good for freezing
Widely available

Greenhouse

DeCicco, Green Comet, Italian Green Sprouting, Spartan Early

broccoli

brussels sprouts

Brassica oleracea
Gemmifera Group
Brassicaceae (Mustard Family)

BRUSSELS SPROUTS are an annual cool-season crop, hardy to frost and light freezes. Plan an average of two to eight plants per person. There are two basic types of Brussels sprouts varieties: (1) the dwarf (e.g., "Jade Cross"), which matures early and is winter hardy but more difficult to harvest; and (2) the taller (e.g., "Long Island Improved"), which is less hardy but easier to harvest. Brussels sprouts have shallow roots, so as they become top heavy you may need to stake them, particularly if exposed to strong winds. As with other brassicas, Brussels sprouts are susceptible to pests and diseases that must be kept under control early in the season. Row covers are one of the easiest pest controls to use, provided no pest eggs are already present. To prevent spreading soilborne diseases, don't compost brassica roots. Rotate at least on a 3-year basis, or optimally, on a 7-year basis.

This vegetable is high in calcium and iron, as well as a good source of vitamins A and C.

Temperature

For germination: 50°F–80°F
For growth: 60°F–65°F

Soil & Water Needs

pH: 6.0–7.5

Fertilizer: Heavy feeder; use compost, or 2–3 bushels of manure per 100 square feet.

Side-dressing: Apply 2 weeks after transplanting, and twice more at monthly intervals.

Water: Medium and evenly moist

Measurements

Planting Depth: 1/4"
Root Depth: 18"–36"

Height: 24"–48"
Breadth: 24"

Space between Plants:
In beds: 16"–18"
In rows: 18"–24"
Space between Rows:
24"–40"

Pests

Aphid, cabbage butterfly, cabbage looper, cabbage maggot, cutworm, flea beetle, harlequin bug, mite, root fly, slug, thrips, weevil, whitefly

Diseases

Black leg, black rot, club-root, damping off, leaf spot, rhizoctonia, yellows

Allies

Some evidence: Candytuft, clover (white), cover grass, french beans, shepherd's purse, weedy ground cover, wormseed mustard
Uncertain: Celery, chamomile, dill, garlic, mint, onion family, radish, rosemary, sage, savory, tansy, thyme, tomato, wormwood

Companions

Artichoke, beet, peas, potato, spinach

Incompatibles

Kohlrabi, all pole beans, strawberry, tomato (latter may also be an ally)

First Seed-Starting Date

Germinate	+	Transplant	+	Days Before LFD	=	Count Back from LFD
3 to 10 days	+	28 to 49 days	+	14 to 21 days	=	45 to 80 days

Last Seed-Starting Date

Germinate	+	Transplant	+	Maturity	+	SD Factor	+	Frost Tender	=	Count Back from FFD
3 to 10 days	+	21 days	+	80 to 100 days	+	14 days	+	0 days	=	118 to 145 days

Harvest

For the best sprout growth, when a node begins to bulge, remove the leaf below it. Harvest from the bottom of the stalk up. When sprouts are firm and no more than about 1" across, use a sharp knife to cut off the sprouts and remove lower leaves. Leave enough trunk so that new sprouts can grow. As the harvest slows, pinch the top of the plant to direct nutrients to the sprouts. For maximum vitamin C, harvest when the temperature is around freezing. Some say never to harvest unless you've had at least two frosts, because frost improves flavor. It has also been reported that sprouts can be harvested throughout the summer and still be tender if continuously picked when they reach the size of marbles. If you want to harvest all at once instead of continuously, cut or pinch off the stalk top 4–8 weeks before your intended harvest time. After harvest, remove the entire plant from the ground to minimize the chance of disease next season.

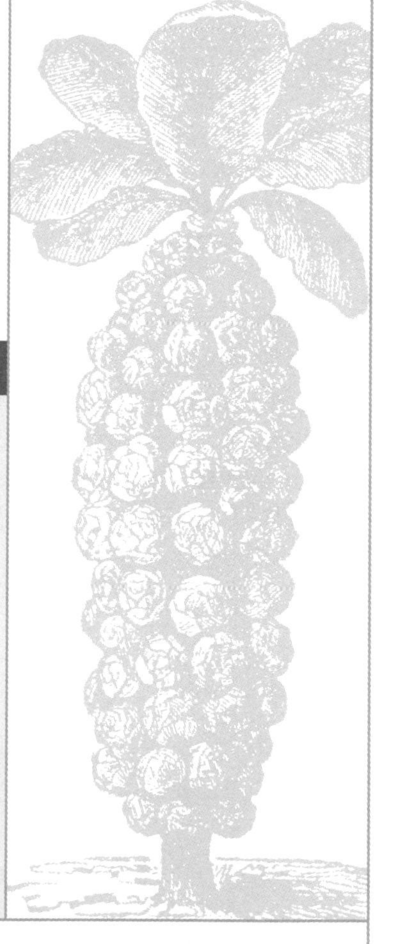

Storage Requirements

Store entire plant in a cool root cellar. Otherwise, leave the stalk in the ground and pick sprouts when ready to eat. Some report harvesting through the snow.

Fresh

Temperature	Humidity	Storage Life
32°F	95%–100%	3–5 weeks

Preserved

Method	Taste	Shelf Life
Canned	fair	12+ months
Frozen	good	12 months
Dried	poor	

Selected Varieties

Bubbles: One of easiest to grow; some tolerance to heat and drought; good disease resistance; bountiful harvest over long period of time
Vermont Bean

Certified Organic Brussels Sprouts Hybrids: Grows 3'; bears uniformly maturing crop; winter hardy to Zone 6
Cook's

Early Dwarf Danish: 95 days; early; open-pollinated; well suited for short seasons; dwarf habit aids mulching; large sprouts
Abundant Life, Garden City

Jade Cross: 80 days; dwarf hybrid; good for short season; sweet and mild (but can be bitter); heat tolerant
Widely available

Long Island Improved (Catskill): 80 days; excellent standard variety; open-pollinated; tall variety; small globe-shaped heads; tender, delicate flavor; good frozen
Nichols, Stokes

Oliver: Hybrid; earliest variety; easy to grow
Burpee, Harris, Johnny's

Ormavon: 117 days; hybrid; grows sprouts up the stem and "cabbage" greens on top; harvest tender greens after sprouts; interesting dual-purpose item
Thompson & Morgan

Peer Gynt: 140 days; dwarf; long maturation and high-quality sprouts
Thompson & Morgan

Prince Marvel: 90 days; tall plant; firm and excellent quality sprouts; tolerates bottom rot; reduced cracking
Widely available

Rubine Red: 80 days; open-pollinated; beautiful red; late variety
Widely available

Silverstar: Hybrid; good flavor; early sprouts; very hardy variety; can overwinter in mild areas or where there is good snow cover
Cook's

Valiant: 110 days; hybrid; 2' plant; cylindrical sprouts; tender and sweet, not bitter; resists rot and cracking; reliable heavy producer
Shepherd's

Greenhouse

Early Dwarf Danish, Jade Cross, Long Island Improved

cabbage

Brassica oleracea
Capitata Group
Brassicaceae (Mustard Family)

CABBAGE is an annual cool-season crop, hardy to frost and light freezes. Plan an average of three to five plants per person. Young plants may bolt if grown at 50°F for a long time; mature plants of late varieties improve flavor in cold weather. A smaller cabbage head has better flavor and can stay in the field longer without splitting. To keep them small, plant close together or, when the head is almost full, give the plant a sharp twist to sever feeder roots. After harvest, continued growth causes cabbage to need additional food reserves. Rapid growers keep poorly, as they use up their food reserves faster. Early varieties are generally the smallest, juiciest, and most tender, but they store poorly and split easily. Midseason varieties keep better in the field. Late varieties, best for sauerkraut, provide the largest and longest-keeping heads. Yellow varieties tend to be hotter than white. To prevent spreading soilborne diseases, don't compost any brassica roots; pull and destroy infected plants. Also, rotate these plants on at least a 3-year basis, or optimally on a 7-year basis.

Temperature

For germination: 45°F–95°F
For growth: 60°F–65°F

Soil & Water Needs

pH: 6.0–7.5 (7.2 deters clubroot)

Fertilizer: Heavy feeder; high N and K; may need to add lime to raise the pH to deter clubroot

Side-dressing: Every 2 weeks

Water: Heavy early and medium late in the season

Measurements

Planting Depth: ¼"–½"
Root Depth: 12"–5'

Height: 12"–15"
Breadth: 24"–40"

Space between Plants:
In beds: 15"
In rows: 18"
Space between Rows: 24"–30"

Pests

Aphid, cabbage butterfly, cabbage looper, cabbage maggot, cabbageworm, cutworm, diamondback moth, flea beetle, green worm, harlequin bug, leaf miner, mite, mole, root fly, seedcorn maggot, slug, stink bug, weevil

Diseases

Black leg, black rot, clubroot, damping off, fusarium wilt, leaf spot, pink rot, rhizoctonia, yellows

Allies

Some evidence: Candytuft, clover (red and white), shepherd's purse, wormseed mustard
Uncertain: Celery, chamomile, dill, garlic, hyssop, mint, nasturtium, onion family, radish, rosemary, sage, savory, southernwood, tansy, thyme, tomato, wormwood

Companions

Artichoke, beet, bush beans, cucumber, lettuce, peas, potato, spinach

Incompatibles

Basil, all pole beans, strawberry, tomato. (*Note:* Some evidence that tomato is also an ally [see page 422].)

First Seed-Starting Date

Germinate	+	Transplant	+	Days Before LFD	=	Count Back from LFD
7 to 12 days	+	42 days	+	14 to 21 days	=	63 to 75 days

Last Seed-Starting Date

Germinate	+	Transplant	+	Maturity	+	SD Factor	+	Frost Tender	=	Count Back from FFD
4 days	+	21 days	+	65 to 95 days	+	14 days	+	0 days	=	104 to 130 days

Harvest

For eating fresh, cut the head at ground level as soon as it feels solid. Smaller heads may grow from the remaining leaves and stems. For the best storage heads, pick when still firm and solid and before the top leaves lose green color. Pull the entire plant and roots from the ground. If left too long in the ground, the cabbage core becomes fibrous and tough, and the head may split.

Storage Requirements

Some recommend curing heads in the sun for a few days before storing for long periods. Such curing requires covering at night. Because of the strong odors emitted, store in either a well-ventilated place or a separate room reserved for brassicas. To store, trim off all loose outer leaves. Hang by its roots, or wrap individually in paper, or layer in straw in an airy bin, or place several inches apart on shelves.

Fresh

Temperature	Humidity	Storage Life
32°F–40°F	80%–90%	about 4 months

Preserved

Method	Taste	Shelf Life
Canned	poor	
Frozen	good	8 months
Dried	fair	12+ months

Selected Varieties

RED

Pierette: Hybrid; midseason; large, round heads; tolerates splitting
Stokes

Red Danish: Open-pollinated; resists thrips, cabbage looper, and moths
Stokes

Ruby Ball: Hybrid; early; red leaves retain color in cooking
Harris, Territorial

WHITE

Danish Ballhead: 100 days; open-pollinated; late keeper; good for kraut; resists thrips; sweet flavor; resists cracking; keeps through winter
Widely available

Early Jersey Wakefield: 63 days; open-pollinated; heirloom; space-saving pointed heads; good flavor; well-adapted to different soils; resists yellows
Widely available

Grenadier: 65 days; 2-pound heads; hybrid; excellent fresh eating; tender; juicy; resists cracking and long holding qualities; prefers cool weather; doesn't tolerate yellows
Shepherd's, Stokes

Lariat: 125 days (from seed); 5–8 pound heads; open-pollinated; very late season; best long-term keeper; still high quality after several months' storage
Johnny's

Selected Varieties *(cont'd)*

Market Prize: 76 days; hybrid; blue-green; good yields under a wide variety of conditions; good holding quality; resists yellows and splitting
Harris

Perfect Ball: 87 days; hybrid; midseason; sweetest and smallest core for midseasons; resists yellows; good holding quality in field; large wrapping leaves
Johnny's

Prime Time: 76 days; hybrid; midseason; especially good for salads if picked early as a loose, round head
Stokes

Savoy Ace: 85 days; AAS Winner; resists heat and frost; holds well for long harvest; resists yellows and insects
Widely available

Savoy Express 2000: AAS Winner; small heads; matures early in about half the time of other savoy types; also sweeter and more tender than other savoys
Widely available

Savoy King: 90 days; very dark green, crinkled (savoyed) leaves; high in vitamin A; semiflat heads resist yellows and insects
Widely available

Stonehead: 67 days; AAS Winner; hybrid; early; popular for firm, solid interior; resists cracking; resists yellows; good for small areas
Widely available

CHINESE

Good in soups, stir-fries, or salads

Jade Pagoda (Michili type): 68 days; hybrid; tall cylinder shape; thick, leafy, and crinkled; mild; sweet; vigorous; high yields; easy to grow; resists bolting
Harris, Park

Joi Choi: 45 days; hybrid between Pak Choi and Lei Choi; slow bolting; stands warm weather; excellent flavor
Harris, Park

Nagoda: 50 days; tolerates heat and cold; standard in stores
Johnny's

Greenhouse

All **Chinese cabbages, Market Prize**

cabbage

carrot

Daucus carota var. *sativum*
Umbelliferae or Apiaceae (Parsley Family)

CARROTS are an annual cool-season crop, half-hardy to frost and light freezes. Plan an average of ten to forty plants per person. One way to break the soil crust for carrot seeds is to plant a few fast-germinating radish seeds in the carrot bed. Carrots produce best in friable soil, so dig well before planting or grow smaller carrots that don't need deep soil. Sow seeds evenly in a very shallow furrow, about ¼-inch deep, and keep seeds moist so they will germinate. When the first leaves emerge, thin to 1 inch apart; when true leaves emerge, thin to 3 inches apart. If you delay the final thinning a bit, you can use the removed roots as baby carrots. The darkest and greenest tops indicate the largest carrots. To prevent greening at the shoulders, hill dirt up around the greens. The sweetest and best textured carrots are the Nantes types, cylindrical and blunt-tipped. The long and tapered characteristics are typical of Imperator varieties. Nantes types absorb more water and therefore have less dry matter, making them more succulent and crisp. They are also lower than other types in terpenoids, which cause a soapy turpentine-like taste; the amount of terpenoids depends entirely on variety, not on the soil. Terpenoids break down in cooking so that carrots taste sweeter when cooked. *Nantes* now describes any carrot with the above traits, not true lineage to the French region where the type originated.

Temperature

For germination: 45°F–85°F
For growth: 60°F–65°F

Soil & Water Needs

pH: 5.5–6.5

Fertilizer: Light feeder; too much top growth may mean too much N

Side-dressing: Apply 3 weeks after germination and again when 6"–8" high

Water: Medium

Measurements

Planting Depth: ¼"–½"
Root Depth: 24"–4'

Height: 12"
Breadth: 12"–24"

Space between Plants:
In beds: 2"–3"
In rows: 6"
Space between Rows:
16"–30"

Pests

Carrot rust fly, carrot weevil, cutworm, flea beetle, leafhopper, nematode, parsleyworm, slug, snail, weevil, wireworm

First Seed-Starting Date

Germinate	+	Transplant	+	Days Before LFD	=	Count Back from LFD
6 to 14 days	+	0 days	+	8 to 14 days	=	14 to 28 days

Last Seed-Starting Date

Germinate	+	Transplant	+	Maturity	+	SD Factor	+	Frost Tender	=	Count Back from FFD
6 to 14 days	+	0 days	+	65 to 70 days	+	14 days	+	0 days	=	85 to 98 days

carrot

Diseases

Alternaria leaf spot, cercospora, damping off, leaf blight, soft rot, yellows

Allies

Some evidence: Onion family
Uncertain: Black salsify, chives, coriander, flax, lettuce, peas, pennyroyal, radish, rosemary, sage, wormwood

Companions

All beans, leek, pepper, tomato

Incompatibles

Celery, dill (retards growth)

Harvest

Gently pull the roots out by their green tops. For most newer varieties, don't let carrots grow fatter than 1½" across or they'll become woody. Some older varieties can still be succulent and delicious when large.

Storage Requirements

Remove the green tops, but do not wash the carrot before storing. Store in sawdust or sand in containers.

Fresh

Temperature	Humidity	Storage Life
32°F	90%–95%	4–5 months
35°F	95%–100%	7–9 months

Preserved

Method	Taste	Shelf Life
Canned	fair	12+ months
Frozen	good	8 months
Dried	fair	12+ months

Selected Varieties

Bolero: Nantes type; bright orange; 6–7", sweet and crunchy
Widely available

Danvers Half Long: 75 days; open-pollinated; good keeper in underground storage; adapted to many soil types
Widely available

Ingot: 68 days; hybrid; Nantes type; high flavor and texture rating by Rodale
Widely available

Mokum: Hybrid; good for early spring crop under row covers
Widely available

Nantes Half Long (aka Coreless and Scarlet Nantes): 70 days; open-pollinated; sweet and tender eating carrot; coreless; standard home gardener variety
Widely available

Nantes Tip Top: 73 days; Dutch; excellent eating quality; for wide range of soils
Shepherd's

Nelson: Earliest and sweetest Nantes; many gardeners rate this as the sweetest early carrot
Widely available

Royal Chantenay: 70 days; excellent for heavy, clay, or shallow soils; popular
Widely available

Thumbelina: AAS Winner; round carrot that will produce in heavy soil
Garden City, Vermont Bean

Touchon: 75 days; French; open-pollinated; Nantes type; fine texture; sweet; flavor suffers in hot summers; good for juicing
Widely available

White Belgium: 75 days; unusual white all the way through; taste may get too strong in hot summers
Nichols

Yellowstone: Adaptable and unique; thin to 4" apart as many yellow roots can reach 12–13"; sweet and bitter-free with strong carrot flavor; alternaria resistant
Cook's

Greenhouse

Thumbelina or other shorter carrots

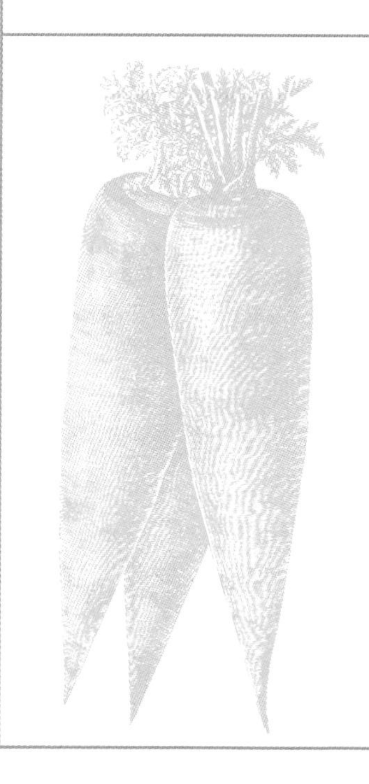

cauliflower

Brassica oleracea
Botrytis Group
Brassicaceae (Mustard Family)

CAULIFLOWER is an annual cool-season crop, half-hardy to frost and light freezes. Plan an average of three to five plants per person. Cauliflower can be difficult to grow as a spring crop because it bolts in the heat. While it is the most sensitive of all brassicas to frost, it is generally easier to grow as a fall crop. Note on the seed-starting dates that cauliflower shouldn't be transplanted outdoors until all danger of frost is past, unless covered. It also needs to mature before hot summer weather arrives. A compromise might be to choose an intermediate starting date and cover the plants when set out to protect them from cold. Purple cauliflower is an easier crop because it's more pest resistant and hardier than white varieties. Early in the season it looks and tastes more like broccoli, but after a cold snap, its flavor is more like cauliflower. Depending on your area, cauliflower might be better grown as a fall crop to reduce the threat of insect damage and bolting.

Plant transplants 1 inch deeper than they were grown in starting pots, and cover with netting to protect from pests. Spacing between plants determines head size: the closer together, the smaller the head. When heads start forming, prevent yellowing by tying several upright leaves loosely together with string, covering the rest of the head from direct exposure to the sun. To prevent spreading clubroot and other soilborne diseases, don't compost any Brassica roots. Pull and destroy all infected plants. Also, rotate Brassica plants on at least a 3-year basis, or optimally on a 7-year basis. Researchers in India report best yields were attained by spraying with a 1.2% concentration zinc sulfate in water, while highest levels of ascorbic acid (vitamin C) were attained with zinc sulfate concentrations of either 0.6% or 0.9%.

First Seed-Starting Date

If covered, transplants can be set out almost 4 weeks earlier (see below).

Germinate	+	Transplant	+	Days Before LFD	=	Count Back from LFD
4 to 10 days	+	35 to 49 days	–	14 days (no cover)	=	25 to 45 days (uncovered at transplanting)
4 to 10 days	+	35 to 49 days	+	14 days (covered)	=	53 to 73 days (covered once transplanted)

Last Seed-Starting Date

Germinate	+	Transplant	+	Maturity	+	SD Factor	+	Frost Tender	=	Count Back from FFD
4 to 10 days	+	21 days	+	50 to 95 days	+	14 days	+	N/A	=	89 to 140 days

Cauliflower

Temperature

For germination:
45°F–85°F
For growth:
Day: 60°F–70°F day
Night: 50°F–60°F

Soil & Water Needs

pH: 6.0–7.5

Fertilizer: Heavy feeder; needs high N and K

Side-dressing: Every 3–4 weeks

Water: Medium; critical early in season and during warm weather

Measurements

Planting Depth: ¼"–½"
Root Depth: 18"–4'

Height: 18"–24"
Breadth: 2' to 2½'

Space between Plants:
In beds: 12"–15"
In rows: 18"
Space between Rows:
24"–46"

Pests

Aphid, cabbage butterfly, cabbage looper, cabbage maggot, cutworm, diamondback moth, harlequin bug, mite, root fly, root maggot, slug, snail, striped flea beetle, weevil

Storage Requirements

Wrap individual plants, head and roots, in plastic. Store in a root cellar or cool place.

Fresh

Temperature	Humidity	Storage Life
32°F	95%–98%	3–4 weeks

Preserved

Method	Taste	Shelf Life
Frozen	good	12 months
Pickled	good	12 months

Diseases

Black leg, black rot, clubroot, damping off, downy mildew, leaf spot, rhizoctonia, seed rot, yellows (*Note:* Two British horticulturists have found hollow stems to be caused by large head size, not boron deficiency as commonly believed.)

Allies

Some evidence: Candytuft, corn spurry (*Spergula arvensis*), lamb's quarters, shepherd's purse, tomato, white or red clover, wormseed mustard
Uncertain: Celery, chamomile, dill, mint, nasturtium, onion, rosemary, sage, savory, thyme, wormwood

Companions

Aromatic plants, artichoke, beet, bush beans, garlic, lettuce, peas, potato, spinach

Incompatibles

Pole bean, strawberry, tomato (latter also might be an ally)

Harvest

When heads are 8"–10" in diameter, harvest by pulling the entire plant from the soil. Cauliflower heads deteriorate quickly, so check periodically and harvest when ready.

Selected Varieties

Amazing: 68 days; large head with long leaves to cover head, tolerant to heat and cold; heads hold longer and can be harvested over longer time
Widely available

Apex: Self-wrapping, self-blanching; for fall crop in Northeast start transplants inside in early June and plant mid-July
Harris

Early Purple Sicilian: Open-pollinated; purple heads, like purple broccoli
Abundant Life

Extra Early Snowball: 50 days; open-pollinated; early variety; pest resistant; resists yellows
Southern Exposure, Stokes

Graffiti: 80 days; hybrid; purple heads; great raw or cooked; unlike Violet Queen, produces a huge head; best for fall harvest
Johnny's

Purple Head: 80–85 days; open-pollinated; easier to grow than white types; turns green when cooked
Burpee, William Dam

Snowball: 68 days; self-blanching; open-pollinated; old white type
Widely available

Selected Varieties *(cont'd)*

Snow Crown: 50–53 days; hybrid; AAS Winner; early and reliable producer; vigorous; tolerates adverse weather
Widely available

Snow King: Hybrid 50 days; AAS Winner; very early; excellent flavor
Widely available

Violet Queen: 54 days; hybrid; earliest purple; considered by some as best purple variety; easy to grow; uniform
Cook's, Harris, Johnny's

White Sails: 68 days; hybrid; good for fall planting; leaves self-wrap to protect inner curds; deep large heads
Southern Exposure, Stokes

Greenhouse

Extra Early Snowball, Purple Head, Snow Crown, Snow King

celeriac

Apium graveolens var. *rapaceum*
Umbelliferae (Parsley Family)

CELERIAC is a cool-season root crop, half-hardy to frost and light freezes. Plan an average of one to five plants per person. It is a pest-, disease-, and problem-free biennial vegetable that is usually grown as an annual. In Zones 5 and north, this root can be overwintered in the ground if mulched heavily with straw to prevent frost penetration. Sow seeds thickly in pots indoors. Thin seedlings to one per pot. Set out when 2 to 2½ inches tall. Remove side shoots from the base of the plant. When transplanting, keep as much as possible of the original potting soil around the roots. After setting in the roots, reach under and gently squeeze the soil around the roots to eliminate air gaps. Too much oxygen exposure during transplanting will cause them to dry out and die. Root quality drops when watered regularly, so water deeply and less frequently. To achieve the deep fertile soil needed for celeriac, try adding 1 bushel of well-rotted manure for 20 feet of row, and spading it in 8 to 10 inches deep. Celeriac has been enjoyed in Europe for ages and only recently has begun to make a showing in the

United States. It is an unsightly root that tends to develop involutions filled with dirt, which cause wastage in food preparation. With a delicious, delicate flavor similar to that of celery heart, celeriac is worth the effort of cutting off all of the skin and the involutions. It can be boiled and added to potatoes, or eaten raw as an addition to salads. We cut it into thin strips, blanch and tenderize it for 30 minutes by adding lemon juice and a little salt, then toss and serve it with a remoulade sauce.

Measurements

Planting Depth: ⅛"
Root Depth: 18"–24"

Height: 12"
Breadth: 12"

Space between Plants:
In beds: 4"–7"
In rows: 12"
Space between Rows:
15"–26"

Pests

Aphid, celeryworm, slug, weevil, wireworm

Diseases

Septoria leaf spot

Allies

None

Temperature

For germination: 70°F
For growth: 60°F–65°F

Soil & Water Needs

pH: 6.0–6.5

Fertilizer: Heavy feeder

Water: Heavy, infrequently, but keep evenly moist, especially during hot spells

First Seed-Starting Date

Usually 8 to 12 weeks before last frost

Germinate	+	Transplant	–	Days After LFD	=	Count Back from LFD
10 to 21 days	+	46 to 60 days	–	0 to 14 days	=	56 to 67 days

Last Seed-Starting Date

Germinate	+	Transplant	+	Maturity	+	SD Factor	+	Frost Tender	=	Count Back from FFD
10 to 21 days	+	50 days	+	110 to 120 days	+	14 days	+	N/A	=	184 to 205 days

Companions

Squash, tomato

Incompatibles

None

Harvest

After the first fall frost, and when the bulbs are 2"–4", carefully dig or pull out the roots. The larger the root (more than 4"), the more it will be contaminated by involutions of the skin.

Storage Requirements

Celeriac is best stored fresh. The results of various preservation methods are unknown. Do not wash the root before storing. Rub off side shoots. Store in boxes of moist peat. When covered with a thick mulch, the roots will keep in the ground for about one month beyond the first frost.

Fresh

Temperature	Humidity	Storage Life
32°F	97%–99%	6–8 months

Selected Varieties

Alabaster: 120 days; open-pollinated; large thick roots
Burpee

Diamant: Takes 4 months from transplant; vigorous grower; will store refrigerated 3–4 months
Kitchen Garden, Cook's

Jose or Brilliant: 110 days; open-pollinated; resists pithiness and hollow heart; relatively smooth; good taste
Cook's, Johnny's

Large Smooth Prague: 110 days; open-pollinated
Widely available

President: 110 days; organic seed; elite strain from one of Europe's top seed houses
Cook's

Greenhouse

Celeriac, as a root vegetable, doesn't grow well in the greenhouse.

celery

Apium graveolens var. *dulce*
Umbelliferae or Apiaceae (Parsley Family)

CELERY is a cool-season crop, half-hardy to frost and light freezes. Plan an average of three to eight plants per person. Soak seeds overnight to help germination. The seedlings need to be transplanted at least once before setting outside. Transplant outside when seedlings are 4 to 6 inches tall and night temperatures don't fall below 40°F. Water plants before they are transplanted. There are two basic types of celery: self-blanching and blanching.

Self-blanching varieties are much easier to grow, as they can be grown in flat soil without trenches. Their harvest, however, is earlier and more limited. For celery that needs blanching, (1) plant in the center of 18-inch-wide trenches, (2) remove suckers in midseason and wrap stalk bunches with brown paper, newspaper, or cardboard to prevent soil from getting between the stalks, (3) fill the trench with soil up to the bottom of the leaves 2 months before harvest, and (4) keep mounding soil around the base of the plant every 3 weeks. Make sure the mound is sloped to help drainage.

Temperature

For germination: 60°F–70°F
For growth: 60°F–65°F, with nights higher than 40°F

Soil & Water Needs

pH: 6.0–7.0

Fertilizer: Heavy feeder; 2–3 weeks before planting, apply compost worked 12" into soil

Side-dressing: Apply every 2 weeks, especially 3 weeks after transplanting; repeat 6 weeks later

Water: Heavy

Measurements

Planting Depth: ¼"–½"
Root Depth: Shallow, upper 6"–12"

Height: 15"–18"
Breadth: 8"–12"

Space between Plants:
In beds: 6"–8"
In rows: 8"–12"
Space between Rows: 18"–36"

Pests

Aphid, cabbage looper, carrot rust fly, carrot weevil, celery leaftier, earwig, flea beetle, leafhopper, mite, nematode, parsleyworm, slug, tarnished plant bug, weevil, wireworm

Diseases

Black heart, celery mosaic, damping off, early and late blight, numerous fusariums, pink rot, yellows

Allies

Uncertain: Black salsify, cabbage, chive, coriander, garlic, nasturtium, pennyroyal

Companions

All beans, all brassicas, spinach, squash, tomato

Incompatibles

Carrot, parsnip

First Seed-Starting Date

Germinate	+	Transplant	–	Days After LFD	=	Count Back from LFD
5 to 7 days	+	70 to 84 days	–	14 to 21 days	=	61 to 70 days

Last Seed-Starting Date

Germinate	+	Transplant	+	Maturity	+	SD Factor	+	Frost Tender	=	Count Back from FFD
6 days	+	30 days	+	90 to 135 days	+	14 days	+	0 days	=	140 to 185 days

celery

Other

Cucumbers provide shade and moisture to other plants such as celery

Harvest

Harvest self-blanching celery before the first frost. Harvest blanched varieties after the first frost. Dig out each plant whenever needed.

Selected Varieties

All varieties listed are open-pollinated.

BLANCHING

Red: Unusual red-bronze stalks; stay red when cooked; hardier than green varieties
Seeds of Change

Summit: 100 days; very dark green; tolerates fusarium wilt; compact; 10"
Stokes

Storage Requirements

Celery is best stored at cold temperatures in a perforated plastic bag. To refresh wilted stalks, simply place them in a tall glass of cold water.

Fresh

Temperature	Humidity	Storage Life
32°F	98%–100%	2–3 months
32°F	80%–90%	4–5 weeks

Preserved

Method	Taste	Shelf Life
Canned	fair	12+ months
Frozen	good	5 months
Dried	good	12+ months

Selected Varieties *(cont'd)*

Tall Utah 52-70R Improved: 98 days; long, dark green, thick stalks; high yields; resists boron deficiency; high quality
Widely available

Tendercrisp: 105 days; early; high yields; more upright than most; Pascal type; ribless; pale green stalks
Field's, Baker's Creek

Ventura: 100 days; tall Utah type; some tolerance of fusarium wilt
Johnny's, Garden City

SELF-BLANCHING

Golden Self-Blanching: 115 days; heirloom; early; crispy; delicate flavor; dwarf
Widely available

SEASONING OR CUTTING CELERY

This has the same general culture as celery but is generally grown only for its leaves. It is easier to grow than other celery due to fewer pest and disease problems and minimal requirements for water and feeding. Cutting celery also can be harvested over a longer period of time.

Amsterdam Fine Seasoning Celery: 12"–18" tall; leaves like a shiny flat-leaved parsley; easy to dry and retains flavor well
Shepherd's

Greenhouse

Try the self-blanching varieties.

sweet corn

Zea Mays var. *rugosa*
Gramineae or Poaceae (Grass Family)

CORN is a warm-season crop, tender to frost and light freezes. Plan an average of twelve to forty plants per person, depending on your needs. For good pollination and full ears, plant in blocks of at least four to six rows and about 15 inches on center. If birds are a problem in your garden, stealing seeds or eating seedlings, cover your corn patch with a floating row cover immediately after planting seeds. Corn easily cross-pollinates, so isolate popcorn and field corn from sweet corn by at least 50 to 100 feet, or plant varieties that have different pollination times, which is when tassels appear. For seed saving, isolate corns by 1000 feet for absolute purity. Once pollinated, corn matures rapidly, usually 15 to 20 days after the first silks appear. Corn has shallow roots, so mulch heavily and avoid cultivating deeper than 1½ inches. In small patches, don't remove suckers; they may bear corn if well side-dressed. White and yellow corn vary in nutrition; white corn contains twice as much potassium, and yellow corn contains about 60 percent more sodium.

Temperature

For germination: 60°F–95°F
For growth: 60°F–75°F

Soil & Water Needs

pH: 5.5–7.0

Fertilizer: Heavy feeder; apply manure in the fall, or compost a few weeks before planting

Side-dressing: Apply every 2 weeks and additionally when stalks are 8"–10" and knee high

Water: Medium; provide more when the stalks flower

Measurements

Planting Depth: 1"–2" sh_2 types, ½"
Root Depth: 18"–6'

Height: 7'–8' (but some flour and field corns grow much taller)
Breadth: 18"–4'

Space between Plants:
In beds: 8"–12"
In rows: 18"
Space between Rows: 30"–42"

Pests

Aphid, birds, corn borer, corn earworm, corn maggot, corn rootworm, cucumber beetle, cutworm, earwig, flea beetle, garden webworm, Japanese beetle, June beetle, leafhopper, sap beetle, seedcorn maggot, thrips, webworm, white grub, wireworm

Diseases

Bacterial wilt, mosaic viruses, rust, smut, southern corn leaf blight, Stewart's wilt

Allies

Some evidence: All beans, chickweed, clover, giant ragweed, peanut, pigweed, shepherd's purse, soybean, sweet potato
Uncertain: Alfalfa, goldenrod, odorless marigold, peas, potato, white geranium

First Seed-Starting Date

Sow every 10 to 14 days for continuous harvest.

Germinate	+	Transplant	–	Days After LFD	=	Count Back from LFD
4 to 21 days	+	0 (direct)	–	0 to 10 days	=	4 to 11 days

Last Seed-Starting Date

Germinate	+	Transplant	+	Maturity	+	SD Factor	+	Frost Tender	=	Count Back from FFD
4 days	+	0 days	+	65 to 95 days	+	14 days	+	14 days	=	97 to 127 days

Companions

Cucumber, melon, pumpkin, squash

Incompatibles

Tomato (attacked by some similar insects)

Harvest

Sweet Corn: About 18 days after silks appear, when they're dark and dry, make a small slit in the husk (don't pull the silks down), and pierce the kernel with a fingernail. If the liquid is (1) clear, wait a few days to pick, (2) milky, pick and eat, or (3) pasty, the ear is past its prime and is best for canning.

Popcorn (var. *praecox*): Pick when the husks are brown and partly dried. Finish drying corn on the husks. A solar drier is the most rapid method, drying the corn in about 5 days. The kernels are ready for storage if they fall off easily when rubbed by a thumb or twisted. Before using, store in bags or jars to even out the moisture content. The ultimate test, of course, is to pop them.
After harvest: Cut stalks and till under or compost immediately.

Storage Requirements

Corn is best eaten immediately. Some gardeners won't even go out to pick ears until the cooking water is already boiling.

Fresh

Temperature	Humidity	Storage Life
40°F–45°F	80%–95%	4–10 days

Preserved

Method	Taste	Shelf Life
Canned	excellent	12+ months
Frozen	good	8 months
Dried	good	12+ months

Selected Varieties

OP = open-pollinated
HL = heirloom

FLOUR VARIETIES

These have soft starch, are easiest to grind, and make flour not meal.

Hopi Blue: OP; HL; drought tolerant; vigorous; colors vary from black to purple; makes pretty blue flour
Abundant Life, Baker's Creed, Native Seed (their strain good for high and low desert). *Note:* Most other sources of Hopi Blue are flint varieties.

White Posole: OP; makes white flour
Plants of the Southwest

DENT OR FIELD VARIETIES

These have hard starch on the side and soft starch on the top of the kernel, are easier to grind than flint types, and are good for fresh eating, roasting, and grinding; versatile.

Bloody Butcher: 110–120 days; 10'–12'; OP; pretty red kernels sometimes interspersed with other colors
Heirloom Seeds, Southern Exposure

Mayo Tuxpeño: 90 days; OP; HL; 10'–12'; with yellow, blue and yellow or pink kernels; short flat ears; staple of region; for low desert
Native Seed

Northstine Dent: 105 days; 7'; OP; HL; excellent for cornmeal and cereal; yellow; good for short seasons; high quality
Johnny's

Reids Yellow Dent: 85–110 days; 7'; OP; HL; very hardy; good in southern heat; high yields; good for tortillas, meal, and hominy
Field's, Southern Exposure

FLINT VARIETIES

These have very hard starch and make cornmeal. Generally, they have good insect resistance and store well.

Longfellow: 117 days; OP; orange kernels; makes a sweet cornbread; good in Northern gardens
Johnny's

Tarahumara Apachito: OP; HL; favorite grown by Tarahumara; kernels light pink to dark rose and pearly; low desert
Native Seed

SWEET VARIETIES

Supersweet corn has the Shrunken-2 (sh_2) gene, which causes shriveled seeds with weak seed coats, resulting in poor germination. Supersweets are often twice as sweet as eh *types. They develop a watery texture when frozen.*

Sugar Enhanced corn has the se *gene, which causes higher sugar levels than* eh *types and also longer retention of sugar*

Selected Varieties *(cont'd)*

and tenderness, for up to 10–14 days.

Everlasting heritage corn has the eh *gene designed for sweetness and texture.*

Cocopah: Sweet white kernels; fast growing; grown by Colorado River tribes
Native Seeds

Country Gentleman or Shoepeg: 96 days; OP; HL; white; tender; resists drought and wilt; kernels are not in rows
Widely available

Honey and Cream: 78 days; *eh* type; white and yellow; tight husks
Widely available

Improved Golden Bantam: 82 days; OP; yellow; midseason; flavorful
Abundant Life, Nichols

Iochief: 86 days; AAS Winner; 7½'; hybrid; yellow; top rated by *Organic Gardening* readers for reliability, productivity, drought tolerance, and taste
Field's, Gurney's

Kandy Korn: 84 days; *eh* type; yellow; good flavor; doesn't require isolation; stays sweet several days after harvest; pretty purplish stalks and husks
Widely available

Northern Extra Sweet: Earliest yellow; high yields and cold tolerant
Johnny's

Polar Vee: 53 days; yellow; produces well in cool regions
Field's, Gurney's, Stokes

Silver Queen: 92 days; standard hybrid; white; late season; top rated for flavor, productivity, drought tolerance, and disease resistance; tolerates Stewart's wilt and leaf blight
Widely available

POPCORN

These have the hardest starch of all corns.

Cherokee Long Ear: Brightly colored long ears
Baker's Creek

Strawberry: 105 days; 4'; HL; OP; stalks bear several ears; red kernels
Widely available

Tom Thumb: 85 days; dwarf 3½'; yellow; can grow closely spaced
Johnny's

Notes

Studies show hybrid varieties have greater ear and kernel growth rates, bigger kernel size, a longer period of kernel-filling, as well as greater redistribution of stalk-stored food to the kernel.

cucumber

Cucumis sativus
Cucurbitaceae (Gourd Family)

CUCUMBERS are a warm-season crop, very tender to frost and light freezing. Plan an average of three to five plants per person, depending on your pickling needs. In warm climates, some recommend planting cucumbers in hills spaced 3 to 5 feet apart, with six to eight seeds per hill. In cooler climates, transplant seedlings on a cloudy day or in the afternoon to minimize transplanting shock.

Most cucumbers are *monoecious* and produce both male and female (fruit-bearing) flowers. *Gynoecious* types bear only female flowers, therefore a few male-flowering pollinators are included in the seed packets. Both require pollination by bees. *Parthenocarpic* types produce few if any seeds and require no pollination, so they can be grown to maturity under row covers. *Bitter-free* are resistant to damage from cucumber beetles. *Dwarfs* are good candidates for intercropping with tomatoes and peppers, but they must have a constant water supply. All types must also be picked daily in warm weather.

Temperature

For germination: 60°F–95°F
For growth: 65°F–75°F

Soil & Water Needs

pH: 5.5–7.0

Fertilizer: Heavy feeder; before planting apply compost

Side-dressing: Every 2–3 weeks

Water: Heavy during fruiting; all other times average and evenly moist; deep watering.

Measurements

Planting Depth: ½"–1"
Root Depth: 12", tap root to 2'–3'

Height: 6'
Breadth:
Trellis: 12"–15"
On ground: 12–20 square feet

Space between Plants:
In beds: 12"
In rows: 24"–48"
Space between Rows: 4'

Support Structures

Use a 6' post, A-frame, tepee (three poles tied together at the top), or a trellis.

Pests

Aphid, cucumber beetle, cutworm (seedlings), flea beetle, garden centipede, mite, pickleworm, root-knot nematode, slug, snail, squash bug, squash vine borer

Diseases

Alternaria leaf spot, anthracnose, bacterial wilt, belly rot, cottony leak, cucumber wilt, downy mildew, leaf spot, mosaic, powdery mildew, scab

Allies

Some evidence: Broccoli, corn
Uncertain: Catnip, goldenrod, marigold, nasturtium, onion, oregano, radish, rue, tansy

Companions

All beans, cabbage, eggplant, kale, melon, peas, sunflower, tomato

First Seed-Starting Date

Germinate	+	Transplant	–	Days After LFD	=	Count Back from LFD
6 to 10 days	+	28 days	–	14 to 21 days	=	17 to 21 days

Last Seed-Starting Date

Germinate	+	Transplant	+	Maturity	+	SD Factor	+	Frost Tender	=	Count Back from FFD
6 to 10 days	+	28 days	+	22 to 52 days	+	14 days	+	14 days	=	84 to 118 days

Incompatibles

Anise, basil, marjoram, potato, quack grass, rosemary, sage, strong herbs, summer savory

Other

Radish can be used as a trap plant; celery is companion plant under the cucumber's A-frame

Harvest

Allow the main stem to grow as high as possible by pinching back some of the lateral shoots and letting others grow into branches. By picking the fruit early you won't have to support heavy fruit or risk arresting plant production.

When the fruit is slightly immature, before seed coats become hard, pick with 1" of stem to minimize water loss. In warm weather, all cucumber plant types should be picked daily. Always pick open-pollinated varieties underripe. Harvest pickling cucumbers at 2"–6" and slicing cucumbers at 6"–10".

Storage Requirements

Keep the short piece of stem on each fruit during storage.

Fresh

Temperature	Humidity	Storage Life
45°F–55°F	85%–95%	10–14 days

Preserved

Method	Taste	Shelf Life
Canned	good (as pickles)	12+ months
Frozen	poor	
Dried	poor	

Selected Varieties

CMV = cucumber mosaic virus
DM = downy mildew
PM = powdery mildew

SALAD/SLICING

Bush Champion: 55 days; short compact vines; 9"–11" fruit; long-season producer in all weather conditions; good for containers and greenhouse
Burpee, Thompson & Morgan

Green Knight Hybrid: 60 days; burpless; heat-resistant; vigorous; thin-skinned
Burpee

Marketmore 80: 68 days; bred for resistance to CMV, DM, PM, and scab; straight slicer, bitter-free
Widely available

Salad Bush: 57 days; AAS Winner; resists PM, DM, CMV, leaf spot, and scab; small 24" vines; 8" fruit; monoecious
Widely available

Selected Varieties *(cont'd)*

Spacemaster: 60 days; open-pollinated; best dwarf; tolerates CMV; resists scab; 7½"; white spine; plant early
Widely available

Surecrop: 60 days; hybrid; AAS Winner; early; 8" long; weather resistant; high yields; good taste
Field's, Gurney's, Jung

Sweet Success: 58 days; hybrid; AAS Winner; early; 12"–14"; burpless; tolerates CMV; resists scab and leafspot; parthenocarpic; provides seedless fruit
Widely available

PICKLERS

Cornichon Vert de Massy: 53 days; very productive; French type; recipe included
Nichols

Edmonson: Open-pollinated; heat and disease resistant
Southern Exposure

Pickalot: 54 days; hybrid; bush; tolerates PM; gynoecious; continuous bearer
Burpee

NOVELTIES

Armenian Long (aka Yard Long): 65 days; open-pollinated; long; good slicer; mild; sweet; bitter-free
Abundant Life, Burpee, Nichols, Shepherd's

China: 75 days; open-pollinated; oriental type; 12"–15"; tolerates CMV; crisp; firm; mild; weather tolerant
Stokes

Crystal Apple: 65 days; open-pollinated; nonbitter skins; good slicing and pickling; 2"–3" globes; creamy white skin and spine
Bountiful Gardens, Thompson & Morgan

Lemon: 60 days; open-pollinated; sweet; size and color of lemon
Widely available

White Wonder: 58 days; open-pollinated; early bearing; all white; heirloom; high yields; crisp; resists fusarium wilt
Garden City, Heirloom Seeds

Yamato: 60 days; oriental type; 1–1½" diameter and 12–16" long; very hardy and reliable in hot humid southeast; flesh is sweet, crisp and succulent
Southern Exposure

Greenhouse

Bush Champion, Holland House (seedless if grown indoors), Pot Luck, Salad Bush, Spacemaster, Sweet Success. See Stokes and Thompson & Morgan catalogs for other varieties specifically for the greenhouse.

eggplant

Solanum melongena var. *esculentum*
Solanaceae (Nightshade Family)

EGGPLANT is a warm-season crop, extremely tender to frost and light freezes. Plan an average of two plants per three people. All parts of the plant except the fruit are poisonous. For indoor seed starting in flats, block out the plants once the seedlings are well-established, by running a knife through the soil midway between the plants, cutting the roots, and leaving each plant with its own soil block. If you start seeds in individual pots, this procedure is unnecessary. Make sure outdoor soil temperature is at least 55 to 60°F before transplanting; otherwise they become stunted, turn yellow, and are slow to bear. Difficulties growing eggplant are often related to cool conditions. Plant them in the hottest, sunniest spot available and cover with plastic jugs (bottom cut out, cap off) until leaves poke through the top. As frost approaches, pinch back new blossoms so that plant nutrients are channeled into the remaining fruits. Eggplant is a versatile fruit often used in Italian dishes such as ratatouille, caponata, and lasagna. It easily absorbs the flavors of whatever sauce it is cooked in.

Temperature

For germination: 75°F–90°F
For growth: 70°F–85°F

Soil & Water Needs

pH: 5.5–6.5

Fertilizer: Heavy feeder; apply manure water or tea every 2 weeks

Side-dressing: Apply after the first fruit appears

Water: Heavy

Measurements

Planting Depth: ¼"
Root Depth: 4'–7'

Height: 24"–30"
Breadth: 3'–4'

Space between Plants:
In beds: 18"
In rows: 18"–30"
Space between Rows:
24"–48"

Pests

Aphid, Colorado potato beetle, cucumber beetle, cutworm, flea beetle, harlequin bug, lace bug, leafhopper, mite, nematode, tomato hornworm, whitefly

Diseases

Anthracnose, bacterial wilt, botrytis fruit rot, phomopsis blight, tobacco mosaic, verticillium wilt

Allies

Some evidence: None
Uncertain: Coriander, goldenrod, green beans, marigold, potato

Companions

All beans, pepper

Incompatibles

None

Other

Potato can be used as a trap plant for eggplant pests

Harvest

Pick when the fruit is no more than 3"–5" long or 4" in diameter, and before the skin loses its luster. Cut the fruit with a small amount of stem. Fruit seeds should be light-colored. Brown seeds indicate the fruit has ripened too long. Eggplant vines are spiny, so be careful to avoid pricking yourself.

First Seed-Starting Date

Germinate	+	Transplant	–	Days After LFD	=	Count Back from LFD
5 to 13 days	+	42 to 56 days	–	14 to 28 days	=	33 to 55 days

Last Seed-Starting Date

Germinate	+	Transplant	+	Maturity	+	SD Factor	+	Frost Tender	=	Count Back from FFD
5 to 13 days	+	42 to 56 days	+	50 to 80 days	+	14 days	+	14 days	=	125 to 177 days

Selected Varieties

Black Beauty: 73–80 days; open-pollinated; almost round fruit; heavy yields; reliable; a common standard-size fruit
Widely available

Burpee Hybrid: 70 days; drought resistant; tall; semispreading; vigorous
Burpee

Dusky: 60 days; popular early variety; pear-shaped; glossy; tolerates tobacco mosaic virus; good for Northern gardens
Widely available

Ghostbuster: 80 days; open-pollinated; introduced in 1989; sweet; excellent for cooking
Harris

Ichiban: 61 days; open-pollinated; slender 12" fruit; 36" plants; an improved variation of this is "Tycoon"
Gurney's, Park; "Tycoon" available through Nichols, Garden City

Little Fingers: 68 days; long and very slim, oriental-type fruit; easier to pick because spineless
Harris, Shepherd's

Orient Express: 58 days; hybrid; slender 8"–10" fruit; excellent early yields; sets fruit in cool weather; also tolerates heat
Johnny's

Osteric: Hybrid; white eggplant; productive; small oval-shaped fruits; good container variety; good for grilling or slicing
Cook's

Rosa Bianca: Open-pollinated; unusual beautiful bright lavender color; high yields; vigorous; mild flavor
Shepherd's

Violette di Firenze: Open-pollinated; oblong-round fruit is rich lavender; beautiful
Cook's

Storage Requirements

Keep a small piece of stem on the eggplant during storage, and don't pierce the skin. Eggplant is best used fresh.

Fresh

Temperature	Humidity	Storage Life
46°F–54°F	90%–95%	1 week
32°F–40°F	80%–90%	6 months

Preserved

Method	Taste	Shelf Life
Canned	fair	12+ months
Frozen	fair	8 months
Dried	fair	12+ months

Greenhouse

Black Beauty, Ichiban

Kale

Brassica oleracea var. *acephela*
Brassicaceae (Mustard Family)

KALE is a cool-season crop, hardy to frost and light freezes. Plan an average of four plants per person. Kale's flavor is reputed to improve and sweeten with frost. An easy vegetable to grow, it is generally more disease and pest resistant than other brassicas, although it can occasionally experience similar problems. Kale also uses less space than other brassicas. Use it as a spinach substitute in a wide variety of dishes. Kale maintains body and crunch much better than spinach and so can be used in dishes where spinach might not be suitable; it's especially delicious in stir-fry dishes. Author and edible land scaper Rosalind Creasy recommends cooking kale over high heat to bring out the best flavor and prevent bitterness. Creasy also noted that "many specialty growers are planting kale in wide beds only ½ to 12 inches apart and harvesting kale small as salad greens." In England, close plantings of kale have been shown to prevent aphid infestations through visual masking. Although kale is usually disease and pest free, some gardeners won't compost any brassica roots to prevent

lettuce

Lactuca sativa
Compositae or Asteraceae (Sunflower Family)

LETTUCE is a cool-season crop, half-hardy to frost and light freezes. Plan an average of ten to twelve plants per person. Closer spacing results in smaller heads, which may be preferable for small families. Specialty growers are spacing lettuce very close for selling baby lettuces, a rapidly growing produce market.

There are two basic categories of lettuce: heading and nonheading. Head lettuces include crisphead (e.g., iceberg) and butterhead (e.g., Bibb and Boston). Nonhead lettuces include leaf and romaine (also known as *cos*). For head lettuce, one source suggests that you strip transplants of outer leaves to help the inner leaves "head up" better. This is not tested, so treat it as experimental. Head lettuces tend to be milder in flavor but are more difficult to grow. Lettuce doesn't do well in very acidic soils, and some say the pH shouldn't be lower than 6.5. During hot weather, sow lettuce in partial shade, as it doesn't do well in the heat, and use heat-resistant varieties.

Temperature

For germination: 40°F–80°F
For growth: 60°F–65°F

Soil & Water Needs

pH: 6.0–7.5

Fertilizer: Heavy feeder

Side-dressing: Every 2 weeks, apply balanced fertilizer or foliar spray

Water: Low to medium; heavy in arid climates; water early in the morning to minimize diseases

Measurements

Planting Depth: ¼"–½"
Root Depth: 18"–36", with a taproot to 5'

Height: 6"–12"
Breadth: 6"–12"

Space between Plants:
In beds:
Head: 10"–12"
Leaf: 6"–8"
Romaine: 10"
In rows: 12"–14"
Space between Rows: 14"

Pests

Aphid, beet leafhopper, cabbage looper, cutworm, earwig, flea beetle, garden centipede, leaf miner, millipede, slug, snail

Diseases

Bacterial soft rot, botrytis rot, damping off, downy mildew, fusarium wilt, lettuce drop, mosaic, pink rot, powdery mildew, tip burn

Allies

Uncertain: Chive, garlic, radish

Companions

Beet (to head lettuce), all brassicas (except broccoli; see Incompatibles below), carrot, cucumber, onion family, pole lima bean, strawberry

First Seed-Starting Date

Germinate	+	Transplant	+	Days Before LFD	=	Count Back from LFD
4 to 10 days	+	14 days (leaf and head)	+	7 to 28 days	=	25 to 46 days

Last Seed-Starting Date

Germinate	+	Transplant	+	Maturity	+	SD Factor	+	Frost Tender	=	Days Back from FFD
4 days	+	14 days	+	60 to 95 days	+	14 days	+	0 days	=	92 to 127 days (head)
4 days	+	14 days	+	45 to 65 days	+	14 days	+	0 days	=	77 to 97 days (leaf)
4 days	+	14 days	+	55 to 80 days	+	14 days	+	0 days	=	87 to 112 days (romaine)

Incompatibles

None; some studies have shown lettuce to be sensitive to plant residues of barley broccoli, broad bean, vetch, wheat, and rye

Harvest

For leaf lettuce, start picking the leaves when there are at least five to six mature leaves of usable size. Usable size means about 2" long for baby lettuce and 5"–6" long for more mature lettuce. Keep picking until a seed stalk appears or the leaves become bitter. For head lettuce, when the head feels firm and mature simply cut it off at the soil surface. Harvest all lettuce in early morning for the maximum carotene and best taste. Refrigerate immediately.

Storage Requirements

Lettuce doesn't store well for long periods and is best eaten fresh.

Fresh

Temperature	Humidity	Storage Life
32°F–40°F	80%–90%	1 month
32°F	98%–100%	2–3 weeks

Selected Varieties

All varieties listed are open-pollinated.

HEAD LETTUCE (Butterhead and Iceberg types)

Note: Butterheads are more resistant to leaf aphids, according to Dutch scientists.

Bibb or Limestone: 55 days; early spring; favored for its delicacy
Cook's, Nichols, Park

Buttercrunch: 50 days; matures midsummer; Bibb; favorite with home gardeners; slow-bolting; tolerates heat; very tender
Widely available

Four Seasons (Merveille des Quatre Saisons): 60 days; spring Bibb; also for late summer and fall; beautiful red outer leaves; pink and cream interior
Cook's, Shepherd's

Ithaca: 72 days; spring iceberg; popular; reliable; the only iceberg good in Northeast; also good in warm climates; resists tip burn in hot weather; flood to avoid "slime"
Cook's, Field's, Garden City, Harris, William Dam

Red Deer Tongue: 60–65 days; heirloom; sweet and crisp tender rosettes with red-tinged, tongue-like leaves; slow to bolt
Baker Creek, Seeds of Change, Southern Seed

Reine des Glaces: 60–75 days; also called Ice Queen; best results from spring plantings but grows in cold when others have stopped; beautiful green heirloom with spiky outer leaves
Cook's, Seeds of Change

Selected Varieties *(cont'd)*

Sangria: 49 days; developed in 1990; beautiful red butterhead; resists bolting and tip burn; superior flavor
Johnny's

Tom Thumb: 65 days; spring; tiny solid butterhead; good for containers
Widely available

NONHEADING, LOOSE-LEAF, OR CUTTING LETTUCE (Romaine, Cos, and other types)

Green Ice: 45 days; spring; very sweet; very crinkled leaves; slow to bolt
Burpee, Park

Little Gem or Sugar Cos: 60 days for baby heads; 80 days to maturity; romaine; 5"–6"; trouble-free
Burpee, Cook's, Shepherd's, Territorial, Thompson & Morgan

Lolla Rossa: 56 days; spring; beautiful crinkly leaves with red margins; mild; keep cut to avoid bolting and bitterness
Widely available

Parris Island Cos: 68–76 days; all seasons; romaine; mild sweet flavor; slow bolting; 8"–10" heads; resists tip burn and mosaic; vigorous
Burpee, Garden City, Harris, Johnny's, Stokes

Red Sails: 45–50 days; spring; AAS Winner; dark red leaf lettuce; resists bolting; little bitterness; fast growing; easy to grow
Widely available

Winter Density: 60 days; winter or year-round in mild climates; romaine like a tall buttercrunch
Cook's, Johnny's

Greenhouse

Parris Island Cos, Tom Thumb
See also Cook's and Johnny's for special greenhouse varieties.

lettuce

muskmelon

Cucumus melo
Reticulatus Group
Cucurbitaceae (Gourd Family)

MELONS are a warm-season crop, very tender to frost and light freezes. Plan an average of two to six plants per person. Muskmelons are often called cantaloupes, but they're not the same botanical variety. True cantaloupes are rarely grown in North America. Winter melons (Inodorus group) include honeydew and casaba. Like all cucurbits, melons need bees for pollination. Melons can be sown directly outside, but some gardeners report better germination with presprouted seeds. If you start melons indoors, use individual cells or peat pots, not flats, as the roots are too succulent to divide. When you direct sow, plant four to five seeds in a hill and then thin to appropriate spacing, depending on whether you train them on a trellis or let them spread on the ground. For direct sowing and transplants, cover seedlings with cloches or hot caps to protect from frost, speed growth, and keep out pests. To encourage side shoots, when seedlings have three leaves pinch out the growing end. When new side shoots have three leaves, pinch out the central growing area again. When fruits begin to form, pinch back the vine to two leaves beyond the fruit. Make sure fruits on a trellis are supported by netting or pantyhose, and fruits on ground vines are elevated by empty pots to prevent disease and encourage ripening. Troughs near the plants can be flooded for effective watering. Melon rinds are good for compost; they decompose rapidly and are high in phosphorus and potassium.

Temperature

For germination: 75°F–95°F
For growth: 65°F–75°F

Soil & Water Needs

pH: 6.0–6.5

Fertilizer: Heavy feeder; before planting, work in compost or rotted manure.

Side-dressing: Apply balanced fertilizer or compost when vines are 12"–18" long and again when fruits form.

Water: Medium. Apply deep watering; withhold water when fruits begin to ripen

Measurements

Planting Depth: ½"
Root Depth: Shallow; some to 4'

First Seed-Starting Date

Germinate	+	Transplant	–	Days After LFD	=	Count Back from LFD
4 to 10 days	+	21 to 28 days	–	14 to 21 days	=	18 days

Last Seed-Starting Date

Germinate	+	Transplant	+	Maturity	+	SD Factor	+	Frost Tender	=	Count Back from FFD
4 days	+	21 days	+	59 to 91 days	+	14 days	+	14 days	=	112 to 151 days

Height: 24"
Breadth:
Bush: 36"–48"
Vine: 30–40 square feet on ground

Space between Plants:
In beds: 2'
In rows: 4'–8'
Space between Rows: 5'–7'

Support Structures

Use an A-frame or trellis to grow vines vertically.

Pests

Aphid, cucumber beetle, cutworm, flea beetle, mite, pickleworm, slug, snail, squash bug, vine borer, whitefly

Diseases

Alternaria leaf spot, anthracnose, bacterial wilt, cucumber wilt, curly top, downy mildew, fusarium wilt, mosaic, powdery mildew, scab

Allies

Uncertain: Chamomile, corn, goldenrod, nasturtium, onion, savory

Companions

Pumpkin, radish, squash

Incompatibles

None

Storage Requirements

Store fruits in a cool area.

Fresh

Temperature	Humidity	Storage Life
35°F–55°F	80%–90%	about 1 month

Preserved

Method	Taste	Shelf Life
Frozen	good	3 months

Other

Morning glories, radish, and zucchini are succession trap plants for cucumber beetles.

Harvest

Melon is ready for harvest as soon as it is at "full slip," the ends are soft (i.e., separate easily from the stem), a crack develops around the stem, and it smells "musky." The skin netting should be cordlike, grayish, and prominent. Winter melons don't "slip" but should be soft. Dip muskmelons in hot water (136°–140°F) for 3 minutes to prevent surface mold and decay during storage. Store in polyethylene bags to reduce water loss and associated softening of the flesh.

Selected Varieties

OP = open-pollinated
HL = heirloom

MUSKMELON AND "CANTALOUPE"

Ambrosia Hybrid: Very popular; juicy; sweet; firm salmon flesh; some resistance to downy and powdery mildews
Burpee, Park, Stokes

Charmel: Early ripening; gourmet French Charantais type; deep orange flesh; resists powdery mildew and cucumber beetles; high yields
Nichols

Edisto 47: 88 days; OP; excellent disease resistance; said to exceed disease resistance of many hybrids; resists or tolerates alternaria leaf spot, powdery mildew, and downy mildew
Southern Exposure, Baker's Creek

Goldstar Hybrid: 87 days; early and long bearing; deep orange flesh; high yields; excellent quality; a standard for mid-Atlantic states
Harris

Green Nutmeg (or Extra Early Nutmeg): 84 days; OP; HL; green flesh with salmon center; small fruits; unusual
Widely available

Honeyloupe: 75 days; unusual cross of honeydew and cantaloupe; skin is smooth and creamy white, interior is salmon; resists verticillium wilt
Stokes

Jenny Lind: OP; HL; green flesh; excellent flavor; rediscovered and offered in 1987
Widely available

Minnesota Midget: OP; uses only 3' of space; earliest maturing melon; 4" fruit; orange flesh; sweet; spicy
Widely available

Old Time Tennessee: 100 days; OP; HL; elliptical fruit 12"–16" long; flavor is outstanding if picked at peak; salmon flesh
Southern Exposure, Baker's Creek

Sugar Nut: 2 lb melons with yellow skin and green flesh; very sweet and juicy
Cook's

HONEYDEW (*C. melo,* Inodorous group)

Melon de Castillo: Sweet with pale yellow, smooth skin; for high and low desert; favorite of Native Seed staff
Native Seed

Sweet 'n' Early: 75 days; best tasting early melon; good for North
Burpee

Venus Hybrid: 88 days; green flesh; medium fruit; very juicy; sweet; aromatic
Burpee

Selected Varieties *(cont'd)*

CRENSHAW

Crenshaw: OP; very sweet; perishable; needs a long season
Heirloom Seeds, Seeds of Change

Early Crenshaw Hybrid: 90 days; early enough to be grown in the North; very sweet; yellow-green flesh; each up to 12–14 pounds
Burpee

WATERMELON *(Citrullus lanatus)*

These require more space, usually 8'×8' minimum. They should be harvested when tendrils shrivel and brown, the bottom turns green, and the skin hardens.

Charleston Gray: 85 days; OP; resists anthracnose, fusarium wilt, and sunburn; red and crisp flesh; 24" long; 28–35 pounds; good in the North or South
Widely available

Moon and Stars: 100 days; from Russia; this heirloom is a green melon with yellow moons and tiny yellow stars; 25 pound fruit is sweet and best for longer hot summer areas
Shepard's

Sugar Baby: 75 days; OP; very popular; one of the easiest watermelons to grow; small and sweet; 8" diameter; thin and hard rinds; 8–12 pounds
Widely available

Sugar Baby Bush: 80–85 days; space-saver; bush-type vines grow to 3½'; bears 2 oval-round fruits; 8"–10"; 8–12 pounds
Burpee, Gurney's; see also Park's disease-resistant Bush Baby II

Yellow Doll or Yellow Baby Hybrid: 65–70 days; AAS Winner; one of earliest watermelons; juicy yellow flesh; sweet; 7" diameter; grows to 10 pounds
Widely available

Greenhouse

Minnesota Midget

muskmelon

onion

Allium cepa
Liliaceae (Lily Family)

ONIONS are a cool-season crop, hardy to frost and light freezes, although certain varieties are exceptions. Plan an average of forty plants per person. Onions are actually easy to grow, although the daylight requirements and numerous varieties for flavor and storage can be confusing. To start, the best strategy is to plant sets of a variety known to do well locally about 4 weeks before the last frost date. Onions started from seeds generally grow larger and store longer, while sets are easier and faster to grow but are more subject to bolting and rot. Multiplier onions (*A. aggregatum*), such as shallots and perennial potato onions, reproduce vegetatively and are usually started by sets. Similarly, bunching onions (*A. fistulosum*), such as scallions, Welsh, and Japanese, don't form full bulbs and are usually started by sets. Sets should be started with small bulbs no larger than 3⁄8 to 7⁄8 inches. Sweet onions are best started from seed, as is the common or regular onion. Sweet onions generally store poorly, whereas pungent varieties store well because of a high content of aromatics, which act as preservatives. For an easy perennial onion patch, grow potato onions. Almost a lost variety, with a flavor stronger than shallots, they can substitute for regular onions. Buy them once, plant in fall or spring, and enjoy harvests for decades. (For details, consult catalogs.)

Temperature

For germination: 50°F–95°F
For growth: 55°F–75°F

Soil & Water Needs

pH: 6.0–7.5 (multiplier types: 6.5–7.0)

Fertilizer: Light feeder; use compost

Water: Medium; dry soil will cause the onion to form two bulbs instead of one; don't water one week before harvest

Measurements

Planting Depth:
Seed: ½"
Sets: 1"
Root Depth: 18"–3'

Height: 15"–36"
Breadth: 6"–18"

Space between Plants:
To grow scallions: 1"
To grow bulbs: 3"

Pests

Japanese beetle, onion eelworm, onion maggot, slug, thrips, vole (storage), white grub, wireworm. (Try radish as trap crop for onion root maggot; when infested, pull and destroy.)

Diseases

Botrytis, damping off, downy mildew, pink root, smut, storage rot, sunscald, white rot

Allies

Some evidence: Carrot
Uncertain: Beet, caraway, chamomile, flax, summer savory

First Seed-Starting Date

The average time to maturity is 100–160 days for spring starts. (See Notes, page 111.)

Germinate	+	Transplant	+	Days Before LFD	=	Count Back from LFD
4 to 10 days	+	28 to 42 days	+	14 to 40 days	=	46 to 92 days

Companions

Lettuce, pepper, spinach, strawberry, tomato

Incompatibles

All beans, asparagus, peas, sage

Harvest

Wait until tops fall over; pushing them can shorten storage life. When bulbs pull out very easily, rest them on the ground to dry and cure. Treat gently as they bruise easily. Turn once or twice in the next few days; cover if it rains. When completely brown, they're ready for further curing. For regular onions, clip tops 1" from the bulb. Do not clip tops of multipliers or separate bulbs. Spread onions no more than 3" deep on wire screens in a shady, warm, dry, well-ventilated area. Cure for up to 2 months before storing for the winter. The flavor and quality of multipliers keeps improving. After 2 months, check for spoilage and remove bad or marginal onions. Separate multiplier bulbs by cleaning and cutting off dried tops about 1" above the bulbs. Keep the smallest bulbs for spring planting.

onion

Storage Requirements

Onions sprout in the presence of ethylene gas, so never store with apples, apricots, avocados, bananas, figs, kiwis, melons, peaches, pears, plums, or tomatoes. Eat the largest first — they're most likely to sprout.

Fresh

Temperature	Humidity	Storage Life
36°F–40°F	65%–70%	1–8 months (dry)

Preserved

Method	Taste	Shelf Life
Canned	good	12+ months
Frozen	fair	3 months
Dried	good	12 months

Selected Varieties

LONG DAY
(need 15+ daylight hours)

Plant these in early spring in Virginia and northward to obtain large bulbs; can plant later in regions south of Virginia

Copra: 104 days; long storage; mild flavor; early maturity
Widely available

Early Yellow Globe: 102 days; early; high yields; keeps 6–12 months; moderately strong
Widely available

Sweet Sandwich: 100 days, stores well; the longer stored the sweeter they become
Garden City, Vermont Bean

Sweet Spanish: 110 days; some are hybrid and some open-pollinated; large globe-shaped yellow bulbs; sometimes known as "hamburger onion"; mild; sweet
Widely available

Walla Walla Sweet: 125 days spring-seeded; 300 days late-summer seeded and overwintered; open-pollinated; large flattened bulbs; mild; plant in August–September for next summer harvest and sweetest onions; short keeper
Widely available

INTERMEDIATE DAY
(need 12–14 daylight hours)

These are best for intermediate latitudes.

Red Hamburger or Red Mac: Large, semiflat bulb; red skin; red and white flesh; best in salads; doesn't store for long periods
Burpee, Gurney's, Park

SHORT DAY
(12 daylight hours)

This is the best type for fall planting south of Virginia, and spring or fall planting in mid-Atlantic; not good in North, where long days force them to bulb too fast.

Granex Hybrid or Vidalia: 170 days; large, flat, yellow bulbs; very mild flesh; especially good for mid-Atlantic and South; good for overwintering in the South
Widely available

Texas Supersweet: Long keeper for short-day variety; uniformly sweet; best in far South
Garden City

PEARL ONIONS

Cippolina Borretana: 110 days; use for boiling or creamed onions
Cook's

Selected Varieties *(cont'd)*

MULTIPLIER (perennial)

These include potato onions, Welsh onions, and shallots.

He-shi-ko: 70–80 days; Welsh scallion; resists bulbing; good in North and South
Field's, Gurney's, Nichols

Shallots: Open-pollinated; French; red-pink bulbs known for culinary uses; tops can be used for scallions
Widely available

Yellow Potato Onion: Open-pollinated; very popular; good drought resistance; resists pink root; good keeping quality; heirloom; flavorful but not too strong
Southern Exposure

BUNCHING

Beltsville Bunching: 65 days; stands dry summer heat; best for August harvest; crisp and mild
Stokes

Crimson Forest: Brilliant red stalks
Baker's Creek

Evergreen Hardy White: 60 days for scallions; open-pollinated; little or no bulbing; hardiest bunching type can be perennial; protect in severe winters
Widely available

Fukagawa: Japanese bunching; slim straight nonbulbing with sweet taste
Kitchen Garden

Greenhouse

Any variety.

Notes

Onions are usually easier to start from sets, which are planted in spring an average of about 4 weeks before the LFD in order to receive enough daylight growing hours for bulb maturation before harvest in mid- to late summer. Seeds for summer harvest should be started inside in early spring for the same reason. For small storage onions, seed can be sown outside shortly after the last frost date. For overwintering onions, sow seed in mid-summer to early fall. For more planting advice, see specific day requirements and variety requirements.

pea

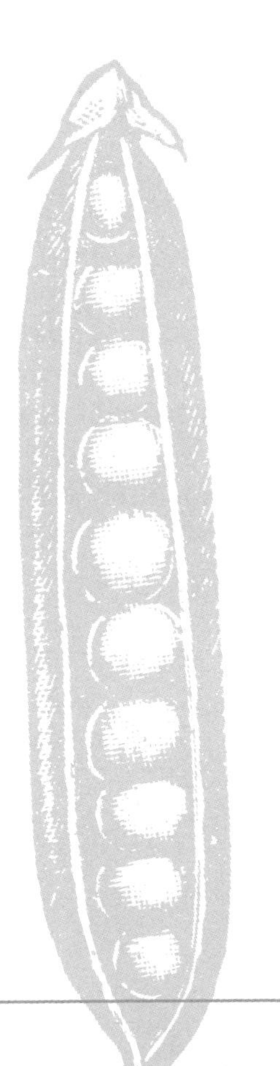

Pisum sativum
Leguminosae or Fabaceae
(Pea Family)

PEAS are a cool-season crop, hardy to frost and light freezes. Plan an average of twenty-five to sixty plants per person, depending on how much you want to freeze, dry, or can for winter. Add organic matter to the beds in fall; in spring when the soil is thawing, gently rake the soil surface. Gardeners with mild winters can plant peas in both spring and fall. Peas have fragile roots and don't transplant well. While some gardeners recommend presoaking seeds, research indicates that presoaked legume seeds absorb water too quickly, split their outer coatings, and spill out essential nutrients, which encourages damping-off seed rot. Yields can increase 50 to 100 percent by inoculating with Rhizobium bacteria. Peas can cross-pollinate, so for seed-saving, space different varieties at least 150 feet apart. Dwarf varieties don't need a trellis if you plant them close together. Pole and climbing peas produce over a longer period and up to five times more than dwarf bush varieties. After the harvest, turn under the plant residues to improve the soil.

Temperature

For germination: 40°F–70°F
For growth: 60°F–65°F

Soil & Water Needs

pH: 6.0–7.5

Fertilizer: Light feeder. When inoculated, peas are N-fixing and need low N. Apply liquid seaweed 2–3 times per season.

Side-dressing: When vines are about 6" tall, apply compost or an amendment high in P and K, and light in N.

Water: Low initially; heavy after bloom; shallow watering is said to increase germination

Measurements

Planting Depth: 1" or ½"–¾"
Root Depth: Shallow to 3'

Height:
Garden peas: 21"–4'
Snap peas: 4'–6'
Breadth: 6"–10"

Space between Plants:
In beds: 2"–4"
In rows: 1"–3"
Space between Rows:
18"–48"

Support Structures

A 6' post, A-frame, or trellis

Pests

Most problems affect seedlings: aphid, cabbage looper, cabbage maggot, corn earworm, corn maggot, cucumber beetle, cutworm, garden webworm, pale-striped flea beetle, seedcorn maggot, slug, snail, thrips, webworm, weevil, wireworm

Diseases

Bacterial blight, downy mildew, enation mosaic, fusarium wilt, leaf curl, powdery mildew, root rot, seed rot

First Seed-Starting Date

Plant every 10 days in case of poor germination.

Germinate	+	Transplant	+	Days Before LFD	=	Count Back from LFD
7 to 14 days	+	0 (direct)	+	28 to 42 days	=	35 to 56 days

Last Seed-Starting Date

Germinate	+	Transplant	+	Maturity	+	SD Factor	+	Frost Tender	=	Count Back from FFD
6 days	+	0 (direct)	+	50 to 80 days	+	14 days	+	0 days	=	70 to 100 days

Allies

Some evidence: Tomato
Uncertain: Brassicas, caraway, carrot, chive, goldenrod, mint, turnip

Companions

All beans, coriander, corn, cucumber, radish, spinach

Incompatibles

Garlic, onion, potato

Harvest

If a plant has only a few peas on it, pinch back the growing tip to encourage further fruiting. When pea pods are plump, crisp, and before they begin to harden or fade in color, harvest them with one clean cut. Sugar snaps are best picked when plump and filled out. Harvest snow peas when the pods are young and peas undeveloped. Pick peas every day for continuous production. Pea shoots, the last 4"–6" of the vine, can also be harvested for stir-fry dishes and salads.

Storage Requirements

Blanch shelled regular peas and whole snap and snow peas before freezing.

Fresh

Temperature	Humidity	Storage Life
32°F	95%–98%	1–2 weeks

Preserved

Method	Taste	Shelf Life
Canned	good	12+ months
Frozen	excellent	12+ months
Dried	good	12+ months

Selected Varieties

BUSH GREEN PEAS

Alaska: 55 days; 18"–36" tall; plump; good dried as split peas
Widely available

Frosty: 65 days; sturdy 18"–24"; excellent for freezing; high yields
Harris

Green Arrow: 70 days; 28"; resists downy mildew; fusarium wilt; excellent freezing quality; high yields; long pods
Widely available

Laxton's Progress No. 9: 62 days; 14"–20", resists fusarium wilt; good flavor and quality
Field's, Gurney's, Jung, William Dam

Little Marvel: 63 days; 18"; very early; summer; good fresh or frozen
Widely available

Selected Varieties *(cont'd)*

Novella: 65 days; 18"–24" bush; fewer leaves provides better air circulation and lower disease and insect damage; resists powdery mildew
Widely available

Waverex: 65 days; 15" bush needs staking; true French petit pois; good in any cool climate; good frozen
Bountiful Gardens, Cook's, Thompson & Morgan

VINE GREEN PEAS

Alderman or Tall Telephone: 68 days; 5'–6' tall; extra sweet; high yields and quality; sow in early warm weather; late maturing
Bountiful Gardens, Gurney's, Harris, Stokes, Territorial

Tarahumara "Chicharos": Peas are good fresh or dried; not heat adapted; plant in spring in cool climates; for low and high desert planting
Native Seed

Wando: 68 days; 30" tall; tolerates heat and drought; good yields; best pea variety for late sowing; good for southern areas
Widely available

EDIBLE POD PEAS (Snow Peas)

Burpee Sweetpod (Mammoth Melting Sugar): 68 days; 4' vines; resists wilt; can be trellised inside greenhouse
Burpee

Dwarf Grey Sugar: 65 days; 18" bush; no staking needed; high yields
Widely available

Norli: 58 days; 4'–5' vine; very early; sweet; good producer
Shepherd's, William Dam

Oregon Sugar Pod II: 64 days; 28"–6' vines; good fresh or frozen
Widely available

SNAP PEAS

Introduced in the 1980s, these peas have sweet, edible pods that are excellent raw, cooked, or frozen.

Sugar Bon: 56 days; 18"–24" vines; matures 2–3 weeks earlier than Sugar Snap; compact 18" plant; resists powdery mildew and pea leafroller; high yields
Widely available

Sugar Mel: 68 days; 30" vines; very resistant to heat and powdery mildew
Park, Southern Exposure

Sugar Snap: 70 days; AAS Winner; sweet; good raw any size; resists wilt
Widely available

Greenhouse

Alaska, Burpee Sweetpod, Dwarf Grey Sugar, Frosty, Green Arrow, Little Marvel, Sugar Bon

peanut

Arachis hypogaea
Leguminosae or Fabaceae (Pea Family)

Also known as goober peas and groundnuts, peanuts are a warm-season crop, very tender to frost and light freezes. Plan an average of ten to twenty plants per person. Peanuts require full sun and can be grown wherever melon grows, as far north as Canada, although commercial production is generally limited to the South. In short-season areas, you may want to start seeds (which are the nuts) inside. For higher yields, inoculate seeds with special peanut inoculant available from nurseries. Plant nuts that are not split and still have their papery skin. Nuts sprout more easily without their shell but filled shells can also be planted. Transplant seedlings outside into soil warmed with plastic, choosing a sheltered, south-facing site. They need loose, enriched, sandy soil. For succession planting, peanuts are good planted after early crops of lettuce or spinach. The plant produces peanuts after the stem blossoms; its lower leaves drop and in their place peduncles grow. The peduncles eventually bend over and root in the nearby soil, where clusters of peanuts then grow. When the plants reach 6 inches, begin to cultivate the rows to control weeds and keep the soil aerated. When about 1 foot high, hill the plants in the same manner as potatoes, mounding soil high around each plant. Hilling is important to help the peduncle root quickly. Mulch between rows with 8 inches of grass clippings or straw.

Research at the Alabama Agricultural Experiment Station has shown that extra soil calcium can minimize the fungal-produced aflatoxins in peanuts; calcium applied at 10 pounds per 1000 square feet can be a justifiable "insurance" against cancer-causing aflatoxins in peanuts.

Peanut hulls are good for mulching and composting as they're rich in nitrogen. For rotation planning, follow peanuts with nitrogen-loving plants.

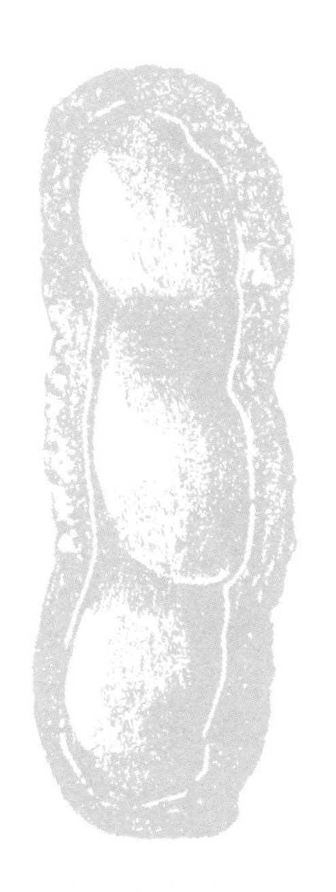

First Seed-Starting Date

Try presprouting extra-large peanuts, which germinate poorly in wet, cool soil.

Germinate	+	Transplant	+	Days Before LFD	=	Count Back from LFD
7 to 14 days	+	0 days	+	15 to 28 days	=	22 to 42 days

Last Seed-Starting Date

Germinate	+	Transplant	+	Maturity	+	SD Factor	+	Frost Tender	=	Count Back from FFD
7 to 14 days	+	0 (direct)	+	110 to 120 days	+	14 days	+	14 days	=	145 to 162 days

Temperature

For germination: 60°F–80°F
For growth: 70°F–85°F

Soil & Water Needs

pH: 5.0–6.0

Fertilizer: Add rotted manure in fall so its decomposition won't affect peanut seeds; at blossom time, add calcium (calcium sulfate or limestone); additional potassium may also be required.

Water: Average. When the plant begins to blossom, stop all watering.

Measurements

Planting Depth: 1"–1½"
Root Depth: Shallow

Height: 12"–18"
Breadth: 15"–20"

Space between Plants: 3"–6" (in 4" high ridge), thin to 12"
Space between Rows: 30"–36"

Pests

Weeds are the most significant; beyond that backyard gardeners may experience only occasional corn earworm, cutworm, pale-striped flea beetle, potato leafhopper, spider mite, thrips, and different caterpillars

Diseases

No significant diseases, except occasional leaf spot and southern blight

Allies

Some evidence: Corn

Incompatibles

None

Harvest

As the first frost approaches, when the leaves turn yellow-white, kernels drop, and pod veins darken, dig up the entire plant. In short-season areas, you may delay harvest until after the first few light frosts; although top growth may be killed, the pods continue to mature. Shake off all loose dirt. Dry roots in the sun for a few days to facilitate separation of the pods, or hang in a dry, airy place. Keep out of reach of small animals.

Storage Requirements

Preserved (Cured or Dried)

Temperature	Humidity	Storage Life
32°F	low	12 months

Spread peanuts on shallow trays or hang the entire plant from rafters in a garage or attic. Cure this way in a warm, dry place for a minimum of 3 weeks and no more than 3 months. Peanuts are best stored shelled, in airtight containers in the refrigerator for short periods, or in the freezer for long periods — they are very susceptible to a fungus that produces a highly toxic substance called aflatoxin. To be safe, don't eat any moldy peanuts. Roast nuts at 300°F for 20 minutes before eating.

Selected Varieties

Peanuts are good container plants.

Spanish or Early Spanish: 100–120 days; dwarf bushes; rich flavor; small kernels; can be grown as far north as Canada; provide light, sandy soil and southern exposure
Widely available

Valencia Tennessee Red: 120 days; large and sweet kernels; southern warm-season type; can be grown as far north as New York
Park

Virginia or Jumbo Virginia: 120 days; 3½' spreading vines; large kernels; grows in the North and Corn Belt, too; good rich flavor fresh or in peanut butter
Burpee, Field's, Gurney's, Park

Greenhouse

The smaller dwarf bushes (Spanish) might be grown in a greenhouse.

pepper

Capsicum annuum
Solanaceae
(Nightshade Family)

PEPPERS are a warm-season crop, very tender to frost and light freezes. Plan an average of five to six plants per person. All parts of the plant except the fruit are poisonous. To start indoors use pots at least 1½-inches wide to minimize transplant shock, make a stockier plant, and encourage earlier production. Growers report that the following cold treatment of seedlings significantly improves yields and early growth: (1) when the first leaves appear, lower the soil temperature to 70°F and ensure 16 hours of light with grow lamps; (2) when the first true leaf appears, thin seedlings to 2 to 3 inches apart or transplant into 4-inch pots; (3) when the third true leaf appears, move the plants to a location with night temperatures of 53 to 55°F; keep there for 4 weeks; (4) return the seedlings to a location with average temperature of 70°F; (5) transplant into the garden 2 to 3 weeks after all danger of frost has passed. Soil temperature should be at least 55 to 60°F for transplanting, or the plants turn yellow, become stunted, and are slow to bear. Some recommend feeding seedlings weekly with half-strength liquid fertilizer until transplanted. Peppers do better planted close together. Except in the West, where peppers may be mostly pest-free, use row covers immediately because pepper pests will be out. If the temperature rises over 95°F, sprinkle plants with water in the afternoon to try to prevent blossom drop.

Temperature

For germination: 65°F–95°F
For growth: 70°F–85°F

Soil & Water Needs

pH: 5.5–7.0

Fertilizer: Medium-heavy feeder; high N; rotted manure or compost; some soils may need calcium

Side-dressing: Apply at blossom time and 3 weeks later. Apply liquid seaweed 2–3 times per season. At blossom time, try spraying leaves with a weak epsom salt mixture (1 teaspoon per quart) to promote fruiting.

Water: Medium-heavy

Measurements

Planting Depth: ¼"
Root Depth: 8", some to 4'

Height: 2'–3'
Breadth: 24"

Space between Plants:
In beds: 12"
In rows: 12"–24"
Space between Rows:
18"–36"

First Seed-Starting Date

Germinate	+	Transplant	–	Days After LFD	=	Count Back from LFD
10 to 12 days	+	32 to 44 days	–	14 to 21 days	=	28 to 35 days

Last Seed-Starting Date

Germinate	+	Transplant	+	Maturity	+	SD Factor	+	Frost Tender	=	Count Back from FFD
6 to 9 days	+	21 days	+	60 to 90 days	+	14 days	+	14 days	=	115 to 148 days

Pests

Aphid, Colorado potato beetle, corn borer, corn earworm, cutworm, flea beetle, leaf miner, mite, slug, snail, tomato hornworm, weevil

Diseases

Anthracnose, bacterial spot, cercospora, mosaic, soft rot, southern blight, tobacco mosaic
Environmental disorders: Blossom end rot, sunscald

Allies

Uncertain: Caraway, catnip, nasturtium, tansy

Companions

Basil, carrot, eggplant, onion, parsley, tomato

Incompatibles

Fennel, kohlrabi

Harvest

For sweet peppers, pick the first fruits as soon as they're usable in order to hasten growth of others. For storage peppers, cut the fruit with 1" or more of stem. For maximum vitamin C content, wait until peppers have matured to red or yellow colors.

Storage Requirements

Hot varieties are best stored dried or pickled. Pull the entire plant from the ground and hang it upside down until dried. Alternately, harvest the peppers and string them on a line to dry. For sweet peppers, refrigeration is too cold and encourages decay.

Fresh

Temperature	Humidity	Storage Life
45°F–55°F	90%–95%	2–3 weeks

Preserved

Method	Taste	Shelf Life
Canned	good	12 months
Frozen	fair	3 months
Dried	excellent	12 months
Pickled	excellent	12+ months

Selected Varieties

SWEET

Ace Hybrid: 50 days; green stuffing pepper; medium thick flesh; short, sturdy plants reliable even in adverse conditions; ripens to red
Widely available

California Wonder: 75 days; open-pollinated; very mild flavor; good stuffing pepper; blocky deep green fruit
Widely available

Early Sunsation: 70 days; our favorite large yellow bell pepper; thick walls good for stuffing; good disease resistance
Nichols, Park, Shepherd's, Tomato Growers

Golden Bell or Golden Summer Hybrid: 70 days; unusual bright yellow when ripe; excellent flavor; open growing habit; good yields
Widely available

Gypsy Hybrid: 62 days; AAS Winner; early; prolific; thin; yellow-red; mild flavor; good for salads and frying; 3"–4" fruits; resists tobacco mosaic virus
Widely available

Large Sweet Cherry: 70 days; open-pollinated; red; small fruit, 1½" across; excellent in salads
Nichols, Southern Exposure

Pepperonicini (Italian): 65 days; open-pollinated; shrubby 3' plant; red pencil-thin fruit; prolific; good fresh or pickled for antipasto
Heirloom Seeds, Nichols, Shepherd's, Stokes

Staddon's Select: 72 days; open-pollinated; early; meaty; prolific even under adverse conditions; resists mosaic; popular in the North
Garden City, Stokes

Sweet Banana or Sweet Hungarian: 75 days; open-pollinated; pretty light yellow; slender 6"–8" fruit; thin flesh; good for pickling or salads; high yields; sturdy plants
Widely available

Sweet Pimento: 65 days; open-pollinated; heart-shaped peppers; bright red; eat fresh in salads or out of hand, roasted, peeled, or canned
Widely available

Yolo Wonder: 76 days; open-pollinated; thick flesh; bell pepper; ripens green to red; resists mosaic; thick foliage protects against sunscald
Bountiful Gardens, Jung, Stokes

HOT

Anaheim: 74 days; open-pollinated; chili type; relatively mild; 7"–8"; thick flesh; good fresh in salsa and chili rellenos, canned, frozen, or dried; may resist tobacco mosaic virus
Widely available

Selected Varieties *(cont'd)*

Early Jalapeño: 70 days; open-pollinated; very hot Mexican type; compact plant; thick flesh; good fresh and in jelly and as pickles
Garden City, Johnny's, Jung, Shepherd's

Hungarian Yellow Wax: 70 days; open-pollinated; pretty yellow; red when ripe; 6"–7"; strong upright plants; medium hot; considered best hot pepper for cool climates
Widely available

Large Red Cherry: 80 days; open-pollinated; red 1¼" hot fruit
Widely available

Sandia: 6"–9"; good for rellenos, enchilada sauce, and stews; for low and high desert
Native Seeds

Serrano: 75 days; popular in Southwest; extremely hot
Widely available

Super Cayenne: 72 days; AAS Winner; 20" plants; more compact and manageable than other cayennes; hot and spicy pepper
Park

Thai Hot: Extremely hot; pretty 8" plant; small 1½" fruit; good for containers
Park

NOVELTY

Ariane: 70 days; unusual orange; first sweet orange bell; beautiful; good flavor when green or orange; early; high yields; resists tobacco mosaic virus
Widely available

Paprika: 80 days; open-pollinated; mildly hot and sweet; thick peppers; excellent dried and ground
Widely available

Greenhouse

Ace Hybrid, Anaheim, Early Jalapeño, Hungarian Yellow Wax, Super Cayenne, Staddon's Select, Thai Hot

pepper

potato

Solanum tuberosum
Solanaceae (Nightshade Family)

POTATOES are a warm-season crop in the North, tender to frost and light freezes, and a cool-season crop in the South and West. Plan an average of ten to thirty plants per person. All plant parts except the tubers are poisonous. Start potatoes with *seed potatoes,* each containing one to three "eyes," or small indentations that sprout foliage. To prepare seed potatoes for planting: (1) Cut the potato into 2-inch pieces 2 days before planting, and cure indoors at about 70°F in high humidity to help retain moisture and resist rot; or (2) plant small whole potatoes, which are less apt to rot, have more eyes, and don't need curing prior to planting. If desired, presprout seed potatoes by refrigerating at 40 to 50°F for 2 weeks before planting to break dormancy. Place in trenches 6 inches deep by 6 inches wide, spaced 10 to 12 inches apart, and cover with 3 to 4 inches of soil. One week after shoots emerge, mound soil around base, leaving a few inches exposed. This "hilling" prevents greening. Side-dress and "hill" again 2 to 3 weeks later. Cover plants if a hard frost is expected.

Temperature

For germination: 65°F–70°F
For growth: 60°F–65°F

Soil & Water Needs

pH: 5.0–6.0

Fertilizer: Light feeder; apply compost at planting.

Side-dressing: 2–3 weeks after first hilling, apply fertilizer 6" away from plant and hill again

Water: Medium; heavy when potatoes are forming, from blossom time to harvest

Measurements

Planting Depth: 3"–4"
Root Depth: 18"–24"

Height: 23"–30"
Breadth: 24"

Space between Plants:
In beds: 9"–12" (seed potatoes)
In rows: 10"–12"
Space between Rows: 24"

Pests

Aphid, cabbage looper, Colorado potato beetle, corn borer, corn earworm, cucumber beetle, cutworm, earwig, flea beetle, Japanese beetle, June beetle, lace bug, leaf-footed bug, leafhopper, leaf miner, nematode, slug, snail, tomato hornworm, white grub, wireworm

Diseases

Black leg, early blight, fusarium wilt, late blight, mosaic, powdery mildew, psyllid yellows, rhizoctonia, ring rot, scab, scurf, verticillium wilt
Environmental disorders: Black heart

Allies

Uncertain: All beans, catnip, coriander, "dead" nettle, eggplant, flax, goldenrod, horseradish, nasturtium, onion, tansy

Companions

All brassicas, corn, marigold, pigweed

Incompatibles

Cucumber, pea, pumpkin, raspberry, spinach, squash, sunflower, tomato

Harvest

For small "new" potatoes, harvest during blossoming; for varieties that don't blos-

First Seed-Starting Date

2–4 or 6–8 weeks before LFD. In the South and West, potatoes are usually started in February or March and harvested in June and July.

Last Seed-Starting Date

90–120 days (average days to maturity) before FFD

som, harvest about 10 weeks after planting. Harvest regular potatoes when the vines have died back halfway, about 17 weeks after planting. Gently pull or dig out tubers with a garden fork. If not large enough, pack the soil back and try again at 2–3 week intervals. If you have many plants, remove the entire plant when harvesting to make room for another crop. For storage potatoes, dig near the first frost when plant tops have died back. To minimize tuber injury, always dig when the soil is dry.

Storage Requirements

Spring- or summer-harvested potatoes aren't usually stored, but keep for 4–5 months if cured first at 60°–70°F for at least 4 days and stored at 40°F. Dry fall-harvested potatoes for 1–2 days on the ground, then cure at 50°–60°F and a high relative humidity for 10–14 days. Don't cure potatoes in the sun; they turn green. Once cured, store in total darkness in a single layer. Never layer or pile potatoes more than 6"–8" deep.

Laboratory experiments conducted in Greece have shown that several aromatic herbs and their essential oils can suppress sprouting of potatoes in storage and have antimicrobial activity against potato pathogens. English lavender, pennyroyal, spearmint, rosemary, and sage suppressed growth of potato sprouts, but two oreganos did not. English lavender was the most effective sprout inhibitor.

Fresh

Temperature	Humidity	Storage Life
55°F–60°F	90%–95%	5–10 months

Preserved

Method	Taste	Shelf Life
Canned	fair	12+ months
Frozen	good	8 months
Dried	good	12+ months

Selected Varieties

WHITE

These are good for a variety of purposes.

Elba: 1987 Cornell release; late; white; resists early and late blight, verticillium wilt, and golden nematode; high yields; good boiled or baked
Garden City, Gurney's

Kennebec: Late; white; delicious big tubers; thin skin; resists mosaic and blight; reliable high yields; good storage
Widely available

RUSSET (BROWNISH SKIN, WHITE FLESH)

These are the best baking potatoes.

Burbank: The famous "Idaho potato"; late maturing
Shepherd's, Territorial

Butte: Baking potato; higher in protein and vitamin C than most
Widely available

Selected Varieties *(cont'd)*

White Cobbler: Early season; the standard popular white potato; smooth skin; tastes best baked; dependable yields under wide growing conditions; not good for storing
Field's

RED (RED SKIN AND WHITE FLESH)

Good for boiling.

Red Norland: Very early potato; very large tubers; resists scab; one of best flavors; smooth skin
Abundant Life, Field's, Gurney's

Red Pontiac: Early to midseason; tolerates heat for Southern growers; excellent for boiling; stays firm for potato salads
Widely available

YELLOW FLESH

Lady Finger: Small 1" wide and 4"–5" long fruits; brown skin and yellow flesh; good for baking; excellent boiled or fried
Gurney's

Princess La Ratte: Culinary prize of European haute cuisine; mildly nutty
Garden City

Purple Peruvian: Only purple fingerling, mild and sweet
Widely available

Yellow Fingerling: Salad potato; yellow flesh; long and slender crescent-shaped tubers
Widely available

Yellow Finn: Bakes and boils nicely; prized in Europe; natural butter flavor
Widely available

Yukon Gold: Round; yellow flesh; all-purpose; good flavor; excellent storage
Widely available

NOVELTY

All-Blue: Good yields; unusual blue flesh all the way through; good baked; boiled; roasted; makes beautiful violet vichyssoise
Widely available

Purple Viking: Excellent for frying, baking, and boiling; good yield and storage
William Dam, Garden City

Greenhouse

None; potatoes won't grow well in the greenhouse.

Notes

Potatoes are very disease prone, so use only certified disease-free potatoes as seed potatoes.

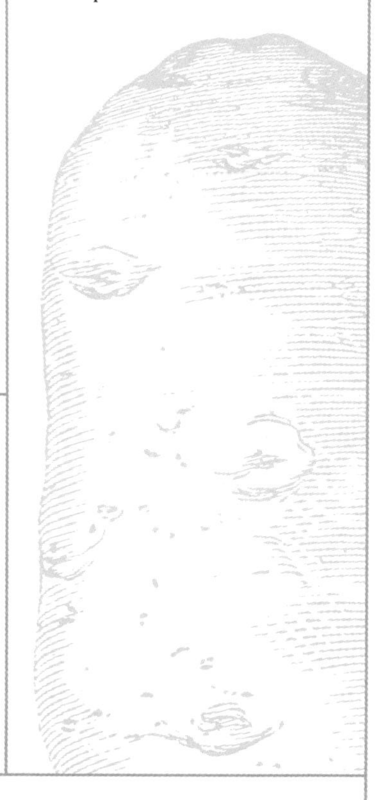

spinach

Spinacia oleracea
Chenopodiaceae
(Goosefoot Family)

SPINACH is a cool-season crop, hardy to light frosts and freezes. Plan an average of ten to twenty plants per person. Spinach can be grown as soon as the soil is workable. After thinning to 4 to 6 inches, cover the plants with row covers to keep pests away. Fall crops usually taste better and suffer no leaf miners or bolting. Also, if you plant a late fall crop and mulch it, a very early crop will come up in spring. Spinach bolts when there's 14 to 16 hours of light, regardless of the temperature, although warmer temperatures will cause it to bolt faster. The exceptions are New Zealand and Basella Malabar "spinach," which thrive in warm weather. They aren't true spinach, but when cooked they taste like the real thing. Malabar is also a pretty ornamental vine that is easily grown on arbors where it provides summer shade and a constant supply of summer greens.

Temperature

For germination: 45°F–75°F
For growth: 60°F–65°F

Soil & Water Needs

pH: 6.0–7.5

Fertilizer: Heavy feeder; before planting apply compost.

Side-dressing: Apply 4 weeks after planting, and thereafter every 2 weeks.

Water: Light but evenly moist

Measurements

Planting Depth: ½"
Root Depth: 1', tap root to 5'

Height: 4"–6"
Breadth: 6"–8"

Space between Plants: 2", thin to 6"–12" as leaves touch
Space between Rows: 12"–14"

Pests

Aphid, beet leafhopper, cabbage looper, cabbageworm, flea beetle, leaf miner, slug, snail

Diseases

Curly top (spread by beet leafhopper), damping off, downy mildew, fusarium wilt, leaf spot, spinach blight (caused by cucumber mosaic virus spread by aphids; see page 336)

Allies

Uncertain: Strawberry

Companions

All beans, all brassicas, celery, onion, peas

Incompatibles

Potato

First Seed-Starting Date

Sow directly every 10 days, starting 4–6 weeks before last frost.

Germinate	+	Transplant	+	Days Before LFD	=	Count Back from LFD
7 to 14 days	+	28 days	+	21 days	=	56 to 64 days

Last Seed-Starting Date

Sow later crops directly, as transplanting encourages bolting.

Germinate	+	Transplant	+	Maturity	+	SD Factor	+	Frost Tender	=	Count Back from FFD
5 days	+	0 days	+	40 to 50 days	+	14 days	+	N/A	=	59 to 69 days

Harvest

Cut individual leaves when they're large enough to eat. Continual harvest prevents bolting. When the weather warms, cut the plant to ground level. Its leaves will grow back. For the best nutrition, harvest leaves in the morning.

Storage Requirements

For freezing and drying, cut the leaves into thick strips. Blanch for 5 minutes before drying, or 2 minutes before freezing. It's best to use only the smallest and most tender leaves for freezing.

Fresh

Temperature	Humidity	Storage Life
32°F	95%–100%	10–14 days

Preserved

Method	Taste	Shelf Life
Canned	good	12+ months
Frozen	good	12 months
Dried	unknown	

Selected Varieties

savoyed = crinkled leaves

Bloomsdale Long Standing: 48 days; open-pollinated; late spring–early summer crop; best flavor for salads; heavy yields; glossy savoyed leaves; bolt resistant; long harvest; can be overwintered
Widely available

Indian Summer: Hybrid; outperforms all others in warm weather; attractive savoyed leaves; high yielding
Cook's, Shepherd's

Melody: 42 days; hybrid; spring or fall crop; quick growing; 40–50 days; resists downy mildew and mosaic; bred for the home garden; upright; easy harvest of dirt-free leaves; good fresh, frozen, or canned
Widely available

Nordic IV: 39–45 days; resists all downy mildews; smooth, thick, 8-inch leaves held off ground for cleaner harvest; resists bolting so good for late spring–summer harvest; delicate flavor and pleasing texture
Kitchen Garden, Territorial

Savoy, Cold-Resistant: 45 days; open-pollinated; late summer or fall crop; good for overwintering for early spring crop; tolerates heat, cold, and blight; good fresh flavor; well-savoyed
Southern Exposure, Stokes

Selected Varieties *(cont'd)*

Space Hybrid: 40 days; excellent smooth-leaf spinach; vigorous; strong, upright fleshy broad leaves; considered good for growing all three seasons; resists heat and mildew
Cook's, Park, William Dam

Tyee: 42 days; hybrid; excellent savoy type; upright habit makes easier harvest; tolerates downy mildew; good for spring, summer, and fall crops; very slow to bolt; stands longer than Melody
Widely available

Wolter: 45 days; Dutch hybrid; rapid growing; high yields; high resistance to downy mildew; very fine flavor
Shepherd's

HOT-WEATHER SPINACH SUBSTITUTES

Basella Malabar Red Stem Summer Spinach: Open-pollinated; a new vegetable introduced in 1987 from Asia; harvest all summer; mild but not as flavorful as spinach; tolerates heat; vigorous; can grow up to 6'; easily trained on a trellis; ornamental red stems are also good in salads; not recommended in northern areas
Bountiful Gardens, Park

New Zealand Everlasting Spinach *(Tetragonia expansa)*: Open-pollinated; perennial; tastes like spinach; tolerates heat and drought; crisp green leaves; continuous harvest
Widely available

Tetragonia (New Zealand Spinach): 50 days; heat and drought tolerant
Johnny's

Greenhouse

Bloomsdale, Long Standing, Melody, Space Hybrid

spinach

squash

Cucurbita pepo and others
Cucurbitaceae
(Cucumber Family)

SQUASH is a warm-season crop, very tender to frost and light freezes. Plan an average of two winter plants per person and two summer plants per four to six people. Winter squash does not transplant well, but can be sown inside in individual pots to minimize root disturbance. Squash is usually planted in small hills. To prepare, dig 18-inch-deep holes, fill partly with compost; complete filling with a mixture of soil and compost. Traditionally, six to eight seeds are placed 1-inch deep in each hole (others recommended only one or two seeds due to high germination rate); when seedlings reach 3 inches, thin to two seedlings. Raise fruits off the ground to prevent rot. Fabric row covers boost and prolong yields. In cooler climates, keep row covers on all season; when the female (fruit) blossoms open, lift the cover for 2 hours in early morning twice a week to ensure bee pollination, which is essential. To keep vines short for row covers, pinch back the end, choose the best blossoms, and permit only four fruits per vine.

Temperature

For germination: 70°F–95°F
For growth: 65°F–75°F

Soil & Water Needs

pH: 6.0–7.5

Fertilizer: Heavy feeder; apply lots of compost; high N requirements

Side-dressing: Apply compost midseason; in boron-deficient soils, apply 1 teaspoon of borax per plant.

Water: Heavy

Measurements

Planting Depth:
In hills: ½"–1"
Vine: 72"–96"
Root Depth: 18"–6'

Height:
Winter: 12"–15"
Summer: 30"–40"
Breadth:
Bush: Up to 4 square feet
Vining: Up to 12–16 square feet

Space between Plants:
In beds: 12"–18"
In rows: 24"–28"
Space between Rows:
Bush: 36"–60"

Support Structures

Use an A-frame or trellis to grow vines upright.

Pests

Aphid, beet leafhopper, corn earworm, cucumber beetle, Mexican bean beetle, pickleworm, slug, snail, squash bug, squash vine borer, thrips, whitefly

Diseases

Alternaria leaf spot, anthracnose, bacterial wilt, belly rot, cottony leak, cucumber wilt, downy mildew, mosaic, powdery mildew, scab

Allies

Some evidence: Corn
Uncertain: Borage, catnip, goldenrod, marigold, mint, nasturtium, onion, oregano, radish, tansy

First Seed-Starting Date

Germinate	+	Transplant	–	Days After LFD	=	Count Back from LFD
7 to 10 days	+	28 to 42 days	–	21 to 28 days	=	14 to 24 days

Last Seed-Starting Date

Germinate	+	Transplant	+	Maturity	+	SD Factor	+	Frost Tender	=	Days Back from FFD
3 days	+	0 (direct)	+	40 to 50 days	+	14 days	+	14 days	=	71 to 81 days (summer)
3 days	+	0 (direct)	+	80 to 110 days	+	14 days	+	14 days	=	111 to 141 days (winter)

squash

Companions

Celeriac, celery, corn, melon

Incompatibles

Potato, pumpkin (cross-pollinates with other pepo plants, which is only important if you are saving seeds; keep pumpkin distant or plant 3 weeks later)

Harvest

Cut all fruit except hubbard-types with a 1" stem. Don't ever lift squash by the stem. Treat even those with hard skins gently to avoid bruising.

Summer: Cut before 8" long, when skin is still soft, and before seeds ripen.

Patty Pans: Cut when 1"–4" in diameter and the skin is soft enough to break with a finger.

Winter: Cut when the skin is hard and not easily punctured, usually after the first frost has killed the leaves and the vine begins to die back but before the first hard frost.

Storage Requirements

Cure winter squash after picking by placing in a well-ventilated, warm or sunny place for 2 weeks. If you cure fruit in the field, raise them off the ground and protect from rain. Or, dip fruit in a weak chlorine bleach solution (9 parts water:1 part chlorine), air dry, and store. Store only the best fruit. Don't allow fruit to touch. Wipe moldy fruit with a vegetable-oiled cloth.

Fresh

Temperature	Humidity	Storage Life
50°F–60°F	60%–70%	4–6 months

Preserved

Method	Taste	Shelf Life
Canned	good	12+ months
Frozen	good	8 months
Dried	good	12+ months

Selected Varieties

OP = open-pollinated
HL = heirloom

PEPO

Almost all bush varieties are pepo, including the most commonly grown such as summer squash, acorn, spaghetti, and pumpkin.

Summer

Aristocrat: 53 days; hybrid zucchini; AAS Winner; upright bush; early; high yields; excellent quality; dark green; smooth; long harvest
Jung, Nichols, Thompson & Morgan

Cocozelle or Italian Vegetable Marrow: 50 days; OP; slim zucchini; striped; very flavorful raw or cooked
Widely available

Eight Ball: AAS Winner; first round zucchini of intense green; pick when fruits are 3"–4"
Widely available

Golden Bush Scallop: OP; HL; long season; space-saving bush
Heirloom Seeds, Southern Exposure

Greyzini: OP; zucchini; high eating quality; long harvest
Stokes

Patty Pan: OP; flat with scalloped edges; excellent flavor; picky when very young, survives well even in English climate
Gardens Bountiful, Heirloom Seeds

Scallopini Hybrid: AAS Winner; best when 3" or less; good raw, boiled, or fried
Jung, Stokes, Territorial

Winter

Delicata or Peanut or Sweet Potato: 100 days; OP; acorn type; oblong 8" fruits with dark green stripes; superb keeper; good taste; compact vines
Widely available

Gold Nugget: 85 days; OP; runnerless bush plants; each plant bears about four slightly flattened orange fruits; 5" across; stores well
Widely available

Jersey Golden Acorn: 80 days; OP; AAS Winner; smaller; good for small gardens; when picked 1–3 days after flowering, the fruit tastes like corn with a sweet nutty flavor; also good when matured for winter storage
Widely available

Spaghetti Squash: 100 days; OP; use pulp like spaghetti
Widely available

Table King: OP; winter acorn; AAS Winner; compact bush; dark green with small seed cavity; improves with storage
Garden City, Stokes, Territorial

Selected Varieties *(cont'd)*

Winter Luxury Pumpkin: OP; best for smooth, tasty pie fillings; ripens early; excellent keeper; high yields; about 10" fruit
Jung

MAXIMA

These are excellent keepers, tolerant of borers, and include the largest fruit, such as buttercup and banana.

Buttercup (Burgess strain): 105 days; OP; turban-shaped with light stripes; deep orange; rich, sweet, and very dry flesh; tastes like sweet potato; excellent keeper
Widely available

Mayo Blusher: 120 days; large fruit turns pink when ripe; very sweet orange flesh; good for pies; keeps well; good for low desert
Abundant Life, Native Seed

Red Kuri: 92 days; beautiful red-orange; teardrop-shaped; good for pies and purees
Johnny's

MOSCHATA

The sweetest squashes, such as butternut, cushaws, and cheese, are all Moschata. They have high pest resistance and also have the highest vitamin content.

Early Butternut: 92 days; hybrid; AAS Winner; compact vines; high quality fruit; stores 2–3 months
Widely available

Magdalena Big Cheese: Large; ribbed; flat pumpkin shape; good for soups or pies; good for low desert
Native Seeds

MIXTA

This is a Southern-growing group like Moschata.

Cushaw: 115 days; OP; green-striped; resists squash vine borer; light yellow flesh; good for pies; excellent canned
Field's, Gurney's, Southern Exposure

Hopi Vatnga: Striped or solid green; thick hard shells (sometimes used for musical instruments); for high and low desert
Native Seeds

Greenhouse

Aristocrat, Gold Nugget, Greyzini, Patty Pan, Scallopini Hybrid, Table King

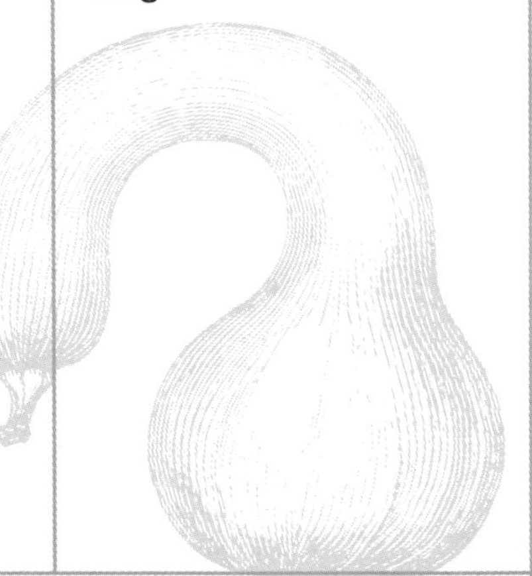

sweet potato

Ipomoea batatas
Convolvulaceae (Morning Glory Family)

The SWEET POTATO is a warm-season crop, very tender to frost and light freezes. Plan an average of five plants per person. Other than extreme sensitivity to frost, sweet potatoes are easy to grow, mostly pest-free, and, once the transplants are anchored, drought hardy. Start slips with a sweet potato cut in half lengthwise. Lay the cut side down in a shallow pan of wetted peat moss or sand. Cover tightly with plastic wrap until sprouts appear, then unwrap. The slip is ready when it has four or five leaves, is 4 to 8 inches tall, and has roots. A second method is to place a whole potato in a jar, cover the bottom inch with water, and keep warm. When leaves form above the roots, twist sprouts off and plant in a deep flat or, if warm enough, outdoors. A third method is to take 6-inch cuttings from vine tips in fall just before frost. Place cuttings in water and, when rooted, plant in 6-inch pots set in a south-facing window for the duration of winter. By late winter you can take more cuttings from these. To prepare the ground in April, fill furrows with 1 to 2 inches of compost. Mound soil over compost to form at least 10-inch-high ridges. This mini-raised bed optimizes both tuber size and quality, because tuber growth is easily hindered by obstructions in the soil. After all danger of frost is past, transplant slips into these ridges. Unlike potatoes, sweets are not true tubers and keep expanding as the vine grows.

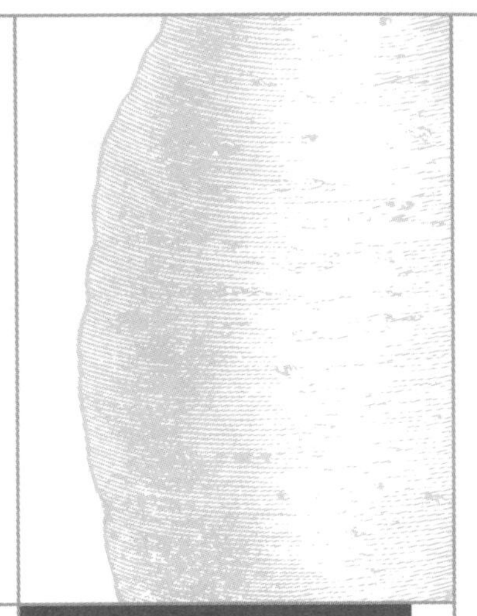

Temperature

For germination: 60F°–85°F
For growth: 70°F–85°F

Soil & Water Needs

pH: 5.0–6.0

Fertilizer: Light feeder. Low N. Before planting, place 1"–2" of compost in furrows.

Side-dressing: Once anchored apply high P fertilizer like bone meal, about 1 cup per 10 row-feet.

Water: Dry to medium. Water well the first few days until anchored, then ease back on water.

Measurements

Planting Depth: 4"–6"
Root Depth: Length of the potato

Height: 12"–15"
Breadth: 4–8 square feet

Space between Plants:
In beds: 10"–12"
In rows: 12"–16"
Space between Rows:
36"–40"

First Seed-Starting Date

Germinate	+	Transplant	–	Days After LFD	=	Count Back from LFD
8 to 12 days	+	42 to 56 days	–	7 to 21 days	=	43 to 57 days (6 to 8 weeks)

Last Seed-Starting Date

Germinate	+	Transplant	+	Maturity	+	SD Factor	+	Frost Tender	=	Days Back from FFD
8 to 12 days	+	42 to 56 days	+	100 to 125 days	+	14 days	+	14 days	=	178 to 221 days

Pests

Flea beetle, nematode, weevil, wireworm. (Problems vary by region, so check with your Extension agent.)

Diseases

Black rot (fungal), fusarium surface rot (storage), rhizoctonia, soil rot or scurf

Allies

Uncertain: Radish, summer savory, tansy

Companions

None

Incompatibles

None

Harvest

Some harvest after the vines are killed by frost, but most warn that frost damages the root. Always harvest on a dry day. Start digging a few feet from plant to avoid damage. Bruises or cuts as small as a broken hair root will shorten the shelf life by serving as an entry point for fusarium surface rot. Dry for 1–3 hours on the ground. Do not wash unless absolutely necessary; never scrub.

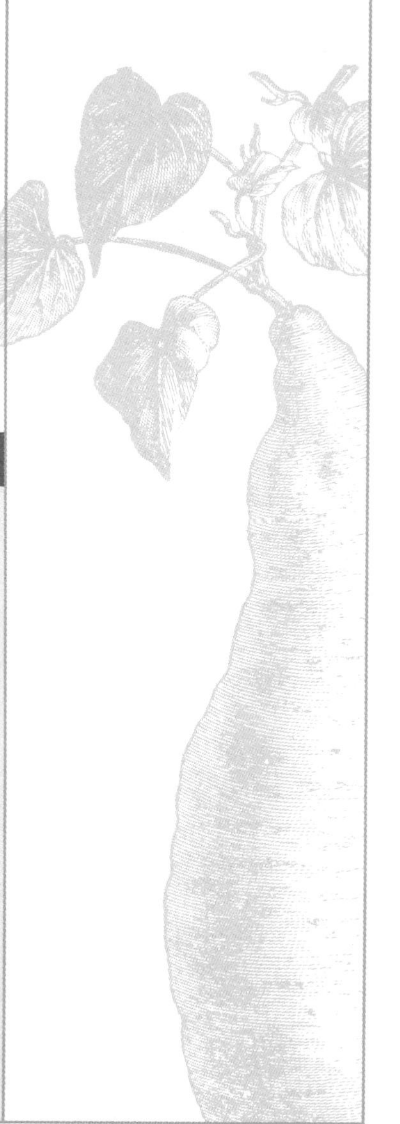

Storage Requirements

Cure sweet potatoes before dry storage to seal off wounds and minimize decay. Place in a warm, dark, well-ventilated area at 85°F–90°F and high humidity for 4–10 days. Store in a cool place, making sure they don't touch. Temperatures below 55°F cause chill injury. Don't touch until ready to use.

Fresh

Temperature	Humidity	Storage Life
55°F–60°F	85%–90%	4–7 months

Preserved

Method	Taste	Shelf Life
Canned	good	12+ months
Frozen	excellent	6–8 months
Dried	good	12+ months

Selected Varieties

WHITE

These white sweet potatoes, not very well known, have less beta-carotene but still more vitamin C than tomatoes. They are a good substitute for white potatoes.

Sumor: Very white sweet potato; an excellent and more nutritious substitute for Irish white potato; some disease and insect resistance; stores well; must boil before removing skin
South Carolina Foundation

White Delight: Heavy yields; unusual white flesh; texture and sweetness resemble orange-flesh varieties
South Carolina Foundation

YELLOW-ORANGE

Centennial: 100 days; very popular; bright copper skins; high yields; keeps well; good for Northern climates; tolerates clay soils; resists wilt
Burpee, Jung, Park

Excel: Excellent for baking or making fried chips; earlier and higher yields than Jewel; good shapes; high level of resistance to internal cork and stem rot (fusarium wilt); low resistance to soil rot (pox); good resistance to southern root-knot nematode
South Carolina Foundation

Jewel: 100 days; leading commercial variety; excellent keeper (up to 50 weeks); bright copper skin; highest yields of all; disease resistant; prefers sandy soil
Field's, Jung, Park, South Carolina Foundation

Regal: Excellent baking quality; high yield potential; high level of resistance to internal cork and stem rot (fusarium wilt); low level of resistance to soil rot (pox); good resistance to southern root-knot nematode; resistance to tobacco and southern potato wireworm, banded cucumber beetle, spotted cucumber beetle, elongate flea beetle, pole striped flea beetle, sweet potato flea beetle and to at least two species of white grubs; stores well but not as long as Jewel
South Carolina Foundation

Southern Delight: Dark orange flesh; high pest resistance, especially root-knot nematodes; resists diseases; excellent baking quality; introduced in 1988
South Carolina Foundation

Vardaman: 110 days, a "bush" with short vines of only 4'–5'; resists fusarium wilt better than longer vined types; worth trying in the greenhouse
Field's, Jung, Park

Greenhouse

Vardaman

tomato

Lycopersicon escutentum
Solanaceae (Nightshade Family)

TOMATOES are a warm-season crop, very tender to frost and light freezes. Plan an average of two to five plants per person. All parts of the plant except the fruit are poisonous. Never plant near the walnut family trees (see Walnut on page 237). To start in flats, sow seeds at least ½-inch apart. Seedlings will be spindly with less than 12 to 14 hours of light per day. When seedlings have four leaves, transfer to a deeper pot and again when 8- to 10-inches tall. Each time, place the uppermost leaves just above the soil line and remove all lower leaves. Transplant into the garden when the stem above the soil has again reached 8- to 10-inches tall. Allow up to 10 days to harden off. Soil temperature should be at least 55 to 60°F to transplant, otherwise plants turn yellow, become stunted, and are slow to bear. To transplant, pinch off the lower leaves again, and lay the plant on its side in a furrow about 2½ inches below the soil surface. This shallow planting speeds up growing since the plant is in warmer soil. Put in stakes on the downwind side of the plants. Some sources suggest that indeterminate and larger semideterminate varieties be pruned of all suckers (tiny leaves and stems in the crotches of larger stems) because they may steal nourishment from the fruits. However, the Erie City Extension Service has demonstrated that removing leaves decreases photosynthetic production. Hand pollinate in greenhouses.

Unlike most crops, you may solarize soil as you grow tomatoes because they're very heat tolerant. Solarizing helps control disease, particularly verticillium wilt. Wet the soil and cover with clear plastic for the entire season for best results.

First Seed-Starting Date

Set out 2 to 4 weeks after the last frost. In Florida, Texas, and southern California, tomatoes can be transplanted in late winter and removed in summer when they stop bearing.

Germinate	+	Transplant	–	Days Before LFD	=	Count Back from LFD
7 to 14 days	+	42 to 70 days	–	14 to 28 days	=	28 to 56 days (avg 6 weeks/42 days)

Last Seed-Starting Date

In Florida, Texas, and southern California, gardeners often plant fall crops.

Germinate	+	Transplant	+	Maturity	+	SD Factor	+	Frost Tender	=	Days Back from FFD
7 to 14 days	+	42 to 70 days	+	55 to 90 days	+	14 days	+	14 days	=	132 to 202 days

Temperature

For germination: 60°F–85°F
For growth: 70°F–75°F

Soil & Water Needs

pH: 5.8–7.0

Fertilizer: Heavy feeder. Fertilize 1 week before and on the day of planting. Avoid high N and K at blossom time. Too much leaf growth may indicate too much N or too much water.

Side-dressing: Every 2–3 weeks apply light supplements of weak fish emulsion or manure tea. When blossoming, side-dress with a calcium source to prevent blossom-end rot.

Water: Medium and deep watering until harvest. Even moisture helps prevent blossom-end rot.

Measurements

Planting Depth: ½"
Root Depth: 8", some to 6'

Height
Determinate: 3'–4'
Indeterminate: 7'–15'
Breadth: 24"–36"

Space between Plants:
In beds: 18"
In rows: 24"–36"
Space between Rows:
3'–6'

Support Structures

Use a wire cage, stake, or trellis; most gardeners prefer cages.

Pests

Aphid, beet leafhopper, cabbage looper, Colorado potato beetle, corn borer, corn earworm, cucumber beetle, cutworm, flea beetle, fruit worm, garden centipede, gopher, Japanese beetle, lace bug, leaf-footed bug, mite, nematode, slug, snail, stinkbug, thrips, tobacco budworm, tomato hornworm, whitefly

Diseases

Alternaria, anthracnose, bacterial canker, bacterial spot, bacterial wilt, botrytis fruit rot, curly top, damping off, early blight, fusarium wilt, late blight, nematode, psyllid yellows, septoria leaf spot, soft rot, southern blight, spotted wilt, sunscald, tobacco mosaic, verticillium wilt.

Environmental disorders: Blossom-end rot, sunscald

Storage Requirements

Wash and dry before storing. Pack no more than two deep.

Fresh

Temperature	Humidity	Storage Life
ripe: 45°F–50°F	90%–95%	4–7 days
green: 55°F–70°F	90%–95%	1–3 weeks

Preserved

Method	Taste	Shelf Life
Canned	excellent	12+ months
Frozen	good	8 months
Dried	good	12+ months

Allies

Some evidence: Cabbage
Uncertain: Asparagus, basil, bee balm, borage, coriander, dill, goldenrod, marigold, mint, parsley, sage

Companions

Brassicas, carrot, celery, chive, cucumber, marigold, melon, nasturtium, onion, pea, pepper

Incompatibles

Corn, dill, fennel, kohlrabi, potato, walnut

Harvest

Pick when fruit is evenly red but still firm. If warmer than 90°F, harvest fruit earlier.

Selected Varieties

IND = indeterminate
DET = determinate
OP = open-pollinated
H = hybrid
V = verticillium
F = fusarium
N = nematodes
T = tobacco mosaic
L = leafspot

STANDARD

Ace 55 VF: 70 days; H; DET; well-suited for hot, dry areas in West
Burpee, Tomato Growers

Beefmaster VFN: 80 days; H; IND; tomatoes up to 2 pounds; need staking
Park, Tomato Growers

Bonny Best: 76 days; OP; IND; early season; heirloom; meaty and good flavor; best grown in a wire cage
Heirloom Seeds, Nichols, Tomato Growers

Carmello: 70 days; IND; this French hybrid is our all-time favorite tomato; heavy bearer of medium to large crack-resistant fruits; disease resistant; exceptional sweet flavor; hands down the best tomato we've ever grown
Kitchen Garden, Nichols, Shepherd's

Coldset: OP; DET; medium fruit; withstands soil temperature of 50°F; can sow directly; doesn't tolerate wet, humid conditions well
Gurney's, Southern Exposure, Stokes

Delicious VFN: 77 days; OP; IND; midseason; very large fruit; good for slicing; little cracking
Widely available

Early Girl VFF: 54 days; H; IND; earliest slicing/canning; sweet-tart; improved variety has more disease resistance
Widely available

Floramerica VFFA: 70 days; H; DET; AAS Winner; developed in the South to resist fifteen diseases; resists cold, heat, and humidity; good fresh or canned
Field's, Jung, Tomato Growers, William Dam

Selected Varieties *(cont'd)*

Orange Strawberry: 80 days; IND; large orange-yellow oxheart-type fruit shaped like a fat strawberry; can range from 8 oz to 1 lb; great yellow tomato flavor; good fresh or for golden tomato sauces
Tomato Growers

Patio Prize VFNT: 68 days; H; DET; bush; no staking; medium-sized fruit (4 ounces); excellent disease resistance
Park, Stokes, Tomato Growers

Pink Ponderosa: 80 days; OP; IND; late; very large; meaty and solid beefsteak fruit; low acid
Widely available

Quick Pick VFFNT: 68 days; H; IND; excellent flavor and texture; high yields
Park, Tomato Growers

Silvery Fir Tree: 55–60 days; heirloom from Russia; 2′ dwarf; produces early; good for containers and grows well in overcast coastal areas; 4–6 oz fruit
Seeds of Change, Tomato Growers

Striped Stuffer: 85–90 days; H; IND; unusual thick-walls like peppers; good for stuffing or grilling; red with golden-orange stripes
Burpee, Thompson & Morgan

Sub-Arctic Maxi: 62 days; OP; DET; early; sparse foliage for quick ripening; vigorous; no staking
Widely available

Yellow Oxheart F: OP; stores 3–6 months; excellent flavor
Southern Exposure, Tomato Growers

CHERRY

Most of these are determinate. Indeterminate are better flavored but need more space.

Chiapas: Sprawling; survived curly top in low desert; prolific; for low desert
Native Seeds

Juliet Hybrid: 60 days; IND; AAS Winner; grapelike clusters of 1" fruit; high yields; crack resistant; tolerates late blight and leaf spot
Widely available

Penn State Cherry: DET; high concentrations of the antioxidant lycopene; resists fungal blights and is well-suited to humid growing conditions in the Mid-Atlantic states
Look for it in coming years.

Sweet 100: 65 days; H; IND; excellent sweet flavor; midseason; plants are tall and need staking; high yields over prolonged period; disease resistant; high vitamin C
Widely available

Tiny Tim: 60 days; OP; DET; early; 15" plants; good for containers; fruit ¾"; not as sweet as large cherry types
Widely available

Selected Varieties *(cont'd)*

Yellow Pear: 75 days; OP; IND; pretty yellow pear-shaped; resists heat; meaty; great for eating fresh and soup; extremely prolific; requires significant staking and pruning; drought tolerant
Widely available

SAUCE AND PASTE

Most of these are determinate but can be staked.

Del Oro VFNA: 72 days; H; DET; best disease resistance of paste types
Harris, Tomato Growers

Milano: 60–65 days; DET; this Italian hybrid is our favorite plum-type tomato; prolific; meaty fruits have a rich sweet taste that is great for sauces and drying
Kitchen Garden, Shepherd's

Roma II VF: 80 days; OP; DET; late season; compact plants; heavy bearer; very solid and meaty with few seeds; good canned whole and good for paste
Widely available

San Marzano: 80 days; OP; IND; popular; rectangular pear-shaped fruit; meaty; excellent paste tomato
Widely available

Super Sarno: 78 days; OP; IND; classic Italian tomato; 5 to 5½ oz; delicious; meaty; for salads and sauces; produces earlier than other San Marzona types
Johnny's

DRYING

Principe Borghese: 75 days; H; DET; bush cherry grown in Italy for drying; meaty and large cherry; excellent flavor dried or fresh; plants bear profusely and hold fruit so that whole plants can be dried in arid climates
Cook's, Southern Exposure, Tomato Growers

Greenhouse

Coldset, Patio Prize VFNT, Sub-Arctic Maxi, Tiny Tim

tomato

Chapter Three

fruits &nuts

apple

Malus pumila
Rosaceae (Rose Family)

THOUSANDS of apple varieties have been grown since ancient times. Many are lost to posterity, but more varieties of apple exist today than of any other fruit. Breeding for disease resistance has focused on the apple more than other fruits, so organic orchardists may have the greatest chance of success with this fruit. After the June drop and when the fruit is no more than 1 inch in diameter, thin to 8 inches apart, or remove about 10 percent of the total fruit. Also thin all clusters to just one fruit; this will help produce large fruit and encourage return bloom the next year. The inclination of the apple branch is thought to determine its fruitfulness — the more horizontal, the more fruit. U.S. researchers have shown that shining red lights on apple trees for 15 minutes each night, beginning 2 weeks before harvest, delays fruit drop for 2 weeks. Other researchers are experimenting with inoculating bare roots with hairy root organism, previously thought to be a problem disease but now shown to promote early root growth and fruiting.

To reduce the need for pest management, some growers have shown that, after early-season management of diseases and pests that attack trees, apples can be covered with small paper bags when they reach ½ to ¾ inch in diameter. Fruit can be thinned at the same time the apples are bagged. Apple bags are now available commercially.

Site

Southeast exposure. Clay loam. To espalier, some suggest siting apples on an eastern wall or slope in very hot summer climates to avoid sunburn.

Soil & Water Needs

pH: 6.5–7.0
(6.0–6.5 for bitter pit)

Fertilizer: Low N for young trees. Appropriate new growth is 6"–14".

Side-dressing: Apply compost in late autumn and work into soil.

Water: Medium

Measurements

Root Depth: 10' or more, with a spread 50 percent beyond drip line

Height:
Dwarf: 6'–12'
Semidwarf: 12'–18'
Standard: 20'–40'

Spacing:
Dwarf: 8'–20'
Semidwarf: 15'–18'
Standard: 30'–40'

Growing & Bearing

Bearing Age:
Dwarf and semidwarf: 2–3 years
Standard: 4–8 years

Chilling Requirement: 900–1000 hours below 45°F, though some require less

Pollination: Most require cross-pollination

Support Structures

Branch may need support when fruiting; branch separators can increase yields.

apple

Shaping

Training
Freestanding: Central leader
Wire-trained: All cordons, espalier, fans, stepovers, palmettes

Pruning: Spur types require little annual pruning because spurs bear for about 8 years, but each spring remove 1 out of 10 spurs and thin fruit by 10 percent. Tip-bearers fruit on 1-year-old wood. For these, prune back some of the long shoots and some of the spurs. For bitter pit, a sign of unbalanced growth, remove the most vigorous shoots at the end of the summer.

Pests

Aphid, apple maggot, cankerworm, codling moth, European apple sawfly, European red mite, flea beetle, fruit worm, gypsy moth, leafhopper, leafroller, mice, oriental fruit moth, pear slug, plum curculio, potato leafhopper, scale, tent caterpillar, weevil, whitefly, white grub, woolly apple aphid

Storage Requirements

Wrapping in oiled paper or shredded paper helps prevent scald. Some apples stored over winter develop a rich flavor that is excellent for pies. Israeli research shows that summer apples keep better if held at high temperatures several days before storage. Do not store apples with potatoes because apples will lose their flavor and potatoes will develop an off flavor.

Fresh

Temperature	Humidity	Storage Life
32°F–40°F	80%–90%	4–8 months

Preserved

Method	Taste	Shelf Life
Canned	good	up to 2 years
Dried	good	6–24 months

Diseases

Apple scab, baldwin spot, canker dieback, cedar apple rust, crown gall, crown rot, cytospora canker, fire blight, powdery mildew, sunscald

Allies

Some evidence: Buckwheat, *Eryngium* genus of herbs (e.g., button snakeroot and sea holly), *Phacelia* genus of herbs (e.g., California bluebells), weedy ground cover

Uncertain: Caraway, coriander, dill, garlic, nasturtium, tansy, vetch, wildflowers, wormwood

Incompatibles

Mature walnut tree, potato

Harvest

For summer apples, pick fruit just before fully ripe, otherwise the apples become mealy. In fall, pick fruit only when fully ripe. Make sure you pick with the stems, or a break in the skin will occur that will permit bacteria to enter and foster rot. Be aware that ripening apples give off small amounts of ethylene gas that may inhibit the growth of neighboring plants and cause early maturing of neighboring flowers and fruits.

Selected Varieties

EARLY (SUMMER APPLES)

Jerseymac: Red fruit; one of the best early apples but bruises easily; susceptible to scab and blight; productive; naturally large tree
Southmeadow

Redfree: Large, bright red fruit; immune to scab and cedar rust; moderate resistance to mildew and blight; excellent flavor; naturally small tree
Adams

Williams Pride (Co-op 23): Very early; large red fruit; 1988 Purdue release; best flavor of disease-resistant apples; immune to scab; resists mildew, cedar rust, and fire blight; stores 1 month refrigerated; naturally moderate-sized tree; tested in Zones 5–6
Raintree, Sonoma

MIDSEASON

Gravenstein: Yellow with red stripes; old favorite; great flavor; vigorous; bears biennially; ideal for sauce and cider; infertile pollen; requires pollinator; loved on the West Coast
Widely available

Honeycrisp: Moderately vigorous tree with spreading growth habit; amazing flavor and storage; very winter-hardy, moderate disease resistance
Widely available

Selected Varieties *(cont'd)*

Jonafree: Red; good for fresh eating; immune to scab; resists fire blight and cedar apple rust; good keeper; high yields; vigorous and spreading tree; Zones 5–8
Miller

Liberty: Red with yellow; McIntosh-type; resists scab, mildew, rust, and fire blight; good dessert apple; naturally large tree; Zones 4–8
Widely available

Pristine: Vigorous tree; yellow fruit with high sugar content; excellent resistance to fire blight and apple scab
Adams

LATE

"Braeburn" and later bearers need about 150 frost-free days for fruit to mature. Late ripeners tolerate temperatures as low as 29°F before internal freezing occurs, according to University of Minnesota Extension. If freezing does occur, let the apples thaw before harvesting, to prevent severe bruising.

Ashmead's Kernel: Historical yellow apple with orange-brown blush; russet; natively large tree; excellent flavor fresh or juiced; excellent keeper; resists mildew; tart when tree ripe; mellows with storage; naturally moderate-sized tree; hardy through Zone 3
Applesource, Bay Laurel, Sonoma Apple, Southmeadow

Braeburn: Yellowish with red blush; new from New Zealand; crisp; juicy; stores 6–12 months; bears young; manageable by homeowner because only moderately vigorous; susceptible to powdery mildew; doesn't appear to be susceptible to fire blight
Widely available

Brown Russet: Very late; russet with patches of green and red; good fresh, stored, or as sweet cider apple; resists scab and mildew; naturally small to moderate-sized tree; hardy through Zone 1
Raintree

Golden Delicious: Universal pollinator; vigorous spreading tree; precocious; heavy crops; resists scab; crops well; susceptible to cedar apple rust; discovered in West Virginia
Widely available

Goldrush: Dessert-quality; spur-bearing; excellent fresh or for baking; resistant to apple scab and powdery mildew; moderately resistant to fire blight; excellent storage; Zones 5–8
Applesource, Stark

Tydeman's Late Orange: Excellent storage apple; reaches full flavor around Christmas; some say flavor is better than Ashmead's Kernel; naturally moderate-sized tree
Bay Laurel, Sonoma Apple, Southmeadow

Selected Varieties *(cont'd)*

CRAB (any apple smaller than 2" in diameter)

Dolgo: Early; best all-purpose crab; excellent pollinator for all apples; excellent ornamental; excellent jelly; high disease resistance; needs to chill 400 hours or less; naturally small to moderate-sized tree; hardy through Zone 1
Widely available

Rootstocks

Most growers now choose dwarf apples; standard trees grow very large, don't bear for years, and are difficult to harvest because of their height. To choose a rootstock you must know your soil type, drainage, and depth; then consider the specific variety's natural growing habit and size. These factors determine how much dwarfing you need in a rootstock. In rich, fertile soil all rootstocks grow more vigorously than predicted and need extra spacing. Buy smaller trees when possible because they suffer less transplanting shock, and are more productive and vigorous.

Most nurseries don't offer a choice of rootstocks for a particular variety, but choice can be found between nurseries. The purpose of the chart on pages 164–165 is not to help you make an independent decision; it is to help you conduct an informed discussion with the nursery. We urge you to seek and follow the nursery growers' advice.

Notes

Stark Brother's Nurseries cautions that extremely dwarfing rootstocks (M27 and P22) won't do well for most gardeners. Miller Nurseries also notes that Malling 9 is very difficult for homeowners; its roots are shallow and brittle. It must be securely staked or the tree will blow over. The British Institute of Horticultural Research showed that wrapping M9 stems in July with black polyethylene film to 6" above ground increases root production.

apple rootstocks

Rootstock Name	Size (percent of standard)	Height / Width	Best Soil	Anchorage	Hardiness	Drought
Malling 27 EMLA 27/M27	Minidwarf 15%–30%	H: 4'–8' W: 2'–8'	Clay loam	Poor/stake	Low hardiness	—
Poland 22 P22	Minidwarf 15%–30%	H: 5'–6' W: 2'–8'	—	Poor/stake	Very hardy	—
Malling 9 EMLA 9/M9	Dwarf 20%–40%	H: 8'–10' W: 8'–10'	Sandy and grainy loam	Poor/stake	—	—
Malling 26 EMLA 26/M26	Dwarf 30%–40%	H: 8'–14' W: 10'–14'	Sandy and grainy loam	Fair/might stake	Very hardy	—
Mark Mac-9	Dwarf 30%–45%	H: 8'–14' W: 8'–12'	Clay	Good/no stake	Very hardy	—
M9/111	Dwarf 25%–50%	H: 10'–15' W: 10'–15'	—	Good/no stake	—	Tolerant
M7a Malling VII EMLA 7	Dwarf 40%–60%	H: 11'–20' W: 12'–16'	Most soils; avoid heavy clay	Good/no stake	Very hardy	Tolerant
MM106 Merton-Malling 106	Dwarf 55%–85%	H: 14'–21' W: 14'–18'	Sandy loam; avoid poor drainage	Fair to good	Low hardiness	Tolerant
MM111 Merton-Malling 111	Dwarf 65%–85%	H: 15'–24' W: 15'–20'	All soils; can also tolerate wet soils	Very good/ no stake	Very hardy	Tolerant

HR = highly resistant; LR = low resistance; MR = moderately resistant; P = precocious (early bearing); R = resistant; S = susceptible; SS = somewhat susceptible; VP = very precocious; VS = very susceptible

From Robert Kourik, *Designing and Maintaining Your Edible Landscape Naturally* (Santa Rosa, Calif.: Metamorphic Press, 1986), 166–67.

Pests and Diseases						
crown rot	woolly aphid	nematodes	fire blight	powdery mildew	Precocious	Other Factors
HR	LR	—	S	MR	P	Remove fruit first 2 years; stops growing when bears fruit; good for espalier when grafted to vigorous varieties
—	—	—	—	—	—	Roots are brittle; union with some types is brittle when young
HR	S–LR	S	VS	MR	VP	Mice love this rootstock, so use tree guard; produces large fruit; defruit or thin fruit in first 2 years to prevent loss of leader
S–MR	S–LR	—	S–VS	MR	VP	Can form root galls at graft union; defruit or thin first 2 years to prevent loss of leader; doesn't sucker much; produces large fruit
R	S–LR	—	S	—	P	Open structure and roots well in stoolbeds; hardy to Zone 4 but not as hardy as M26
R–VR	R–LR	—	S–VS (in early years)	MR	—	Combines benefits of 111 and 9; bury rootstock so MM111 part is underground and M9 part is exposed
L–MR	LR	—	R	MR	P	Susceptible to burr knot; better than M26 on wet soils; remove suckers each year
R	HR	—	HR	S	P	Susceptible to burr knot; good stock for spurs; fumigate site for nematodes, which spread union necrosis and ringspot. Stake in hardpan soil.
R	R	—	SS	SS	No	Smaller harvest than a stock like M106, but still productive; ideal for an interstem or under low vigor varieties

apricot

Prunus armeniaca
Rosaceae (Rose Family)

APRICOTS are good additions to the orchard. They're pretty (with glossy green leaves), easily managed, and one of the most drought-resistant fruit trees. They are, however, vulnerable to winter damage, and their buds are particularly susceptible to late frost damage (see Site below). Apricots grow vigorously and require annual pruning and thinning. After the natural fruit drop in late spring and when the fruit is about 1 inch in diameter, thin fruit to 3 to 4 inches apart. Summer temperatures over 95°F cause pit burn, a browning around the pit. Apricots enjoy long lives of approximately 75 years.

Site

Not too rich or sandy. In the North, plant 12'–15' from the northern side of a building. This delays buds and minimizes late frost injury but ensures full summer sun. Avoid windy locations. Do not plant on former apricot, cherry, or peach tree sites; when waterlogged, their roots release hydrogen cyanide that may linger in the soil and hinder growth.

Soil & Water Needs

pH: 6.0–6.5

Fertilizer: Appropriate new growth on a young tree is 13"–30"; on a bearing tree 10"–18". Because the tree is naturally vigorous, go easy on N.

Side-dressing: Apply compost or well-rotted manure mixed with wood ashes annually in spring, before leaves appear.

Water: Medium

Measurements

Root Depth: 50–100 percent farther than drip line

Height:
Dwarf: 6'–7'
Semidwarf: 12'–15'
Standard: 20'–30'

Spacing:
Dwarf: 8'–12'
Semidwarf: 12'–18'
Standard: 25'–30'

Growing & Bearing

Bearing Age: 3–9 years

Chilling Requirement:
Very low, 350–900 hours, which results in early blooming.

Pollination: Most are self-pollinating, but yields are higher with more than one variety.

Shaping

Training:
Freestanding: Open center, dwarf pyramid. In colder areas use central leader.
Wire-trained: Fan

Pruning: If the tree bears fruit only in alternate years, prune heavily when more than half of the flowers are blooming. Pruning encourages new spurs, each of which bears fruit for about 3 years. Prune yearly to encourage fruiting spurs. Remove wood that is 6 years old or more.

Pests

Aphid, cankerworm, cherry fruit sawfly, codling moth, gopher, gypsy moth caterpillar, mite, peach tree borer, plum curculio, white fly

Diseases

Bacterial canker, bacterial spot, black knot, brown rot, crown gall, cytospora canker, scab, verticillium wilt

Allies

Some evidence: Alder, brambles, buckwheat, rye mulch, sorghum mulch, wheat mulch
Uncertain: Caraway, coriander, dill, garlic, nasturtium, tansy, vetch, wildflowers, wormwood

apricot

Incompatibles

Persian melon, plum. Also, don't plant where any of the following have grown in the previous 3 years: eggplant, oats (roots excrete a substance that inhibits the growth of young apricot trees), pepper, potato, raspberry, strawberry, tomato.

Harvest

When all green color is gone and the fruit is slightly soft, twist and gently pull upward. If possible, harvest apricots when fully ripe. If plagued with animal problems, you may want to pick them slightly green and ripen them at 40°–50°F.

Storage Requirements

For canning, use only unblemished fruits, or all fruits in the container will turn to mush. For drying, split the apricot first and remove pit. If after drying the fruit is still softer than leather, store in the freezer.

Fresh

Temperature	Humidity	Storage Life
60°F–65°F	90–95%	a few days
40°F–50°F	90–95%	3 weeks

Preserved

Method	Taste	Shelf Life
Canned	good	up to 2 years
Frozen (after partially drying)	excellent	12–24 months
Dried	good	6–24 months
Jam	good	up to 18 months

Selected Varieties

Aprium: Late season; apricot-plum hybrid; self-fertile; medium-large, apricot-sized fruit with some plum taste; clear yellow skin; tree can be maintained at 10'; Zones 6–9
Raintree

Dwarf Sungold and Dwarf Moongold: Early midseason; very cold hardy for the North; developed by the University of Minnesota; recommended to pollinate each other; good fresh, canned, or as jam; Zones 3–8
Miller

Goldcot: Midseason; bred deep in Michigan snowbelt; excellent cold-hardiness; vigorous; consistent production; tangy flavor, great for canning; Zones 5–8
Southmeadow, Stark

Harcot: Early ripening; freestone; firm and sweet; cold hardy; late blooming; vigorous; good resistance to perennial canker, bacterial spot, and brown rot; self-fertile; should be well-thinned to avoid over-setting; Zones 5–8
Adams, Bay Laurel, Sonoma Apple

Hargrand: Midseason; very large, firm fruit; freestone; tolerant to brown rot, bacterial spot, and perennial canker; winter hardy and productive
Adams

Harogem: Late season; upright, productive tree; very hardy; fruit is beautifully colored; freestone; good flavor; resistant to brown-rot and perennial canker
Adams

Puget Gold: Mid- to late season; semidwarf; self-fertile; survives frosty Springs where others fail
Raintree, Sonoma Apple

Royal Blenheim: Late season; grown in California; self-fertile; needs warm, dry weather during bloom; medium to large fruit; good fresh, canned, or dried; high yields and intense flavor; fruit is subject to pitburn in warmest springs; good pollinator
Bay Laurel, Sierra Gold, Sonoma Apple

Tomcot: Early-ripening; large, sweet fruit; self-fruitful but bigger crops if cross-pollinated; very reliable in areas with wide day-night temperature ranges, which can cause early blossom drop
Bay Laurel, Raintree, Vanwell

Rootstocks

Generally, don't use peach rootstocks because they're susceptible to peach tree borer, root-knot and lesion nematodes, root winter injury, and uneven growth, which weakens the graft. Try not to use plum rootstocks — those dwarfed on Nanking cherry (*P. tomentosa*) or sand cherry (*P. besseyi*). Although more tolerant of wet soil, these sucker continuously and cause a different fruit flavor. Apricot rootstock offers the best odds for tree survival; it is resistant to nematodes and has some resistance to peach tree borers.

blackberry

Rubus (2 species)
Rosaceae (Rose Family)

BRAMBLE FRUITS are very easy to grow. The keys to good yields are adequate spacing and light, and, because of shallow roots, good weed control and thick mulch. Rather than several short, close rows that limit berry development to only the upper cane parts, plant one long, narrow row that will produce berries to the bottom of the canes.

Blackberries are biennial. *Primocanes,* first-year green stems, bear only leaves. *Floricanes,* second-year brown stems, produce fruit. Upright or erect canes are shorter, whereas trailing varieties (also known as *dewberries* in the South) grow flexible canes as long as 10 feet. Blackberries are usually hardy to Zones 5 through 8; upright varieties are the hardiest. Blackberries have become a desirable crop for small market gardeners because of their ease of culture and the premium price they can command.

Site

Full sun; rich loam. Due to verticillium wilt, avoid planting where nightshade family plants were grown in the last 3 years. Plant at least 300 feet away from wild brambles (which harbor pests and diseases) and from raspberries (to prevent cross-pollination).

Soil & Water Needs

pH: 5.0–6.0

Fertilizer: In spring, apply well-rotted manure or compost before canes break dormancy.

Mulch: In summer, apply 4"–8" of organic mulch; in winter apply 4"–6" of compost.

Water: Medium; drip irrigation is essential to avoid water on the berries, which is absorbed and dilutes flavor. Water regularly because of vulnerability to water stress.

Measurements

Root Depth: More than 12"

Height: 4'–10' when pruned

Spacing:
Erect and Semierect Blackberry: 2'–3' in a row (suckers and new canes fill out row)
Trailing Blackberry: 6'–12' in a row (no suckers, but canes grow very long)

Root Depth: More than 12"

Growing & Bearing

Bearing Age: 2 years

Chilling Requirement: Hours needed depend on the variety

Pollination: Self-pollinating

Propagation

Erect Blackberry: By suckers. When dormant, dig up root suckers no closer than 6" to mother plant.

Thornless and Trailing: By tip layering. Late in the season (August to September), bend and bury the primocane tips 4" deep in loose soil. In spring, cut off the cane 8" from the ground and dig up the new plants.

Shaping

Training: A trellis is critical for disease and pest reduction, quality fruit, and easy harvest. For erect blackberries use a 4' top wire; for trailing blackberries use a 5' top wire. Fan out canes and tie with cloth strips.

Storage Requirements

Freeze within 2 days by spreading out berries on cookie sheets and freezing. When rock hard, store in heavy freezer bags. Refrigerated, they keep 4–7 days.

Preserved

Method	Taste	Shelf Life
Canned	excellent as jam	up to 2 years
Frozen	excellent	12–24 months

blackberry

Pruning: After harvest or in spring, cut out old canes done bearing.

Erect Blackberries: Thin to 5–6 canes per row-foot; to encourage branching, cut off the top 3"–4" when primocanes are 33"–40"; late the following winter, cut the lateral branches back to 8"–12" long.

Trailing Blackberries: Thin to 10–14 canes per hill; don't prune in the first year; in late winter cut canes back to 10'.

Pests

Caneborer, mite, raspberry root borer, strawberry weevil, whitefly, white grub

Diseases

Anthracnose, botrytis fruit rot, cane blight, crown gall, powdery mildew, rust, septoria leaf spot, verticillium wilt

Allies

Some evidence: Grape

Companions

If berries are planted down the center of a 3' bed, plant beans or peas in the first summer to keep the bed in production and to add organic matter and N to the soil.

Incompatibles

Black walnut and all members of the nightshade family transmit verticillium wilt.

Harvest

When berries slide off easily without pressure, harvest into small containers so berries on the bottom won't be crushed. After harvest, cut back floricanes to the ground. For disease control, burn or dispose of all cut canes.

Selected Varieties

TRAILING
(Rubus procerus)

Hull: Midseason; thornless; cane height can reach up to 10'; very sweet; good fresh or in pies; moderately hardy; Zones 5–8
Indiana Berry

Triple Crown: Thornless; large and sweet berries; trailing and semierect; productive and vigorous; firm fruit ships well; cold-hardy
Widely available

UPRIGHT
(Rubus macropetalus)

Apache: Grows 5"–8" tall; thornless; fruits in late summer; suggested "deer-resistant"; Zones 5–8
Widely available

Arapaho: Thornless; tiny seeds; earliest ripening; highly prolific in producing primocanes from roots; some disease resistance; long harvest season; stores well; Zones 6–9
Bay Laurel, Indiana Berry, Raintree, Stark

Illini Hardy: Very hardy; medium-sized berry with good flavor; vigorous; thorny; consistently high-quality fruit
Indiana Berry, Miller, Nourse

Navajo: Huge berries ripen late in season; thornless; released a decade ago by the University of Arkansas, but only recently known for its unusual ability to remain firm for up to 2 or even 3 weeks after harvest; well-adapted in the Pacific Northwest, Southern Plains, and South Atlantic states; Zones 6–10
Park

Siskiyou: Long ripening season; productive; excellent flavor; disease resistant; Zones 7–10
Raintree

blueberry

Vaccinium (3 species)
Ericaceae (Heath Family)

BLUEBERRIES lack abundant root hairs and have shallow, underdeveloped roots concentrated in the top 14 inches of soil. As a result, regular watering and thick mulch are critical to keep down the weeds. A very acid pH is necessary for the plant to extract iron and nitrogen from the soil. Most blueberry problems are caused by stress related to pH, either underfertilization or overfertilization, or underwatering or overwatering. Plant 2-year-old bushes. When your plants arrive, do not put them in water. Follow directions and "heel in" until ready to plant. Try inoculating the roots with the beneficial mycorrhizal fungi, which increases yields significantly. Lowbush varieties are grown primarily in New England, highbush throughout the United States, and rabbiteye only in the South and West. Lowbush and rabbiteye require another variety for cross-pollination; highbush types don't, but yields increase with cross-pollination. To encourage root growth, remove all blossoms for a full 2 to 3 years. The delayed harvest will reward you with higher yields and healthier plants. Blueberries mature about 50 to 60 days from pollination. For areas prone to late spring frosts, blueberries are a good choice, with strong frost resistance. If your soil is sufficiently acid and well draining, consider adding blueberries to your landscape as easy, low-care plants that provide fresh fruit yearly and beautiful red foliage in fall.

Site

Full sun; choose a site where plants won't be disturbed, away from paths, roads, and driveways.

Soil & Water Needs

pH: 4.0–5.6

Fertilizer: Apply 1" of compost under mulch. Avoid high N, aluminum sulfate, or urea. If you must apply ammonium sulfate, use ½ ounce in year 1, and 1 ounce for every additional year thereafter.

Mulch: 3"–6" acid mulches such as pine needles, peat moss, shredded oak leaves, or rotted sawdust.

Water: Heavy and evenly moist

Measurements

Root Depth: Very shallow, top 14" of soil

Height:
Lowbush: 2'–4'
Highbush: 5'–6'
Rabbiteye: 6'–30'

Spacing:
Lowbush: 3'–4'
Highbush: 7'–8'
Rabbiteye: 6'–7'

Rows: 10'

Growing & Bearing

Bearing Age: 3–8 years

Chilling Requirement:
Lowbush and highbush: 650–800 hours below 45°F
Rabbiteye: 200 hours

Pollination: All varieties benefit from cross-pollination, while only lowbush and rabbiteye require cross-pollination by another variety.

Propagation

Layering; bend and bury the tip of a lower branch and cover with soil. Rooting hormone helps plant establish. The following spring, cut off the cane 8" from the ground and dig up the new plant.

Shaping

Pruning: Don't prune until the third year after planting because blueberries fruit near the tips of 2-year and older branches. To prune, cut out diseased tips and, for larger fruit, cut branches back to where buds are widely spaced. Also cut out weak and diseased branches, or *canes,* as they're called by commercial growers. Every 2 or 3 years you may need to cut back to the main stem canes that are 5 years old or older. Don't leave any stubs because suckers will be weaker than new canes growing from the roots. A good rule of thumb is to allow one branch per year of age plus one or two vigorous new branches. If new branch growth on an old bush (15 years) is thinner than ¼ inch, cut out half of the new canes.

Pests

Apple maggot, birds, cherry fruitworm, fruit fly, mite, plum curculio, weevil

Diseases

Bacterial canker, cane gall, crown gall, mummy berry, *Phytophthera cinnomomi,* powdery mildew. Many problems are due to a lack of acidity.

Allies

None

Companions

None

Incompatibles

None

Harvest

Leave berries on the bush 5–10 days after they turn blue. They're fully ripe when slightly soft, come off the bush easily, and are sweet. Pick directly into the storage bowl or containers so that as little as possible of their protective wax is removed.

Storage Requirements

Don't wash the fruit if you're going to freeze it.

Fresh

Temperature	Humidity	Storage Life
35°F–40°F (refrigerated)	80%–90%	7 days

Preserved

Method	Taste	Shelf Life
Canned	fair	up to 2 years
Frozen	good	12–24 months
As preserves	good	up to 18 months

Selected Varieties

Northern and Ornamentals

HIGHBUSH *(Vaccinium acorymbosum)*

Bluecrop: Midseason; large, delicious berry; 10–20 pounds/plant; upright, 4'–6'; very hardy; drought resistant; leading highbush; excellent ornamental; Zones 4–7
Widely available

Blueray: Early midseason; large berry; 10–20 pounds/plant; very sweet; leading U-pick; upright; 4'–6'; excellent ornamental; Zones 4–7
Fall Creek, Miller, Nourse, Sonoma Apple

Bluetta: Very early; small-medium berry; 10–20 pounds/plant; compact; spreading; 3'–5'; good ornamental due to low stature and scarlet foliage; Zones 5–7
Fall Creek, Hartmann's

Elliot: Late; small-medium berry; 10–20 pounds/plant; one of highest yielding blueberries; may need extra pruning due to heavy set; very tart until 60 percent of fruit is ripe; upright; 5'–7'; good ornamental; Zones 4–7
Bay Laurel, Fall Creek, Indiana Berry, Miller

Hardyblue: Midseason; small berry with superior flavor, excellent for baking; adaptable to heavy, clay soils; vigorous
Fall Creek

Northblue: Midseason; large berry; 3–7 pounds/plant; 20"–30"; best with snow protection; good ornamental; very cold hardy; Zones 3–7
Miller, St. Lawrence

Patriot: Early; large berry; 10–20 pounds/plant; upright; 4'–6'; resists *Phytophthera cinnomomi*; good ornamental; adaptable to many soil types; very cold hardy; Zones 3–7
Widely available

Spartan: Early; heavy bearer; prefers well-drained soil; large berry with excellent flavor; mummy-berry resistant; Zones 4–7
Raintree

LOWBUSH *(Vaccinium angustifolium)*

Northsky: 12"–18"; midseason; very cold-hardy but also adapts to warmer zones; small, tasty berries; well-suited for containers or group-plantings; self-fertile; Zones 3–8
Fall Creek, Raintree, St. Lawrence

Selected Varieties *(cont'd)*

Tophat: Very beautiful and one of the best ornamentals; 20" high and 24" breadth; medium berry; profuse blooms; tolerates partial shade; cold hardy; bonsai type; Zones 4–7
Hartmann's, Nourse, Raintree, Stark

Vaccinium angustifolium: Maine wild blueberry; ornamental and commercial uses; 6"–18"; spreads rapidly once established; excellent plant; Zones 3–7
Hartmann's

Wells Delight: Late; creeping; 5"–8"; evergreen similar to holly; excellent low maintenance ground cover; Zones 5–7
Hartmann's

Southern and Ornamentals

HIGHBUSH

All southern highbush are self-pollinating and ripen 20–30 days earlier than rabbiteye. Low-chill varieties can be grown farthest South.

Misty: Ornamental, hot pink spring flowers; tolerant of higher pH soils; hardy to 0°F; Zones 7–9
Raintree, Sonoma, Stark

Ozarkblue: Performs in hottest areas but also very cold hardy; large, delicious berries; extends berry season up to 1 month beyond other varieties
Fall Creek, Stark

Sharpblue: Early; large berry; 8–16 pounds/plant with irrigation; vigorous; highly adaptable; performs well in both heavy and sandy soils; 5'–6'; needs little chill; good ornamental; Zones 7–10
Bay Laurel

Sunshine: Late; medium berry; 5–10 pounds/plant; very hardy but needs little if any chill; will tolerate higher pH soils; excellent patio or pot culture; excellent southern ornamental; Zones 6–10
Bay Laurel, Fall Creek, Raintree

RABBITEYE
(Vaccinium ashei)

Pollinate with another in same maturation group.

Climax: Early midseason; large berry; 8–22 pounds/plant; very good quality; tall and spreading; 6'–10'; leading pollinator; good ornamental; berries tend to ripen uniformly; Zones 7–9
Indiana Berry

Premier: Early; grows 6'–10' tall; light blue, large berries; nice ornamental
Indiana Berry

sour cherry

Prunus dulcis
Rosaceae (Rose Family)

SOUR CHERRIES are the easiest cherries to grow because they're more tolerant than sweet cherries of cold winters, hot and humid summers, and heavy and cool soils. They also are less vigorous and therefore require less pruning. Sour cherries reportedly have a varying productive life, ranging from 15 to 35 years and averaging 20 to 25 years. Sometimes described as tart and known for their use in pies and other desserts, most sour cherries, with the exception of the Morella variety, are sweet enough to be eaten fresh. The key and perhaps the biggest challenge for cherry growers is to keep birds away long enough for the cherries to ripen on the tree. Netting, noisemakers, scare-eye balloons, scarecrows, and "trap crops" of mulberry bushes, which are preferred by birds, are all possible defensive strategies. See Sweet Cherry for more information.

Site

Full sun to partial shade. A southern exposure, or 12' from north-facing walls to delay blooming. Well-drained soil and good air circulation. Likes 4' of topsoil. Don't plant on former apricot, cherry, or peach sites; when waterlogged, their roots release hydrogen cyanide, which may linger in the soil and hinder growth. Also don't plant between other fruit trees, because cherries bear at a time others may need to be sprayed.

Soil & Water Needs

pH: 6.0–6.5

Fertilizer: Low N until 2 years old. Appropriate new growth is 12"–24" when young and 6"–12" bearing.

Water: Heavy and even supply. Cherries are especially sensitive to water stress.

Measurements

Root Depth: 50 percent beyond dripline

Height:
Dwarf: 6'–10'
Semidwarf: 12'–18'
Standard: 15'–20'

Spacing:
Dwarf: 8'–10'
Semidwarf: 18'–20'
Standard: 20'

Growing & Bearing

Bearing Age: 2–7 years

Chilling Requirement: 800–1200 hours at below 45°F

Pollination: All sour types are self-pollinating.

Shaping

Training:
Freestanding: Central leader or open center
Wire-trained: Fan-trained against a south- or north-facing wall, using the same method as for peaches. All sizes of sour cherries are suitable for training because they're not vigorous growers.

Pruning: The lowest branch should be about 2' off the ground and limbs about 8' apart. Both sour and sweet cherries bear on spurs as well as on 1-year-old wood. Sour cherries, however, produce several adjacent fruit blossoms on the year-old wood, which causes future bare spaces that lack foliage. If there are too many of these lateral flower buds, trim 1"–2" off the branch ends in June to stimulate leaf buds. Sour cherry spurs bear for 2–5 years.

Pests

Apple maggot, birds, blackberry fruit fly, cherry fruit fly, cherry fruit sawfly, codling moth, peach leaf curl, peach tree borer, plum curculio

sweet cherry

10–12 years; exercise extreme caution when harvesting or pruning to avoid damaging them.

Pests

See Sour Cherry (page 181).

Diseases

See Sour Cherry (page 182).

Allies

See Sour Cherry (page 182).

Companions

See Sour Cherry (page 182).

Incompatibles

None

Harvest

Wait until the fruit is fully ripe. Gently pull on the stem and twist upward. Be extremely careful not to damage or rip off spurs, as these bear for 10 years or more.

Selected Varieties

RED

Black Gold: Late; self-fertile; very sweet; large crop; Zones 4–9
NYFT, Raintree

Compact Stella: Midseason; genetic dwarf, 8'–14'; self-pollinating; dark red fruit; Zones 6–7
Miller

Lambert: Midseason; flavor and looks of the Bing; later season; more resistant to cracking; excellent fresh or canned; hardy; high chill requirement; pollinated by Rainier
Bay Laurel, Northwoods, Sierra Gold

Lapins Sweet Cherry: Late midseason; self-pollinating; Bing-type fruit; crack resistant; firm, meaty texture; from British Columbia; precocious; one of the best Bing-types for home orchards
Widely available

Selected Varieties *(cont'd)*

Starkrimson: Early; genetic dwarf, 12'–14'; self-pollinating; large fruit; very sweet; high yields; Zones 5–8
Stark

Utah Giant: Midseason; large fruit with excellent flavor; disease resistant; sets large clusters of partially freestone fruit
Bay Laurel, Sierra Gold, Sonoma

Van: Late midseason; bred to be hardy; won't crack in rainiest weather; excellent pollinator; not self-fertile
Bay Laurel, Sierra Gold, Sonoma, Stark

YELLOW

Rainier: Early; large fruit with red blush; like Royal Ann; firm, yellow-white flesh; resists splitting; productive; spreading tree
Widely available

Stark Gold: Late; unusual yellow fruit with unique, tangy flavor; survived –30°F; attracts very few birds; crack resistant; pollinate with any other sweet cherry; excellent for processing; Zones 5–7
Adams, Stark

White Gold: Early–midseason; consistent heavy crops; resistant to cracking and bacterial canker; self-fertile; Zones 4–9
NYFT, Raintree

DUKE CHERRIES

The Duke cherry, not widely grown since the beginning of the twentieth century, might still be available at specialty fruit suppliers. Duke cherries are hardier than sweet cherries and excellent substitutes. Their fruit is yellow-red, soft, and juicy. Pollination is less difficult than with sweet cherries. If you can't grow sweet cherries because the climate is too cold, you might try these.

Rootstocks

Mazzard (*Prunus avium*) rootstock is good for sweet cherries. It is a vigorous grower, produces large trees, is slow to bear, and is tolerant of wet soil. Mazzard is particularly good in the West, Northwest, and areas of the East where moisture is sufficient and winter hardiness is not a problem. GM61, a new rootstock from Belgium, is considered by some to be the best overall dwarfing stock for cherry. Its trees are 50–60 percent of standard size; they can be maintained below 15', are hardy to at least –20°F, are spreading and precocious, do well in heavier soils, and are good for both sweet and sour. Trees propagated on GM61 are available from Northwoods, Raintree, Rocky Meadow, and Stark Nurseries. For information on other rootstocks, see Sour Cherry (page 183).

chinese chestnut

Castanea mollisima
Fagaceae (Beech Family)

The CHINESE CHESTNUT is an attractive, globe-shaped landscape shade tree. It also offers rot-resistant timber and regular annual crops of one of the sweetest nuts. In early summer, the tree is decked with pretty (but odoriferous) yellow catkins, and its glossy, dark green serrated leaves cling late into fall. Chestnut fungal blight swept through North America in the early 1900s, destroying nearly all American (*C. dentata*) and European (*C. sativa*) chestnuts, both of which are now planted on the West Coast, where blight is less of a problem. The Chinese chestnut, introduced in 1853, resists blight and grows well in Zones 5 through 8 in a wide variety of soil and climatic conditions, although it's grown primarily in the East and Northwest. Most seedlings do better than grafted trees in the northern Zones 5–6. Hardy to about –20°F; late blooming permits it to escape spring frosts. For rot-resistant poles, chestnuts can be coppiced: Cut down the tree to stimulate sucker growth, let suckers grow to desired pole thickness, and cut again. Chestnuts can live 50 years or more.

Site

Full sun. Preferred soil is light and sandy, but it can be rocky or silty as long as it isn't alkaline or very dry. Soil must be well drained. Avoid frost pockets, areas that are subject to soil compaction, and sites with potential disruption of the root system.

Soil & Water Needs

pH: 5.0–6.0

Fertilizer: Medium feeder. Use low N for trees under 2 years; you don't want rapid growth when the tree is young. Unlike some nuts, the chestnut likes to be fed throughout its life and produces better quality initially, as the chestnut thrives in acid soil.

Side-dressing: Apply compost in late autumn or rotted manure and leaf mold in early spring.

Water: Medium; however, established trees are fairly drought resistant.

Measurements

Root Depth: Deep

Height: 30'–40'
Breadth: 20'

Spacing: 8' if thinned out in 5 years; 20' if thinned out to 40' in 20 years; 40' if not thinned

Growing & Bearing

Bearing Age: 3–4 years

Chilling Requirement: 250 hours below 45°F are required for development of nut blossoms.

Pollination: All require cross-pollination; plant two of the same or different varieties.

Shaping

Training:
Freestanding: Central leader
Wire-trained: N/A

Storage Requirements

The chestnut can be eaten raw if cured first to maximize the nut's free sugar. To cure, dry deburred nuts in a shady, warm, dry place for 1–3 days until the nut texture is spongy. To roast, cut an X in the shell and cook at 400°F for 15 minutes. If not eaten immediately, store uncured nuts in a cold place that is either dry or has high humidity but no free moisture. A good method is to mix freshly harvested and dehulled nuts with dry peat moss, pack in plastic bags, seal, and refrigerate. You can also dry and grind nuts into a baking flour.

Fresh (uncured)

Temperature	Humidity	Storage Life
32°F	high, but no free moisture	6–12 months

chinese chestnut

Pruning: Prune young trees minimally, just enough to train the tree to a single trunk and basic scaffold; too much pruning stimulates extra vegetative growth and delays bearing.

Pests

Chestnut weevil, gall wasp (in Georgia), mite, nut curculio, squirrel

Diseases

Blossom end rot, chestnut blight, oak wilt

Allies, Companions, and Incompatibles

None

Harvest

Use thick gloves to protect your hands from the spines of the outer chestnut hull. To minimize daily gathering, when the burs begin to crack open in late summer, pick or knock down small bunches of them onto a harvest sheet. For final ripening, store the burs at 55°F–65°F for about 1 week or until they split open. If unable to pick them in bulk, you must gather fallen nuts each day, for chestnuts on the ground are particularly susceptible to rapid degradation by fungi and bacteria.

Selected Varieties

CHINESE AND CHINESE HYBRIDS

Suited for east of the Rocky Mountains, where blight is widespread.

Au-leader: Grafted Chinese variety; large nut; good flavor; hardy; disease-resistant; grows only 25'–35'; Zones 4–9
Nolin River

Colossal: Early-bearing; prolific; high-yield; very large, sweet nuts; tree is large, long-lived, and spreading; Zones 5–9
Bay Laurel, Raintree

Crane: 32 nuts/pound; flavor and storage quality are excellent when cured; good annual bearer; good anywhere from north Florida to the Great Lakes region; hardy to –20°F
Nolin River

Douglass: Chinese-American hybrid; well-filled nuts; good flavor; available as graft or seedling from different nurseries; Zones 5–7
Grimo

Eaton: Ornamental and high-quality nut tree; 30–40 nuts/pound; good texture, flavor, and sweetness; need 2–3 trees for a good crop; originally from Connecticut; good from northern Florida to Michigan and Wisconsin; Zones 5–6
Nolin River

Manchurian: Chinese-American hybrid; flavor and nut size of American; vigorous grower; hardy to –28°F
Grimo, Miller

Sleeping Giant: Good-quality nut; also good for timber; spreading crown; originally from Connecticut; not prone to frost problems
Nolin River

AMERICAN *(C. dentata)*

Suited only for the Northwest, where blight is not a problem.

American seedlings: Excellent; huge timber tree
Grimo, Raintree

EUROPEAN *(C. sativa)*

Suited only for areas west of the Rocky Mountains.

European seedlings: Planted widely on West Coast
Grimo

filbert/hazelnut

Corylus (several species)
Betulaceae (Birch Family)

FILBERTS, also known as hazelnuts, are unusual because they can be grown as shrubs, hedgerows, or trees. An ideal size for home growers, they are generally hardy. Very early blooming renders them particularly susceptible to late frosts, however. In the Northwest, they never go fully dormant. Filberts tend to bear in alternate years, depending on how much new wood was produced and how much was pruned out the prior year. In areas that are subject to temperatures of 5°F or lower, they do not produce nuts consistently. In the first 2 years, protect young tree trunks from sunscald. Filberts produce numerous suckers that are easily used for propagation and good for coppice management (see Chinese Chestnut, page 188). When buying a filbert or hazelnut, make sure it is intended to be a nut-bearing tree; many are grown only as ornamentals and may be poor nut bearers. Nuts from grafted or layered trees are generally higher quality than those from seedling trees. Delicious in baked goods, filberts are also high vitamin and protein snacks.

Site

Full sun in maritime climates; partial shade in very sunny, hot climates. In the East, choose a northern, cold exposure to delay premature bloom. Can adapt to clay and sand, but prefers deep, fertile, well-drained soil. Must avoid low frost pockets and poorly drained areas.

Soil & Water Needs

pH: 6.5

Fertilizer: Heavy feeder, but do not fertilize unless foliage is pale and growth is slow. Appropriate new growth is 6"–9".

Side-dressing: Apply compost in late autumn and organic mulch in early spring.

Water: Medium; water well in times of severe drought. In maritime climates like the Northwest, mature trees rarely need watering. Sawdust mulch helps keep moisture in the soil.

Measurements

Root Depth: Unlike other nut trees, this has no taproot.

Height: 20'
Breadth: 15'

Spacing:
For trees: 15'–20'
For hedges: 3'–5'

Growing & Bearing

Bearing Age: 2–4 years

Chilling Requirement: Medium–high hours (800–1600)

Pollination: All need cross-pollination.

Shaping

Training:
Freestanding: Open center
Wire-trained: N/A

Pruning: To grow a shrub, cut excessive sucker growth yearly. For a tree, prune to establish a central leader and basic scaffold, and remove all suckers. Nuts develop on 1-year-old wood, so prune lightly every year to stimulate new growth. Make thinning-out cuts only where branches are cut back to their base, not cut in half or stubbed off.

Pests

Blue jay, filbert bud mite, filbert weevil, filbertworm, squirrel

Diseases

Crown gall, Eastern filbert blight, filbert bacterial blight, powdery mildew (*Note:* Eastern filbert blight should not be excluded as a possible problem in West Coast growing regions. It was first found on the West Coast in the Willamette Valley of Oregon in 1986.)

Allies

None

Companions

None

filbert/hazelnut

Incompatibles

None

Harvest

Nuts usually turn brown and ripen by late summer, but an immature husk, shaped like a barely opened daffodil blossom, prevents them from dropping for almost another month. The nut is ripe when it readily turns in the husk if pressed. When ripe, you can either hand-harvest the husks or wait until nuts drop to the ground, in which case you risk competing with squirrels and birds.

Storage Requirements

Place hulled nuts in water; remove the rotten and diseased nuts that float to the top. Dry and cure by spreading nuts in one layer and leaving them in a cool, dry, well-ventilated place for several weeks. They're ready for storage when the kernels rattle in the shell, or, when unshelled, the nutmeat snaps when bent. Avoid big piles of nuts, which encourage rot.

Fresh

Temperature	Humidity	Shelf Life
Shelled: 65°F–70°F	low	several weeks
Unshelled: 34°F–40°F	low	several months

Preserved

Method	Taste	Shelf Life
Frozen (shelled)	good	12–24 months

Selected Varieties

EUROPEAN SPECIES (*C. avellana*)

Commonly known as filberts, these are a major commercial crop in the Northwest, grow best in maritime climates, and are the kind of nut usually found in supermarkets. They have occasionally been grown in the East but generally do poorly there due to Eastern filbert blight.

Butler: Good pollinator for Ennis; medium-large nut; good flavor; smooth kernel; moderate bud-mite resistance; very hardy
Bay Laurel, Miller

Casina Late: Low-care; recommended for backyard gardeners; heaviest producer; open pollination; very thin-shelled nuts; Zones 5–8
Bay Laurel, Stark

Halls Giant: Hedge growth, approximately 10'–15' tall; pollinator variety; large, round nut; moderate blight resistance; Zones 5–9
Bay Laurel, Raintree

Lewis and Clark: Companion pollinator varieties; high yield of delicious nuts; heirloom variety, one of Thomas Jefferson's selected cultivars; Zones 6–9
Raintree

Royal Filbert: Large nuts; thin shells; pollinate with Barcelona; Zones 5–8; best in the East
Miller

AMERICAN SPECIES (*C. americana*)

Most often known as hazelnuts, these species grow in native, wild hedgerows throughout the northern U.S. and southern Canada. American hazelnuts are usually smaller than European hazelnuts. The American species serves as a host for Eastern filbert blight fungus but is very tolerant of its attack. Check with suppliers about availability.

HAZELBERTS

A cross between the European filbert and American hazelnut, these nuts combine the large European nut size with American hardiness and early ripening.

Gellatly: Early bearing; hardy; moderate blight resistance
Grimo

Grimo: Productive tree; available grafted, layered, or seedling; high blight resistance
Grimo

Northern: Very cold hardy; possibly even to Zone 3
Grimo

Turkish Trazel *(C. colurna* x *C. avellana):* Excellent small shade tree and nut producer; shapely; good flavor; only 20'–30' tall; hardy to –25°F
Grimo

Vitis labrusca (American)
Vitacae (Grape Family)

GRAPE vines need well-drained soil and should have at least 20 inches of topsoil. They don't need especially fertile soil because of vigorous, extensive roots; vines that grow more slowly also develop more "character." Plant vines at the same depth as grown in the nursery. Mound soil over the crown to prevent wind damage, except in the West because of possible crown rot. Cut out all but one or two stems (with two to three buds each) for the central trunk. Grape roots prefer warm soil, so mulch with stones or black plastic to raise soil temperature. If vines overbear, thin flowers before berries form.

Grape seeds secrete growth hormones within the berry, so commercial growers spray seedless grapes with growth hormones. Homegrown, unsprayed, seedless grapes will be smaller.

Site

Best on a 15-degree south-facing (southeast or southwest) slope

Soil & Water Needs

pH: 6.5–7.0

Fertilizer: Apply compost only at the beginning of the growing season or during blooming. When applied late in the season, N delays ripening, inhibits coloring, and subjects vines to winter injury if they keep growing too long into fall. American grapes and hybrids are especially sensitive to N deficiency in the early spring and during blooming.

Water: Low to dry. To harden off the vines for winter, don't water much after August.

Measurements

Root Depth: In deep soil they can easily extend 12'–40'.

Height:
Pruned: 12'–20'
Unpruned: 50'–100'

Spacing: 8' is best; 7' in shallow soil and 10' in deep soil

Growing & Bearing

Bearing Age: 3–4 years

Chilling Requirement: None

Pollination: All are self-fertile with the exception of a few muscadine vines.

Shaping

Training: For vigorous vines, the less common Geneva double-curtain method provides the best aeration, most sun, and highest yields. Plant vines down the center, prune each to two trunks, and grow trunks to long, 6'–8' cordons on upper wire. Train the first vine to the front wire, the second to the back wire, and so on. The more common four-arm Kniffen method provides an attractive privacy screen but shades the lower vine parts. For both methods, bury 9' end-posts 3' into the ground. Use heavy galvanized #11 wire.

Pruning: To spur-prune European vines, consult other sources. Cane-prune all others as follows. (1) Cut out water spouts, which are shoots on more than 2-year-old wood; (2) remove winter-damaged wood; (3) cut out last year's fruiting cane; (4) identify canes receiving the most sun by their darker wood and closely spaced nodes; select the thickest to bear this season's fruit, and cut back to only eight to fifteen nodes; (5) for each selected fruiting cane, choose a cane nearby as a renewal spur for next year's fruit; cut back to two buds. Each year, replace the fruiting arm with a cane from a renewal spur, and select a new renewal spur for the following year.

grape

Pests

Grape berry moth, Japanese beetle, leafhopper, mealybug, mite, phylloxera root aphid, plum curculio, rose chafer

Diseases

Anthracnose, black rot, botrytis fruit rot, crown gall, downy mildew, leaf spot, Pierce's disease (spread by leafhoppers), powdery mildew

Allies

Some evidence: Blackberries (*Rubus* sp.), Johnson and Sudan grass (see caution on Johnson Grass in chart on page 429)

Uncertain: Chives, hyssop

Harvest

Cut bunches when fruits are fully colored, sweet, slide off easily, and stems and seeds are brown. Grapes don't ripen further once picked. For raisins, use a hydrometer and harvest when the grapes reach 20 percent soluble solids. A lower percentage significantly decreases raisin weight and quality.

Storage Requirements

Cool to 50°F soon after picking and spread fruit in single layers. Dry in the sun under clear polyethylene until stems shrivel slightly; this shortens drying time and yields raisins with higher sugar and vitamin C levels than raisins dried in an oven. Store no more than 4" deep in trays.

Fresh

Temperature	Humidity	Storage Life
40°F	slightly humid	2–3 months

Preserved

Method	Taste	Shelf Life
Jelly	good	up to 18 months
Dried	excellent	6–24 months
Juice	good	12+ months

Selected Varieties

AMERICAN

Pure strains are rich in pectin, best for jelly, and renowned for a "foxy" flavor (cloyingly sweet, like the Concord) that, if possible, winemakers avoid. Newer hybrids are good fresh and for wine.

Red

Canadice: Early; seedless; very good flavor; high yields; keeps on vine a long time; very disease resistant; hardy to –15°F without protection; for table, jelly, or wine
Widely available

Catawba: Very late; high vigor; susceptible to mildew; keeps for up to 2 months; for wine, juice, jelly and also table; hardy to –10°F; Zones 5–8
Miller, Southmeadow

Einset: Midseason; high yields; stores well; resists botrytis; flavor has a hint of strawberry; good fresh or dried as raisins; hardy to –5°F; Zones 5–8
Raintree

Reliance: Early; seedless; fruity; high yields; stores to 3 months; hardy to –10°F without protection; resists anthracnose, black rot, powdery and downy mildew; good fresh; Zones 4–8
Widely available

Saturn: Midseason; excellent flavor; high yields; good fresh or as dessert wine; disease resistant; hardy to -5°F without protection; Zones 5–8
Indiana Berry, Miller

Swensen: Very early; good for Northern states; hardy to –40°F with occasional winter injury; very disease resistant; for table and wine; good for cold storage; Zones 3–7
Raintree, St. Lawrence, Sonoma

Vanessa: Early; seedless; high quality; vinifera character; hangs in compact clusters; from Vineland Ontario Station; has survived –25°F
Miller, Raintree, Southmeadow

White

Cayuga: Late; high vigor; moderately susceptible to mildew; for juice and jelly; very high quality wine; hardy to –5°F without protection; Zones 4–7
Miller

Himrod: Early; low yields; vigorous; great for home garden and arbors; stores well; excellent fresh flavor; good for raisins; hardy to 0°F without protection; Zones 5–8
Widely available

Interlaken: Early; small berries; high yields; excellent flavor fresh; excellent raisins; hardy to 5°F without protection; resistant to powdery mildew; Zones 5–8
Bay Laurel, Miller, Raintree, Sonoma, Southmeadow

Kay Gray: Extremely hardy; medium berry; mild fruity flavor; disease resistant; vigorous; hardy to –15°F without protection
St. Lawrence

Selected Varieties *(cont'd)*

MUSCADINE (var. *rotundifolia*)

Indigenous to the American Southeast, these require a warm, moist climate and are good for jelly, juice, and wine. They require cross-pollination, so plant two varieties.

Cowart: Good for pollinating Fry; blue-black fruit; vigorous and productive vines; disease resistant; Zones 7–9
Stark

Fry: Gold-yellow; large berry; very sweet; vigorous; high yields; good fresh; Zones 7–9
Stark

Tara: Early ripening; bronze-colored fruit; sweet mild flavor; a good pollinator for other muscadines; Zones 7–9
Stark

HYBRIDS *(French Hybrids, American-European Hybrids)*

These represent an attempt to obtain the best of both worlds: European taste and American hardiness and disease resistance.

Aurore (Seibel 5279): Gold-pink berry; early; moderate vigor; high yields; for wine, juice, jelly, and table; hardy to –10°F without protection; Zones 6–7
Widely available

Seibel 9110: Gray-yellow berry; crisp; meaty; delicious table grapes; adherent skins; needs protection from severe winters
Southmeadow

EUROPEAN AND CALIFORNIA (var. *vinifera*)

Low in pectin, these require a frost-free season of 170–180 days and are not hardy below 10°F without protection. Most wines are made from Vinifera or the American-European hybrids. Among the many varieties, these are hardiest.

Cabernet Sauvignon: Black; very seedy; for a cool climate and long growing season; red wine; cane prune; Zones 5–9
Bay Laurel

grape

Actinidia (3 species)
Actinidiaceae (Chinese Gooseberry Family)

KIWI is a productive and tasty candidate for the home grower—but only if proper attention is paid to site, water, support, and pruning. The buds, young shoots, and fruit of all species, regardless of hardiness, are particularly frost tender and need to be protected when temperatures fall below 30°F for any length of time, in spring or fall. Cover at night to protect from frost damage, or use overhead sprinkling until temperatures surmount 32°F. Frost-damaged fruit emits ethylene gas in storage, thereby hastening the softening of other fruit. Kiwi vines can bear for 40 to 50 years. Light pruning and fruit thinning, rather than heavy pruning and no thinning, optimize yield and fruit size. Thin to about sixty fruits per square meter before flowers open.

One male can fertilize up to eight female vines; tag them to be sure you always know which is which. Try to buy larger plants because they have a higher survival rate. Otherwise, you may want to start small rooted cuttings in 5-gallon containers.

Site

Full sun (minimum 6 hours), except for the *A. kolomikta,* which likes partial shade in hot climates. Wind protection is important. Prefers rich, fertile soil, but will tolerate heavy soil. In any soil, good drainage is vital. Avoid soggy, low areas. Best spot is to the north of a building or tree to delay bud break. Cover vines if frost threatens in spring or fall.

Soil & Water Needs

pH: 6.0–6.5

Fertilizer: Heavy feeder. Apply slow-acting organic fertilizers, very thick compost, or well-rotted manure in late winter and spring, several inches away from the crown. Don't fertilize past mid-June. Kiwi needs high K, and also Mg to prevent K-induced deficiency. Never apply Boron, as above-optimum levels can be severely toxic.

Side-dressing: Apply twice during the growing season; use only mildly nitrogenous substances.

Water: Heavy. Drip irrigation is best. Overhead sprinkling can protect fruit and foliage from frosts.

Measurements

Root Depth: Shallow; they're susceptible to crown and root rot in wet areas

Height: Up to 30' long vines

Spacing:
Between plants: 10'–20'
Between rows: 15'–20'

(*Note:* Male and female plants must be within at least 100' of each other.)

Growing & Bearing

Bearing Age: 3–5 years, except for self-pollinating types, which can bear 1 year after planting

Chilling Requirement: Vines benefit from 400–600 hours below 45°F.

Pollination: Cross-pollination between male and female vines required; 1 male can pollinate 8 female vines.

Storage Requirements

On a regular basis, remove from fresh storage soft, rotten, or shriveled fruit. Freeze whole kiwis in plastic bags; or freeze ¼" unpeeled slices and then pack in plastic bags. Cut kiwi in half, then dry at 120°F for 2 hours and repeat the next day.

Fresh (whole)

Temperature	Humidity	Storage Life
32°F	85%–95%	2 months (hardy) 4 months (fuzzy)

Shaping

Training: Requires trellis, arbor, T-bar fence, pergola, wall, or chain-link fence to support fruiting vines. Supports should be 6' tall for females and 7' tall for males. Stake when planting.

Pruning

Similar to grapes. Kiwis fruit on the lower 6 buds of this season's fruiting shoot, off of 1-year-old wood. On planting, prune the main stem back to 4–5 buds. The first summer allow the vine to grow freely. Cut back females to 6' and males to 7', and remove all but 2 or 4 of the strongest cordons for each. The second winter, head back female cordons to 8–10 buds and males to half that length (4–5 buds).

General Female Winter-Pruning Guidelines: Cut out damaged wood, curled or twining growth, and all wood that has fruited for two seasons. Select fruiting arms that have short internodes (less than 2" apart), are 10"–14" apart, and are well-exposed to sunlight; cut these back to 8 buds.

General Female Summer-Pruning Guidelines: Select next season's fruiting arms, remove other shoots, cut out erect water shoots, and shorten curled or tangled growth. Make sure enough light passes through to cast patterns on the ground below.

General Male-Pruning Guidelines: After July bloom cut back flowering arms to 20"–24"; if necessary, trim again in August or September to 29"–31". Trim vines in winter.

Pests

No significant pests reported in North America.

Diseases

No significant diseases reported in North America.

Allies

None

Companions

None

Incompatibles

None

Harvest

Allow fruit to ripen on the vine until the first signs of softening; it should give with a little finger pressure. Clip hardy kiwi with some stem. Snap off fuzzy kiwi, leaving the stem on the vine. Even minor damage causes ethylene production, which prematurely softens other fruit. In dry climates, you can let the kiwi dry on the vine; they will become intensely sweet and keep for about 6 weeks.

Selected Varieties

VERY HARDY *(A. kolomikta)*

Hardy to Zone 3 (–40°F), these are beautiful plants. The male has variegated leaves, and females display some variegation, too. Northwoods Nursery brought seven "Arctic Beauty" varieties back from the Soviet Union in 1987, only one of which is listed below. Fruits are smaller than fuzzy kiwi and are smooth-skinned.

"Arctic Beauty": Ornamental male with no fruit; tolerates poor soils; quickly covers trellis or arbor; good for Zones 3–7
Hartmann's, Miller, Raintree, Northwoods

Raisin Kiwi: Small fruit; delicious eaten fresh; beautiful ornamental with unique leaf coloring
Hartmann's

September Sun: Large fruit; good flavor; variegated foliage
Northwoods

HARDY *(A. arguta)*

Hardy to Zone 4 (–25°F) and native to northern China. The fruit is smaller than fuzzy kiwi and its smooth, grapelike skin can be eaten.

Ananasnaja: Means *pineapple-like* in Russian; very sweet fruit with a hint of mint flavor; brought from Belgium to the Northwest; actually a hybrid of *A. arguta* and *A. kolomikta*
Indiana, Northwoods, Raintree

Issai: Self-fertile Japanese variety; ideal for limited spaces; fruit will be larger with pollination, although if pollinated it will have seeds; Zones 5–9
Bay Laurel, Northwoods, Sonoma

Miller's Kiwi: Sweet; vigorous; hardy; fully rooted plants
Miller

74–49: Proven vigorous and reliable in Virginia; large fruit; aromatic; sweet
Northwoods

TENDER OR FUZZY KIWI *(A. deliciosa)*

Native to China and first commercially grown in New Zealand. This fuzzy fruit is hardy to Zones 8–10 (5°–10°F) and can be grown as far north as the Pacific Northwest and British Columbia or cultivated in greenhouses.

Hayward: The variety usually found in supermarkets; good for warm areas of the maritime Northwest
Northwoods

Saanichton: Large fruit; the hardiest fuzzy variety; same flavor and size of store kiwi; grown on Vancouver Island, British Columbia, for 30+ years
Northwoods, Raintree

peach

Prunus persica
Rosaceae (Rose Family)

PEACHES are considered one of the most difficult fruits to grow, particularly for growers who don't use chemical sprays, because of multiple pests and an early bloom period that makes them extremely susceptible to frost damage. Spraying flat-white interior latex paint on the trees, buds and all, in mid-January can help delay bloom by up to 5 days; this is often enough time to make a significant difference in flower survival and fruit set, according to Rutgers University. Plant at about the same depth that it was grown in the nursery; its upper roots should be only a few inches below the soil surface. Thinning is crucial to a good harvest. After the June drop, but before the fruit is 1¼ inch in diameter, thin to 1 fruit per 30 to 40 leaves, *or* 1 fruit per 10 inches on early-ripening varieties, *or* 1 fruit per 6 to 8 inches on late-ripening varieties. "Cling" varieties are firm and best for canning but are rarely available to home growers. "Freestone" varieties don't can well. Peach trees live a mere 8 years in the South and up to 18 years in the North; poor drainage renders even shorter life cycles.

Site

Full south or southeast exposure, whether or not espaliered. Will not survive on heavy clay soils. Do not plant on former apricot, cherry, or peach tree sites; when waterlogged, their roots release hydrogen cyanide that may linger in the soil and hinder growth.

Soil & Water Needs

pH: 6.0–6.5

Fertilizer: Low N when under 3 years. Appropriate new growth is 12"–15" when young, and 8"–18" when bearing.

Water: Keeping the peach's shallow roots evenly moist is critical.

Measurements

Root Depth: Shallow, over 90 percent in top 18". Roots won't branch out if planted too deeply.

Height:
Dwarf: 4'–10'
Standard: 15'–20'

Spacing:
Dwarf: 12'–15'
Standard: 15'–25'

Growing & Bearing

Bearing Age: 2–3 years

Chilling Requirement: Most need 600–900 hours below 45°F.

Pollination: Most self-pollinate, but yields will be higher with cross-pollination.

Shaping

Training:
Freestanding: Open center
Wire-trained: Fan, against a south-facing wall

Pruning: Unlike apple, peach bears fruit on 1-year-old wood only and should be pruned to encourage new growth. Cut out branches that shade or cross each other, intrude on the center, or are winter-damaged. Remove at least one-third of the previous year's growth or too much fruit will be set. Every few years, cut out some older wood. Cut back upright-growing shoots to the outward-pointing buds.

Pests

Aphid, birds, cherry fruit sawfly, codling moth, gopher, gypsy moth caterpillar, Japanese beetle, mite, oriental fruit moth, peach tree borers, peach twig borer, plum curculio, root lesion nematode, tarnished plant bug, tent caterpillar, weevil, whitefly

Diseases

Bacterial canker, bacterial spot, brown rot, crown gall, cytospora canker, peach leaf curl, scab, verticillium wilt

Allies

Some evidence: Alder, brambles, buckwheat, goldenrod, lamb's-quarters, ragweed, rye mulch, smartweed, sorghum mulch, strawberry, wheat mulch

peach

Uncertain: Caraway, coriander, dill, garlic, nasturtium, tansy, vetch, wildflowers, wormwood
Bird control: A border of dogwood, mulberry, or other aromatic fruit, all of which birds prefer

Incompatibles

None

Harvest

Pick when fruit is firm, almost ready to eat, and easily slides off the stem by tipping or twisting. Never pull it off directly or you'll bruise the peach and hasten spoilage. If fruit has a mild case of brown rot, harvest only those peaches that are infected and dip them in hot water for 7 minutes at 120°F (or alternatively, 3 minutes at 130°F or 2 minutes at 140°F). This kills the fungi without harming the fruit and prepares it to be stored for further ripening.

Storage Requirements

To freeze the fruit, peel, pit, and cut it in halves or slices. Pack it with some honey mixed with lemon or a pectin pack. Peaches don't store well in a root cellar.

Fresh

Temperature	Humidity	Storage Life
50°F–70°F	low	3–14 days

Preserved

Method	Taste	Shelf Life
Canned	fair	up to 2 years
Frozen	good	12–24 months
Dried	fair	12+ months

Selected Varieties

YELLOW FLESH

Avalon Pride: Midseason; semifreestone; reliable; self-fertile; disease resistant; Zones 5–9
Raintree

Canadian Harmony: Midseason; large fruit; good flavor; some resistance to bacterial spot; good for California and Northwest; 850 hours chill
Adams

Compact Redhaven: Same as Redhaven but semi-dwarfed to 10'; tolerates bacterial spot and leaf curl; 850 hours chill
Van Well

Frost: Midseason; medium fruit; semifreestone; good flavor; highly resistant to leaf curl; good in Pacific Northwest
Bay Laurel, Sonoma, Raintree

Harrow Beauty: Midseason; beautiful tree and fruit; good fresh; excellent flavor; hardy; medium vigor; very open; spreading; productive; tolerates bacterial spot and brown rot; 850 hours chill
Adams

Jerseyglo: Late; large fruit; hardy and easy to manage; very resistant to bacterial spot
Adams

Madison: Late midseason; good fresh and canned; sweet; juicy; thin skins; very hardy (hardier than Redhaven); vigorous; tolerates frost during blooming; good bud hardiness; 850 hours chill
Adams, Stark

Redhaven: Early; good fresh, frozen, and canned; almost fuzzless; requires thinning early; hardy wood and buds; disease resistant; not good for warm winters; 950 hours chill; Zones 5–8
Widely available

Reliance: Late; very hardy; produces heavily after even the coldest winters; beautiful pink flowers; Zones 4–8
Adams, Miller, Southmeadow, Stark

Rio Oslo Gem: Late; large fruit; good fresh, frozen, and in pies; reported nonbrowning; large showy flowers; naturally smaller tree; vigorous; 850 hours chill
Sierra, Sonoma

Sunhigh: Midseason; large oblong fruit; very good flavor; sweet and melting flesh; susceptible to bacterial spot; 750 hours chill; best in East
Adams, Southmeadow

WHITE FLESH

Babcock: Early; small to medium fruit; very sweet; low acid fruit; requires heavy thinning; good in California
Sonoma

Selected Varieties *(cont'd)*		Rootstocks
Blushing Star: Midseason; ships and stores very well; does not brown when cut; hardy; good resistance to bacterial spot; freestone Adams, Sierra, Vanwell	GENETIC DWARF PEACHES AND NECTARINES **Arctic Glow:** Early midseason; white flesh; vigorous and productive; very hardy; resistant to bacterial spot Adams	See Plum Rootstocks chart on page 226, as most are the same for peaches. Peaches do best on a peach-seedling rootstock, which produces standard trees. Standard trees can be kept small by pruning. Elberta rootstocks are not hardy for the North.
Champion: Midseason; old variety; medium fruit; excellent fresh; extremely tender and juicy flesh; very hardy; vigorous Miller, Sonoma, Southmeadow	**Honey Babe:** Early midseason; 3'–5'; yellow flesh; firm; sweet; delicious; for container growing; keep rain off to avoid leaf curl; good in Northwest Bay Laurel, Raintree	
Raritan Rose: Early; large fruit; excellent fresh; firm flesh; tender; juicy; honeysweet; hardy buds; very vigorous; susceptible to brown rot; 950 hours chill; best in East Adams, Southmeadow	**Super Hardy Chinese Dwarf:** Mid to late season; extremely hardy; large, delicious crop; self-fertile but cross with Moorpark or similar variety for biggest crops Miller	
Stark White Giant: Early; huge fruit; requires minimal care; tolerant to bacterial leaf spot; semifreestone Stark		

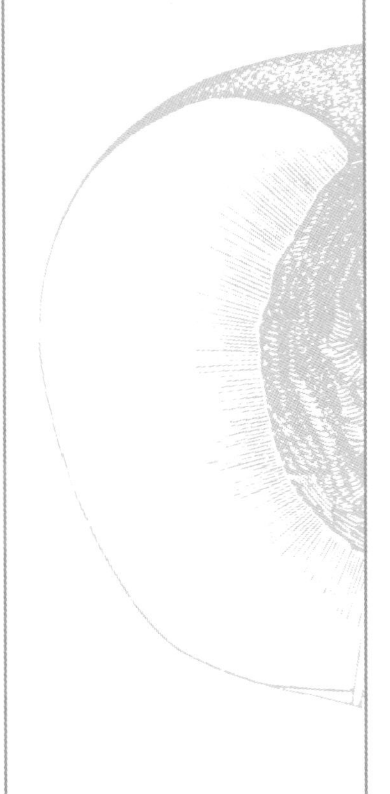

pear

Pyrus communis
Rosaceae (Rose)

PEARS are a good choice for backyard gardeners because they require less care than apples. Although they are more tolerant of heavy clay soils than are other fruits and prefer heavy organic mulches, they do not generally do well in light, sandy soils. They grow best in cool, moist, overcast conditions, which, unfortunately, are also perfect for fireblight, the most serious pear problem. Pear blossoms are fairly frost resistant and can withstand temperatures as low as 28°F. European pears don't usually need thinning because of a low "set ratio" (the number of blooms producing fruit). Asian pears should be thinned to one fruit per cluster to avoid overbearing. The higher the temperature immediately after blooming, the sooner the fruit will mature, generally in 106 to 124 days. Pears are among the longest-lived fruit trees, surviving to 200 to 300 years and producing for 100 years or more.

pear

Site

Full south or southeast exposure

Soil & Water Needs

pH: 6.0–6.5

Fertilizer: Avoid high N; pears need a lot of boron, so periodically check soil for deficiency.

Water: Heavy and constant supply; ground irrigation is especially important to minimize fire blight

Measurements

Root Depth: Deep

Height:
Dwarf: 8'–15'
Semidwarf: 15'–20'
Standard: 30'–40'

Spacing:
Dwarf
  Cordon: 3'
  Espalier: 15'
  Fan: 15'
  Pyramid: 5'
Semidwarf: 20'
Standard: 30'

Growing & Bearing

Bearing Age: 2–4 years

Chilling Requirement: Most need 600–900 hours below 45°F

Pollination: All require cross-pollination.

Support Structures

Limb spreaders strengthen joints and encourage earlier blossoming and higher yields. More than other fruit trees, pear branches sag with fruit and may need to be tied up.

Shaping

Training:
Freestanding: Central leader for Comice and Anjou; open center for Bartlett, Bosc, those with flexible limbs, and those susceptible to fire blight
Wire-trained: All cordons, espalier, stepover, palmettes

Pruning: Cut as little as possible because each cut exposes tissue to fire blight.

Pests

Aphid, apple maggot, cherry fruit fly, codling moth, European apple sawfly, flea beetle, gypsy moth, mite, oriental fruit moth, pear psylla, pear slug, plum curculio, tarnished plant bug, tent caterpillar, thrips, weevil, whitefly

Diseases

Bitter pit, blossom blast (boron deficiency), cedar apple rust, crown gall, crown rot, cytospora canker, fire blight, pear curl, pear decline, scab

Allies

Some evidence: Alder, brambles, buckwheat, rye or sorghum or wheat mulch

Uncertain: Caraway, coriander, dill, garlic, nasturtium, tansy, vetch, wildflowers, wormwood

Incompatibles

None

Harvest

Pick when pears are at least 2" in diameter. Don't allow European varieties to ripen on the tree, as they'll become mealy and coarse. Handle carefully; although the fruit appears hard, it bruises easily. Asian pears should ripen on the tree.

Storage Requirements

Some pears require lengthy storage before they begin to ripen, but if left in cold storage too long some will never ripen. After cold storage, ideal ripening temperature is 60°F–70°F; some pears won't ripen after cold storage if the home temperature is too high. If you wish to avoid the need for cold storage, place pears in a paper bag with ripe apples or pears; these emit ethylene gas, which stimulates the final stages of fruit ripening.

Pear	Minimum Storage Life at 30°F–32°F	Maximum Storage Life at 30°F–32°F	Maximum Storage Life at 40°F–42°F
Anjou	2 months	4–6 months	2–3 months
Bartlett	none	1½ months	2–3 weeks
Bosc	none	3–3½ months	2–2½ months
Comice	1 month	2½–3 months	1½–2 months
Seckel	none	3–3½ months	none

From Diane E. Bilderback and Dorothy Hinshaw Patent, *Backyard Fruit and Berries* (Emmaus, Penn.: Rodale, 1984), 218.

Selected Varieties

EUROPEAN
(Pyrus communis)

Beurre Bosc: Late; large pear; brown russet over green or yellow; long-necked; perfumed and melting flesh; excellent keeper; large and vigorous tree can grow up to 25′; does well in heavy clay soils; can withstand hot summer temperatures and resists cold winters
Widely available

Collette: Early midseason; very high-quality fruit; long bearing for several months; good fresh and canned; hardy to –10°F to –15°F
Miller

Comice: Late; Oregon-grown winter pear; large fruit; top flavor; vigorous; some fire blight resistance (or moderate susceptibility); erratic crops; requires storage before ripening; good storage pear; a low chill requirement makes it suitable also for southern California — it prefers milder climates with less summer heat but is grown commercially in southern California; Zones 5–9
Widely available

Harrow Delight: Very early; medium-large fruit; juicy; melting; hardy; productive; good fire blight and scab resistance; Zones 5–7
Raintree

Magness: Late; needs pollinator; Comice-Seckel seedling; medium-large fruit; delicious flavor and melting, juicy flesh; slightly russeted; more spreading than most pears; cross-pollinates with Asian varieties; resists fire blight; Zones 5–8
Adams, Sonoma Apple, Southmeadow

Moonglow: Early; Comice seedling; Bartlett type; good fresh, in pies, or canned; upright; heavy spurs; hardy; strong pollinator of other pears; vigorous; wide climate tolerance; good fire blight resistance; Zones 5–8
Bay Laurel, Miller, Southmeadow, Stark

Seckel: Midseason; self-fertile; very small fruit; yellow-green with russet cover; "sugar pear"; fresh or canning; keeps well; productive; very hardy; some fire blight resistance; Zones 5–8
Widely available

Warren: Midseason, self-fertile; extremely fire-blight resistant and cold-hardy; sweet and juicy fruit; Zones 4–9
Bay Laurel, Raintree, Sonoma

ASIAN
(Pyrus serotina)

All are attractive ornamentals

Chojuro: Late midseason; medium oblong fruit; brown and russeted; rich aromatic flavor; crisp like apples; can keep until March; medium-sized tree; spreading and vigorous; will pollinate with Seckel and Bartlett
Miller, Sonoma Apple, Southmeadow

pecan

Site

Full sun and rich, well-drained soil

Soil & Water Needs

pH: 5.8–7.5

Fertilizer: Low N when young

Side dressing: Apply compost in late winter to early spring.

Mulch: Avoid sawdust

Water: Medium, but during drought periods water up to 3"–4" per week

Measurements

Root Depth: Very deep

Height: 75'–100'

Spacing: 25' × 25' to 70' × 70' (depending on the variety)

Growing & Bearing

Bearing Age: 3–4 years in the South; 8–10 years in the North

Pollination: Self-fertile but nuts will be higher quality with cross-pollination

Chilling Requirement: Most require some chilling but the hours depend on the variety (200–1600).

Shaping

Training:

Freestanding: Central leader

Wire-trained: N/A

Pruning: At planting, cut back one-third to one-half of the top, to 2"–3" above a bud facing the prevailing wind. Cut out crowded and crossed branches.

Pests

Aphid, birds, fall webworm, hickory shuckworm, pecan casebearer, pecan weevil, scale, squirrel, walnut caterpillar

Diseases

Canker dieback, crown gall, liver spot, pecan bunch, root rot, scab, sunscald, walnut bunch

Allies

None

Companions

None

Incompatibles

None

Harvest

Gather nuts from October to January. As the pecan matures, the outer hull splits to expose the inner nut; as the hull dries, the nut falls to the ground. For most varieties, the limbs must be shaken to encourage nut drop. Spread a canvas sheet on the ground, before shaking, to collect the nuts. Don't begin harvesting until the hulls of the nuts on the inner part of the tree have split open; they will mature last.

Storage Requirements

Place hulled nuts in water, and remove the rotten and diseased nuts that float to the top. Dry and cure by spreading a single layer of nuts in a cool, dry, well-ventilated area. They're ready for storage when the kernels rattle in the shell or when unshelled nutmeats snap when bent. Store unshelled nuts in attics over winter or in cool underground cellars, where they'll keep for a year. Store shelled nuts in ventilated plastic bags or in tightly sealed, paper-lined tins that have a hole punctured in the side beneath the lid. Shelled nuts can also be stored in the refrigerator or freezer for up to a year.

Fresh

Temperature	Humidity	Storage Life
34°F–40°F	dry	12+ months

Selected Varieties

SOUTHWEST REGION (ARID CLIMATES)

Kanza: Early; new variety; large nuts; high yield; disease resistant and cold-hardy; regular annual bearing

Nolin River

Mohawk: Early; large nuts; good flavor; vigorous; upright and spreading; partially self-fertile; replaces Mahan; very good producer; may be good for northern California but may not fill well there; best in Southwest, Zones 7–9, and southern edge of Zone 6

Bay Laurel, Nolin River, Sonoma

Pawnee: Early; large well-filled nut; vigorous with long unbranched limbs that later branch at strong angles; good producer; bears throughout tree; more scab resistant for growing in more northern areas; plant with Posey

Bay Laurel, Nolin River

Selected Varieties *(cont'd)*

Posey: Early; good for more northern areas; plant with Pawnee
Nolin River

SOUTHEAST (HUMID CLIMATES)

Candy: Vigorous; prolific; scab resistant
Widely available in Southeast

Cape Fear: High-quality nut; disease resistant; kernels won't break when shelled; Zones 7–9
Stark

Stuart: High yields; thin-shelled nuts; spreading branches; vigorous; prone to disease; also grows in the West; Zones 7–10
Stark

NORTH (SHORT SEASON)

Here, North *refers to the northern part of the pecan-growing region such as Indiana, Illinois, Iowa, Missouri, Kentucky, Tennessee, and Kansas. Pecans can be grown farther north — through southern New England and the Northwest — but they will never produce filled nuts there.*

Fisher: Good producer as far north as Scranton, Pennsylvania; good flavor; medium nut; good cracker; Zones 5–7; grow with Lucas
Nolin River

Greenriver: Late; very good producer; medium nut; good south of Ohio River and as far north as southern Indiana, Illinois, and Missouri; scab resistant; pollinates with Major
Bay Laurel, Nolin River

Lucas: Good medium nut; cracks and fills well; high yields; good as far north as Scranton, Pennsylvania; grow with Fisher
Grimo, Nolin River

Major: Medium-large nut; large yields; thin shell; easy cracker; scab resistant; pollinates with Greenriver; good in Tennessee, Kentucky, Virginia, and as far north as southern Indiana, Illinois, and Missouri; Zones 6–9
Bay Laurel, Nolin River

HICAN

This cross between shellbark hickory and pecan is a good ornamental and is reputedly easier to grow than pecan in Northern areas. Its nuts generally favor the hickory flavor. Plant several varieties for full crops. Nolin River carries numerous varieties. You may not want to grow Hicans if you have a lot of hickory trees nearby, as hickory weevils are said to enjoy Hicans.

Burton: Self-pollinating; medium nut; good producer and bears young; thin shell; good for South and Midwest as far north as northern Indiana, Illinois, and Ohio
Nolin River

Henke: Early; self-pollinating; small nut with high quality kernels that fill well; good crop at young age
Nolin River

plum

Prunus domestica (European)
Rosaceae (Rose Family)

Plums are naturally smaller than apple, pear, and peach trees and, as a rule, grow well wherever pears do. For a stone fruit, plums offer an unusual range of flavors, sizes, and shapes. Japanese plums, now common in stores, are extremely juicy and soft and make excellent dessert fruits. European plums are much sweeter and dry well as prunes. Plum curculio greatly damages plums. Thin fruit no later than 2 months after full bloom. For both European and Japanese, allow only one fruit per spur or cluster. While usually self-thinning, European plums may need to be thinned to 2 to 3 inches apart. Japanese, a more vigorous tree, should be thinned to 4 to 5 inches. When thinning, destroy any fruit having the curculio's crescent-shaped, egg-laying scar.

Selected Varieties

EUROPEAN

These are the sweetest plums; pollinate them with other Europeans, Damsons, or Americans.

Count Althann's Gage: Late; cling; leading European dessert plum; sweet; juicy; golden flesh resists bacterial spot; moderate resistance to black knot; Zones 5–7
Southmeadow

Green Gage (Reine Claude): Late; self-fertile; yellow-green fruit; juicy; very sweet; compact tree; not for warm winters; susceptible to brown rot; good fresh, frozen, or canned; Zones 5–9
Widely available

Pearl: Late; red-speckled yellow fruit; yellow flesh; extremely sweet; tender and melting flesh; tree is moderately upright; thought to be unsurpassed by Southmeadow
Southmeadow

Stanley: Late midseason; self-fertile; freestone; vigorous; productive; resists bacterial spot; susceptible to black knot; good fresh, canned, cooked, or dried; Zones 5–7
Widely available

Starking Delicious: Late midseason; clingstone; consistent bearer of dessert plums; disease resistant; Zones 5–9
Stark

DAMSON *(P. institia)*

Very small and tart, these are best suited for preserves and cooking.

Blue Damson: Midseason; self-fertile; cold-hardy; heavy-bearing; dependable; great for jams and jellies
Bay Laurel, Sonoma

JAPANESE *(P. salicina)*

Very juicy and soft; excellent fresh; some can be canned and cooked; generally prefer warmer climates. Cross with other Japanese or Americans.

Burbank: Midseason; red-purple firm fruit; dwarf; somewhat drooping; needs thinning; hardy; prolific; resists bacterial spot and black knot; susceptible to bacterial canker, brown rot, and especially to leaf scald; pollinate with Shiro; Zones 6–9
Bay Laurel, Miller, Stark

Elephant Heart: Late; self-fruitful; self-pollinating; red; very juicy; good flavor and quality; vigorous; hardy; prolific; cross with Redheart
Bay Laurel, Sonoma, Southmeadow

Hollywood: Midseason; self-fertile; beautiful ornamental; disease resistant; delicious; Zone 5–9
Raintree

Methley: Very early; self-fertile; ripen in 10 days; fruit very sweet but does not keep well; Zones 4–9
Adams, Bay Laurel, Raintree

Selected Varieties *(cont'd)*

Ozark Premier: Early midseason; red skin; yellow flesh; very tasty; sweet; large fruit; resists bacterial spot, canker, and black knot; susceptible to leaf scald and brown rot; Zones 5–9
Adams, Stark

Redheart: Early midseason; red; juicy; firm flesh; excellent flavor; holds quality when canned or frozen; good pollinator; cross with Elephant Heart; Zones 5–9
Bay Laurel, Stark

Shiro: Early; yellow fruit; hardy; prolific; low tree; moderately resistant to bacterial spot and black knot; susceptible to bacterial canker, leaf scald, and brown rot; extremely juicy; excellent sweet flavor; clingstone; good fresh, canned, or cooked; Zones 5–9
Widely available

Weeping Santa Rosa: Early; very ornamental; self-fertile; easy to keep at 8′ or espaliered; small quantity of large fruits; rich sweet-tart flavor; good pollinator; low chill
Adams, Bay Laurel, Raintree, Sierra Gold, Sonoma Apple

BEACH *(P. maritima)*

These plums are good in poor soil and are extremely hardy. Pollinate with other Beach.

Miller's Beach: Small shrub or tree; once established can withstand long droughts, extreme cold, and most diseases; great for jam and jelly; good windbreak; Zones 4–7
Miller

OTHER VARIETIES

Pluot: Early; a cross of plum and apricot; reputed best-tasting new fruit; ripens midsummer, cross with Japanese, Plumcot, or Aprium (see Apricot, page 169); Zones 6–9
Raintree, Sonoma Apple, Stark

Rootstocks

Unlike apple trees, standard-sized plum trees are naturally smaller and can be kept small by pruning. Because of this, a particularly dwarfing rootstock is not essential for espalier work. Pay attention to variety characteristics, also, as each has a different growing habit. For example, a Japanese plum, which is naturally more vigorous, might be better on a more dwarfing stock than would a European plum. Buy smaller trees when possible because they suffer less transplanting shock; in several years they're more productive and vigorous than a larger transplant.

Most nurseries don't offer an in-house choice of rootstocks. The purpose of the chart on pages 226–227 is not to help you make an independent decision but to help you engage the nursery grower with informed questions about what will grow best in your garden.

plum rootstocks

Rootstock Name	Size (percent of standard)	Height / Width	Best Soil	Anchorage	Hardiness	Precocious
Citation	Dwarf 70%–85%	H: 14' W: 13'	Sandy loam	Good	—	—
Marianna 2624	Semidwarf 90%	H: 16' W: 14'	Clay loam	Good	—	—
Myrobolan	Standard 100%	H: 18' W: 16'	Clay loam	Excellent	—	—
Myrobolan 29c	Standard 100%	H: 18' W: 16'	Clay loam	Fair	—	—
***P. americana* (Bailey)**	Standard 100%	H: 18' W: 16'	Sandy loam	—	Hardy	—
***P. americana* (St. Julien)**	Dwarf 70%–80%	H: 10'–15' W: 13'	Sandy loam	Good	Hardy	—
***P. besseyi* (Western Sand Cherry)**	Dwarf 20%–35%	H: 3'–5' W: 3'–5'	Loam	Poor	Hardy	—
***P. domestica* (Pixie)**	Dwarf 30%–40%	H: 6' W: 5'	Loam	Good	—	Precocious
***P. persica* (Nemagaard)**	Augmented 110%	H: 20' W: 18'	Clay loam	Very good	—	—
***P. tomentosa* (Nanking Cherry)**	Minidwarf 15%–25%	H: 2'–4' W: 2'–3'	Loam	Poor	Hardy	Very precocious

From Robert Kourik, *Designing and Maintaining Your Edible Landscape Naturally.* (Santa Rosa, Calif.: Metamorphic Press, 1986), 175–76.

Pests and Diseases							
crown rot	oak root fungus	nematodes	crown gall	bacterial canker	mice	borer	**Other Factors**
Low resistance	——	Moderately resistant	Susceptible	——	——	——	Good disease and pest resistance
Moderately resistant	Moderately resistant	Resistant	Moderately resistant	Highly susceptible	Moderately resistant	——	
Moderately resistant	Moderately susceptible	Resistant	Moderately resistant	Susceptible	Susceptible	Susceptible	
Moderately resistant	Moderately susceptible	Susceptible	Highly susceptible	Susceptible	Susceptible	Susceptible	
——	——	——	——	Resistant	——	——	
Low resistance	Susceptible	——	——	Resistant	Susceptible	——	Good disease resistance; Citation can replace this
Highly resistant	——	——	——	Susceptible	——	——	Lives less than 8 years
Low resistance	——	Moderately resistant	——	Moderately resistant	——	——	Rootstock promotes small fruit
Low resistance	Susceptible	Resistant	Susceptible	Susceptible	Moderately resistant	Highly susceptible	
Highly resistant	——	——	——	Susceptible	——	——	Lives less than 10 years

raspberry

Rubus (3 species)
Rosaceae (Rose Family)

Red RASPBERRIES are one of the most delicate and prized berry types and are considered the hardiest bramble fruit. Sufficient light and adequate spacing are two critical factors in raspberry production; without these, berries may be produced only on the upper half of the canes rather than all the way to the bottom. Yield also depends on the length of the cane after pruning, winterkill effects on blossoms, and the thickness of the canes. (See Blackberry, page 170, for yields, rows, floricanes, primocanes.) The everbearing raspberry cane bears annually: its floricanes bear a summer crop, and its primocanes bear on their tips in fall and on their bottom the following summer. For continuous harvest, try growing biennials for a spring crop and everbearers for a bumper fall crop (see Pruning, page 229). Always select certified disease-free stock.

Site

Full sun; rich loam; good drainage; east-facing spot sheltered from late afternoon sun. Plant at least 300 feet away from any wild brambles, which harbor insect and disease pests. Also keep at a distance from different types of berries to minimize cross-pollination.

Soil & Water Needs

pH: 5.5–7.0

Fertilizer: See Blackberry, page 171.

Mulch: See Blackberry, page 171. One gardener, who has a 60-year-old blackberry patch, layers 8"–10" of leaf mulch in autumn, and some wood ashes in winter to counter leaf acidity. He rarely needs to water.

Water: See Blackberry, page 171.

Measurements

Root Depth: 12"

Height: 5'–10'

Spacing:
Red and Yellow: 2'–4' in a row (suckers and new canes will fill out row)
Black and Purple: 3'–4' in hills (no suckers, but need room for branching canes)

Storage Requirements

To freeze, spread out berries on cookie sheets and freeze within 2 days of harvest. When rock hard, store in heavy freezer bags. Refrigerated, berries keep 4–7 days.

Preserved

Method	Taste	Shelf Life
Canned (as jam)	excellent	up to 2 years
Frozen	excellent	12–24 months
Dried	good	6–24 months

Growing & Bearing

Bearing Age: 2 years

Chilling Requirement: Most require some chilling, but the amount depends on variety.

Pollination: Self-pollinating

Propagation

Red and yellow by suckers; black and purple by layering. See Blackberry, page 171.

Shaping

Training: A trellis is important for disease and pest reduction, quality fruit, and easy harvest. Place bottom wire at 2', middle wire at 3', and top wire at 4' for black and purple, or 5' for red and yellow. Fan out canes and tie with cloth strips.

Pruning: After harvest or in the spring, cut out thin, weak, spindly, or sick canes, and ones finished bearing.

Red and yellow: Thin to 8 canes per 3 row-foot, or

4"–6" soil per cane. "Topping off" canes lowers yields, so don't do it unless the cane is taller than 6′. Cut out most sucker growth.

Everbearers: Cut to 5–7 canes per hill. To encourage branching, when primocanes are 18"–24" cut off the top 3"–4" of the cane ("topping off"). In late winter, cut lateral branches back to 8"–12" long.

Pests

Aphid, birds, caneborer, flea beetle, fungus beetle (in ripe fruits), Japanese beetle, mite, raspberry root borer, sap beetle, strawberry weevil, tarnished plant bug, weevil, whitefly

Diseases

Anthracnose, botrytis fruit rot, cane blight, cane gall, crown gall, leaf curl, leaf spot, mosaic, powdery mildew, rust, verticillium wilt

Allies

Uncertain: Garlic, rue, tansy

Companions

If berries are planted down the center of a 3' bed, plant beans or peas in the first summer to keep the bed in production and to add organic matter and N to the soil.

Incompatibles

Because of verticillium wilt, don't plant where nightshade family was grown in last 3 years.

Other

Raspberries, unlike most other plants, do tolerate the juglone in black walnut roots.

Harvest

When berries slide off easily without pressure, harvest into very small containers so berries on the bottom won't be crushed. After harvesting, cut floricanes to the ground; on everbearers cut the tips of primocanes. Burn or destroy all canes for disease control.

Selected Varieties

RED *(R. idaeus)*

Amnity: Everbearing; dark red berries; withstands heavier soils; aphid resistant; strong, self-supporting canes; Zones 3–8
Sonoma

Chilliwack: Midseason; new from British Columbia; excellent for wet sites; resists root rot; large firm berry; very sweet; productive; suited to Northwest; Zones 6–9
Raintree

Heritage: Everbearing; tall canes but usually sturdy enough to grow without stakes; medium berry; excellent quality; very hardy; Zones (4) 5–8
Widely available

Killearney: Early; large and firm; disease resistant; one of hardiest raspberries; Zones 3–8
Miller, Nourse

Selected Varieties *(cont'd)*

Latham: Midseason; large firm berry; excellent flavor; good fresh, frozen, or canned; prone to virus, so buy virus-free; Zones (3) 4–8
Indiana, Miller, Nourse

Summit: Everbearing; productive; will fruit first season; resistant to soil rot; survives wetter sites; Zones 4–9
Stark

Taylor: Midseason; conical large berry; very firm; excellent flavor; good quality; vigorous; hardy
Miller, Nourse

Titan: Early; thornless; firm; sweet and juicy; mild flavor; trellis recommended; high yields; resists diseases and pests; Zones 5–7
Indiana, Miller, Nourse, Stark

YELLOW *(R. idaeus)*

Fallgold: Everbearing; sweet and juicy berry; vigorous; hardy; Zones 4–8
Bay Laurel, Indiana, Miller, Sonoma

Kiwi Gold: Discovered in New Zealand; disease-resistant; great flavor; excellent shelf life; won't crumble; Zones 5–8
Stark

PURPLE (cross of red and black raspberry)

Royalty: Late midseason; juicier than black raspberry; sweeter than Brandywine purple raspberry; high yields; very hardy; resists insects and immune to raspberry aphid that carries mosaic; can be tip-layered but this one is best propagated by suckers; gourmet; good fresh and for jam; Zones 4–8
Hartmann's, Miller, Nourse, Raintree

BLACK *(R. occidentalis)*

Bristol: Very early; large firm berry; vigorous upright canes; no staking needed; leading variety in Finger Lakes area; tolerant to powdery mildew; Zones 5–8
Indiana, Miller, Nourse

Cumberland: Midseason; large glossy fruit; excellent flavor; long picking season into fall; very hardy; Zones 5–8
Bay Laurel

Jewel: Very early; large glossy berry; high quality; high yields; not susceptible to any serious disease; only mildly susceptible to mildew; hardy; Zones 5–8
Indiana, Hartmann's, Nourse, NYFT, Raintree

Munger: Late midseason; large, sweet fruit; small seeds; more disease resistant that other blacks; best for preserving and freezing; Zones 4–8
Stark

strawberry

Fragaria x *ananassa*
Rosaceae (Rose Family)

Plan about twenty-four plants to feed a family of four strawberry lovers. Junebearer strawberries yield a single crop in June or July and make lots of runners. Everbearing strawberries yield a first crop in June or July and a second in late summer or early fall; they require 15 hours of daylight or more during summer (unless it is a newer "day-neutral" type). If your area suffers late spring frosts, choose varieties that flower late and tolerate high humidity. Try to buy certified virus-free plants. When planting, cover all roots, but keep the upper two-thirds of the crown above the soil line. To promote a healthier, more productive plant, deflower all Junebearers the first year and everbearers only in their first late-spring flowering. For a bumper fall crop, however, you can deflower everbearers yearly in June. Most strawberry plants reproduce by runners. These can be pinched off to produce larger berries but lower yields, or trained to 7 to 10 inches for smaller berries and higher yields.

Strawberries tend to become less productive each year, so beds are often renewed by various methods. For a perennial bed, try the spaced-runner system. Space plants in rows 2 to 4 feet apart; train runners to 7 to 8 inches apart by pinning them down with clothespins or hairpins. Cut off excess runners to maintain spacing. In year 2, train runners into the central paths. Immediately after harvest in years 2 or 3, mow down the original plants, till several times, spread lots of compost, till a final time, and mulch with chopped leaves or compost every 2 weeks until fall. Every 2 or 3 years the pathways and growing beds trade places. Also, University of Wisconsin studies show enhanced yields when "Regal" perennial rye grass is grown as a living mulch and mowed to 2 to 3 inches two or three times per year.

Site

Good drainage. North-facing, sunny slope to delay blossoms in areas susceptible to late frost. The previous fall, prepare beds with 5 pounds of manure per 10 square feet; repeat whenever renewing.

Soil & Water Needs

pH: 5.5–6.8

Mulch: Apply 3"–4" after ground is frozen hard, remove in spring, and replace in hot weather. Do not cover crowns. University studies in New Hampshire show that row covers through winter, rather than mulch, produce higher yields. Pull off covers during April bloom (replace if frost is expected).

Fertilizer: Low N; apply very lightly and often rather than in two heavy feedings.

Water: Medium

strawberry

Measurements

Root Depth: Shallow; up to 8"

Height: 8"–12"
Breadth: 6"–12"

Space between Plants: 8"–15"
Space between Rows: 2'–5½'

Growing & Bearing

Bearing Age:
Everbearing (including newer day-neutrals): At end of first summer planted.
Junebearing: 1 year after planting; in the deep South, fall plantings may bear in early spring.

Chilling Requirement: Most require chilling hours; amount depends on variety.

Pollination: Self-fertile, self-pollinating

Shaping

Thinning: When runners get too prolific for a solid 7"–8" spacing, cut off excess runners.

Pests

Aphid, birds, crownborer, earwig, flea beetle, garden webworm, Japanese and June beetles, leafroller, mice, mite, nematode, pill bug, root weevil, sap beetle, slug, snail, strawberry beetle, tarnished plant bug, weevil, white grub, wireworm

Storage Requirements

Do not wash or remove green caps.

Preserved

Method	Taste	Shelf Life
Canned	not good	up to 2 years
Frozen	excellent	12–24 months
Dried (as fruit leather)	good	6–24 months
As jam	excellent	up to 18 months

Diseases

Anthracnose, botrytis fruit rot, leaf blight, leaf scorch, leaf spot, powdery mildew, red stele, root rot, septoria leaf spot, verticillium wilt, walnut bunch, yellows (virus)

Allies

Uncertain: Borage, thyme

Companions

Bush beans, lettuce, onion family, sage, spinach

Incompatibles

All brassicas

Harvest

A few days before harvest you may want to apply foliar calcium chloride. Canadian research shows that this can lengthen storage life by slowing ripening and delaying development of gray mold. Pick all berries as soon as ripe — whether damaged or not — to prevent disease. Handle gently to avoid bruising. Berries picked with green caps last longer in storage.

Selected Varieties

JUNEBEARERS

These are usually the highest-quality berries, yielding one crop per year in late spring or early summer.

Allstar: Midseason; easy to grow; good fresh, frozen, and U-pick; high quality; very hardy; consistently high yields; resists leaf scorch, powdery mildew, red stele, and verticillium wilt (somewhat); good in Northeast, mid-Atlantic, and west to Missouri
Hartmann's, Indiana, Nourse

Earliglow: Early; this sets the standard for all others; medium-large berry; best flavor of all; good fresh, canned, and frozen; resistant to red stele, verticillium, leaf spot, and leaf scorch; best for North Carolina to New England to Missouri
Hartmann's, Indiana, Jung, Miller, Nourse

Jewel: Late; introduced in 1985; very high-quality, large fruit; good fresh and frozen; hardy; not resistant to red stele or verticillium wilt, but has low incidence of fruit rot and foliar diseases; good from Nebraska to Maine, and south to Maryland and Ohio
Indiana, Jung, Nourse

Lateglow: Late; medium berry; firm; sweet and aromatic; high yields; strong resistance to red stele and verticillium wilt; tolerates leaf spot, leaf scorch, gray mold, and powdery mildew
Indiana

Mignonette Alpine: Late; heirloom variety; very cold-hardy; small, sweet berry; tolerates partial shade; will reseed to form dense ground cover; Zones 3–9
Burpee, Raintree

Selected Varieties *(cont'd)*

Sparkle: Late; medium berry; good frozen or for preserves; resists red stele and harsh winters; good in Northeast, west to Wisconsin, and in the Rocky Mountain region
Indiana, Jung, Miller, Nourse

Surecrop: Early midseason; good choice for beginners; large and firm; good fresh, canned, and frozen; very verticillium resistant; resists drought, leaf spot, red stele, and scorch; vigorous even in poor soil; good coast to coast; Zones 4–8
Miller, Stark

Winona: Late midseason; hardy and adaptable; resistance to red stele and black root rot; tolerance to leaf diseases
Indiana, Jung, Nourse

EVERBEARERS

These plants produce two main crops per year; for one large and late crop, just pick off all early-summer blossoms. The standard everbearer needs 15 hours or more of daylight, but a newer breed — day-neutral everbearers — are reportedly not affected by day length.

Fort Laramie: Large berry; good flavor; tolerates -30°F without mulch; pretty in hanging baskets; hardy in mountain states and high plains, and also good in Southern states
Indiana

Ogalla: Strong and vigorous; very winter-hardy, also reliable in dry conditions; sends more runners than other everbearers
Jung

Ozark Beauty: Standard everbearer; large berry; need to keep runners to 2 or 3 to ensure large fruit; good fresh or frozen; extremely productive; good Maine to Missouri and on the West Coast; wide climate adaptability; Zones 4–8
Jung, Miller, Stark

Tristar: Day-neutral; early and continuous crop through fall at about 6-week intervals; heaviest crop in fall; medium and firm; good fresh and frozen; very sweet; resists red stele and verticillium; tolerates leaf spot and scorch; can be grown in towers or hanging baskets because runners bloom and bear before rooting; good in Northeast, Midwest, and throughout West; Zones 5–8
Indiana, Miller, Nourse, Raintree, Rayner

Walnut

Juglans regia (Carpathian)
Juglandaceae (Walnut Family)

Black WALNUT and butternut roots excrete an acid (juglone) that inhibits the growth of many plants, so plant them far away from vegetables and flowers. Weed control is vital to the initial years of walnut tree growth. All walnuts benefit from mulching. Walnut trees grown for nut crops require large crowns; those grown for veneers need a straight trunk. To grow for both, favor veneer requirements and prune for straight growth.

Carpathian walnuts grow rapidly, about 4 to 5 feet per year. Black walnuts, the largest *Juglans*, have a strong flavor, so use sparingly in cooking. Grafted trees usually yield three to four times higher kernel filling than seedlings. Butternuts, or white walnuts, are the hardiest *Juglans*, don't require as much water, and have a rich buttery taste. Buy the smallest tree possible to minimize injury to its straight and long taproot. True nut flavor and quality may not emerge for 2 to 3 years after bearing, so don't judge trees prematurely. Walnuts do better in mixed plantings than in a monoculture walnut grove.

Site

Full sun; deep, well-drained loam. Don't plant anywhere near vegetable garden or orchard.

Soil & Water Needs

pH: 5.5–7.0; for Persian: 6.0–7.0

Fertilizer: Don't fertilize the first year. In following years, apply compost or well-rotted manure. When 10 years old, apply ½ pound of boron (borax) in deep bar holes for steady nut production.

Water: Medium

Measurements

Root Depth: Very deep

Height: 20'–80'
Breadth: 20'–40'

Space between Plants: 10'; when 10" in diameter (about 25 years), thin to 22'–50'.

Growing & Bearing

Bearing Age: 3–5 years for nuts; 10–20 years to produce veneer for market

Chilling Requirement: All require some chilling hours (400–1500).

Pollination: All walnuts require cross-pollination.

Shaping

Training:
Freestanding: Central leader
Wire-trained: N/A

Pruning: Unlike most fruit and nut trees, prune walnuts in fall. Don't start until 4–5 years old. Cut out dead or diseased wood, and crossed or competing branches.

Storage Requirements

Rinse off hulled nutshells and place in water. Remove any rotten and diseased nuts that float to the top. Dry and cure by spreading in a single layer in a cool, dry, well-ventilated area. They're ready for storage when the kernels rattle in the shell — generally in 1–3 weeks — or when unshelled kernels snap when bent. Store shelled nuts in plastic bags with holes, or in tightly sealed, paper-lined tins (that have a hole punctured in the side beneath the lid), or freeze. Store unshelled nuts in an unheated shed during winter; in spring move them to a cool place.

Fresh

Temperature	Humidity	Storage Life
32°F–36°F	60%–70%	1 year

Preserved

Method	Taste	Shelf Life
Frozen (shelled)	excellent	12–24 months

Pests

Aphid, bluejay, codling moth, fall webworm, mouse, navel orange worm, squirrel, walnut caterpillar, walnut husk fly, walnut maggot

Diseases

Crown gall, walnut anthracnose, walnut blight, walnut bunch

Allies

Some evidence: Weedy ground cover

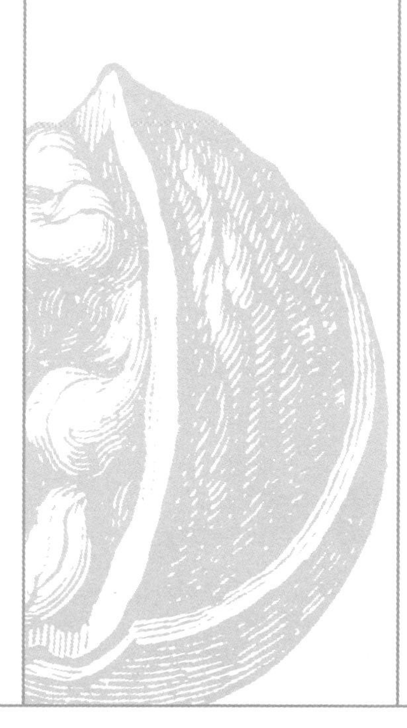

Companions

Autumn olive, black locust, European alder, popcorn (for the first few years), raspberry, soybean.

Incompatibles

Apple, azalea, potato, tomato (which is particularly sensitive to walnut root acid)

Harvest

Due to the indelible stains imparted by the husks, always wear gloves when harvesting walnuts. Remove the outer husks within 1 week of harvest; let dry for several days, after which the husks can be easily removed. There are three methods to dehusk a nut: (1) Use a corn sheller equipped with a flywheel and pulley, driven by ¼ hp motor; (2) spread a single layer in a wooden trough on a driveway and run a car or tractor over them, which splits the husk easily; (3) take them to a professional huller.

Selected Varieties

BLACK *(J. nigra)*

Bicentennial: Excellent timber type; nuts as hard as Thomas; large nut; precocious; vigorous; one of the hardiest black walnuts; good for far North
Grimo

Emma Kay: An excellent variety; high kernel filling; thin shell; excellent cracking; good flavor; heavy yields; good in the Midwest
Grimo, Nolin River

Purdue No. 1: Superior tree; the only known patented hardwood; from Purdue University; very thin shell; extra-large meat size; also excellent for timber.
American Forest Technology, Inc.; AFT offers a nut buy-back program.

Ridgeway: Very large nut; heavy bearing; anthracnose resistant
Nolin River

Thomas: Oldest grafted black walnut; one of largest nuts; well filled; high quality; very hard, thin shells; good cracking; early and heavy yields; good in Northern and Western regions where anthracnose is not a problem; Zones 5–9
Bay Laurel, Grimo, Gurney's, Stark

Thomas Myers: Good straight tree for timber and good nuts; heavy bearer; large nuts crack well; anthracnose resistant; good through Zone 8 and as far north as Massachusetts, Southern Wisconsin, and Michigan
Nolin River

Weshcke: Good flavor; high yields; one of hardiest black walnuts; pollinates Bicentennial; good for the far North
Nolin River

PERSIAN (ALSO CALLED ENGLISH OR CARPATHIAN) *(J. regia)*

Because they leaf early, Persian walnuts grow best where peach trees don't suffer frost-kill. Carpathian *refers to the mountain range in Poland, a source of many imported Persian walnuts; now it usually refers just to "hardy" cultivars.*

Champion: Natural dwarf, growing to only 35'; thin shell; tasty nut; hardy; Zones 5–9
Stark

Colby: Early maturing; medium nut; thin shell; good flavor; hardy; especially good for Great Lakes area
Nolin River

Selected Varieties *(cont'd)*

Franquette: Common cultivar in Northwest; tight nut resists worms; medium-large elongated nuts; high quality; thin smooth shells; tree leafs out and blooms late; slow maturing; not hardy in colder parts of the Northwest; good for the West and Northwest
Bay Laurel

Hansen: Bears young; self-fruitful; self-pollinating thin shell; good flavor; small-to-medium nut; resists anthracnose and husk maggot; naturally small; hardy; widely planted in the East; considered one of the best; especially good for Great Lakes area
Grimo, Nolin River

Himalaya: Very hardy; a good-quality nut; good for the far North
Grimo

Reda: Medium-sized, thin-shelled nut; anthracnose and blight resistant
Nolin River

Spurgeon: Large, flavorful nuts; reliable producer; productive; leafs out late; best cultivar for pockets where late spring frosts are a problem; benefits from pollination by Franquette; good for Northwest
Raintree

BUTTERNUT *(J. cinerea)*

For location in the Ohio Valley and southward, choose varieties with "clean foliage" for anthracnose resistance. Increasingly scarce, these offer high-quality lumber. Hardy in Zones 3–7.

Chamberlin: Extremely hardy; large nuts; medium shell thickness
Grimo, Nolin River

Creighton: Good cracking qualities; medium nut; late vegetating; clean foliage; vigorous
Nolin River

Kenworthy: Large nut; good flavor; good cracking; vigorous; hardy
Grimo

HEARTNUT *(J. sieboldiana)*

A Japanese ornamental walnut, these grow fast and bear nuts that hang in strings or clumps of 10–15. The heart-shaped nutmeat is the sweetest of all walnuts, with no bitterness; it is easy to extract and high in protein. The tree bears even in poor soils.

Heartnut: Grows 40'–60'; cross-pollinate with Butternut and Buarnut; Zones 5–9
Nolin River, Grimo

OTHER VARIETIES

Buarnut or Butterheart (butternut x heartnut): Disease resistant; nuts are similar to butternut; Zones 4–7
Grimo, Miller, St. Lawrence

Chapter Four

herbs

sweet basil

Basilicum
Labiatae (Mint Family)

BASIL is an annual warm-season herb that is particularly tender to frost and light freezes. It transplants easily and also can be grown easily in a greenhouse. Continuous harvest benefits this herb because pruning encourages new growth. A study by the University of California at Davis advises against storing basil in the refrigerator, for it lasts longer when kept in a glass of water at room temperature. Basil, fresh or dried, is a popular seasoning. Some culinary experts suggest that dried basil simply cannot compare with the flavor of fresh basil, but few true basil lovers will pass up either. Basil comes in a range of varieties, from purple to lime green, curly to ruffled-edged leaves, and smooth to hairy leaves. An ingredient in the liqueur Chartreuse, basil may be minty or hint of clove or cinnamon.

Site

Full sun; protected

Temperature

For germination: 75°F–86°F

For growth: Hot

Soil & Water Needs

pH: 5.5–7.0

Fertilizer: Light feeder

Side-dressing: Not necessary

Water: Low, but evenly moist

Measurements

Planting Depth: ¼"

Root Depth: 8"–12"
Height: 18"–24"
Breadth: 20"–30"

Space between Plants:
In beds: 10"–12"
In rows: 12"–18"
Space between Rows: 15"–25"

Propagation

By seed

Pests

Japanese beetle, slug, snail

Diseases

Botrytis rot, damping off

Companions

Pepper, tomato

Other

Basil is said to repel flies and mosquitoes, and to improve growth and flavor of vegetables.

Incompatibles

Cucumber, rue, snap beans. Basil is also alleged to lower cabbage yields and cause a higher incidence of white flies in snap beans.

Harvest

Pick continuously before flower buds open, up to 6" below the flower buds or ends to encourage continuous bushy growth. Cut in the morning after the dew has dried. Do not wash the leaves or aromatic oils will be lost.

First Seed-Starting Date

Germinate	+	Transplant	–	Days After LFD	=	Count Back from LFD
3 to 9 days	+	7 to 14 days	–	14 days	=	9 days before LFD to 4 days after LFD

Last Seed-Starting Date

Germinate	+	Transplant	+	Maturity	+	SD Factor	+	Frost Tender	=	Days Back from FFD
3 to 9 days	+	7 to 14 days	+	30 to 50 days	+	14 days	+	14 days	=	68 to 101 days

Storage Requirements

Leaves can be used fresh, dried, or preserved in oil (must be refrigerated) or vinegar. To dry, find a warm, dry, dark place and hang bunches of snipped stems with leaves, or spread leaves on a wire mesh. When thoroughly dry, strip leaves off stems. Do not crush or grind leaves until you're ready to use them. Store in airtight containers or freezer bags in a dark place. Some people believe that basil stored in oil or vinegar is more flavorful than dried. If storing frozen pesto, don't add garlic until you are ready to serve because garlic can become bitter in the freezer.

Method	Taste
Frozen	excellent (particularly for pesto)
Dried	fair–good (relative to fresh basil)
Fresh	excellent

Selected Varieties

Cinnamon (*O.* var.): Similar to sweet basil; cinnamon flavor; from Mexico
Widely available

Dark Opal Basil: 18"; purple-bronze foliage; dark purple flowers; highly ornamental; good for seasoning; colors vinegars purple
Fox Hill, Nichols, Richters, Sandy Mush

Greek Columnar Basil (*O. basilicum*): Up to 5' tall and 10" wide; leaves have wonderful flavor and can be harvested year-round, as the plant does not flower; best variety for wintering indoors in cold climates
Sheperd's, Park

Green Ruffles (*O. basilicum* "Green Ruffles"): 24"; pretty curly and serrated large leaves
Burpee, Dabney, Richters

Holy (*O. sanctum,* purple Tulsi form): Whole plant has reddish-purple tint; deep, spicy clove scent
Companion Plants, Fox Hill, Sandy Mush

Lemon (*O. citriodorum*): 12"; intense lemon flavor; ideal for tea; compact bush-type
Widely available

Selected Varieties *(cont'd)*

Lettuce Leaf *(O. basilicum crispum)*: 18"; large broad crinkled leaves to 4" long; rich flavor, not as strong as sweet basil
Companion Plants, Nichols, Richters, Sandy Mush, Seeds Blum

Mammoth Basil *(O. basilicum)*: Extremely large leaves; sweet fragrance and tasty
Richters

Mayo/Yaqui Basil *(O. basilicum)*: Do not grow with other basils if saving seed; good for low desert
Native Seeds

Mrs. Burns Famous Lemon Basil *(O. basilicum)*: Pure strain; readily self-seeds; plant spring and fall; for low and high desert; some consider it superior to other sweet basils
Native Seeds

Piccolo Verde Fino: Small leaves; intense true basil flavor; slightly minty; favored for pesto because it maintains flavor better after flowering than others
Casa Yerba, Companion Plants, Fox Hill, Nichols, Richters

Purple Ruffles: AAS Winner; large, very ruffled and fringed dark-purple leaves; pink flowers; beautiful ornamental; also good culinary
Widely available

Siam Green Basil: Purple on leaf undersides; pink blossoms lure beneficial insects; Vietnamese use this variety to garnish food; licorice-, clove-scented leaves
Widely available

Spicy Globe: Great fragrant edging plant and for pots; maintains small mound shape all season; white flowers; good for culinary use
Dabney, Fox Hill, Richters

Sweet Basil *(O. basilicum)*: 16"; a fast grower; tasty and one of best basil flavors; classic pesto basil
Widely available

Thai: Green leaves with purple flowers and stems; good for Southeast Asian cuisine
Nichols

Greenhouse

All varieties work well.

sweetbasil

caraway

Carum carvi
Umbelliferae (Parsley Family)

CARAWAY is a biennial warm-season herb, tender to frost and light freezes. It is commonly grown as an annual in most areas of the United States, except for northern California where it reportedly doesn't grow well. Sow it directly, as it doesn't transplant well, or transplant it while it's still small. It flowers in spring and produces seed in its second summer — or in its first summer if sown in fall. Caraway roots, which are long taproots, can be eaten like carrots. The leaves can be used in salads, soups, or stews. Some trace the use of caraway back to the Stone Age; it has a long history of medicinal use to aid digestion and lactation, relieve flatulence, and as an antispasmodic and antiseptic. Its a popular ingredient in German and Austrian cuisine, cheeses, and breads, and its oil is used to flavor Aquavit and the liqueur Kümmel.

Site

Full sun to light shade

Temperature

For germination: 70°F
For growth: Warm

Soil & Water Needs

pH: 5.5–7.0

Fertilizer: Light feeder

Side-dressing: Not necessary

Water: Low, but evenly moist

Measurements

Planting Depth: ¼"–½"
Root Depth: Long taproot

Height: 2'–3'
Breadth: 12"–18"

Space between Plants:
In beds: 6"
In rows: 12"–18"
Space between Rows:
18"–24"

Propagation

By seed and cuttings

Pests

Carrot rust fly

Diseases

None

Companions

Coriander, fruit trees, peas

Incompatibles

Fennel

Other

Caraway is said to be good for loosening the soil and is reputed to attract beneficial insects to fruit trees.

Harvest

When the seeds are brown and before they begin to fall, snip the stalks. Tie in bundles and hang upside down in a warm, dry, airy place. Place paper-lined trays under the stalks to collect falling seeds, or cover them with a paper bag and let the seeds drop into the bag. Shaking the stalks can dislodge the seeds. After a few weeks, when the fallen seeds are thoroughly dry, store them in an airtight jar.

First Seed-Starting Date

Germinate	+	Transplant	+	Days Before LFD	=	Count Back from LFD
17 days	+	0 (direct)	+	0 to 7 days	=	17 to 24 days

Last Seed-Starting Date

Germinate	+	Transplant	+	Maturity	+	SD Factor	+	Frost Tender	=	Count Back from FFD
17 days	+	0 (direct)	+	55 days	+	14 days	+	N/A	=	86 days

Storage Requirements

Store in airtight jars in a cool, dark place.

Method	Taste
Frozen	Caraway leaf doesn't freeze well
Dried	Seed is excellent dried

Selected Varieties

The single cultivar of caraway is widely available.

Greenhouse

Caraway should grow well in the greenhouse if grown in a pot deep enough to accommodate its taproot.

caraway

chives

Allium schoenoprasum
Liliaceae (Lily Family)

CHIVES are a perennial warm-season herb, hardy to frost and light freezes, and the earliest to appear in spring. This herb likes rich and well-drained soil but can be found growing wild in dry, rocky places in northern Europe and in the northeastern United States and Canada. They also thrive in a cool greenhouse or on a kitchen windowsill. Chives are virtually foolproof because they suffer no diseases or pests. After several years, you can divide them for expansion or renewal and in autumn dig up a clump to pot indoors for continuous winter cutting. Chives blossom midsummer and are an attractive ornamental addition to the garden; if allowed to bloom, cut them back after flowering so new shoots will come up in spring. With a milder flavor than onion, chives are usually snipped raw as a finishing touch for salads, soups, sauces, and vegetable and fish dishes. Chives also work well in egg dishes such as quiche and omelettes.

chives

Site

Full sun to light shade

Temperature

For germination: 60°F–70°F
For growth: Hot

Soil & Water Needs

pH: 5.5–7.0

Fertilizer: Light feeder

Side-dressing: None

Water: Average

Measurements

Planting Depth: ¼"–½"
Root Depth: Bulb clumps

Height: 6"–18"
Breadth: 6"–8"

Space between Plants
In beds: 6"
In rows: 5"–8"

Space between Rows: 12"

Propagation

Division or seed; every 3 years, in mid-May, divide plant into clumps of 6 bulbs

Pests

None

Diseases

None

Companions

Carrot, celery, grape, peas, rose, and tomato

Incompatibles

Beans, peas

Other

Chives are a putative deterrent to aphids on celery, lettuce, and peas; black spot on roses;

First Seed-Starting Date

Chives mature in 50 days, whether or not transplanted.

Germinate	+	Transplant	+	Days Before LFD	=	Count Back from LFD
10 to 14 days	+	21 to 42 days	+	0 days	=	31 to 56 days

Japanese beetles; mildew on cucurbits; and scab on apples

Harvest

After the plant is 6" tall, cut some of the blades down to 2" above the ground to encourage plant production. Herbs should be cut in the morning after the dew has dried. Cut near the base of the greenery, not the chive tips, so new, tender shoots will emerge. Do not wash the cuttings or aromatic oils will be lost.

Storage Requirements

Chives are best fresh or frozen but can also be dried. To dry, tie them in small bunches and hang them upside down in a warm, dry, dark place. Do not crush or cut until ready to use. Store the stem whole, if possible. If harvested with the flower, chives can be stored whole in white vinegar to make a pretty, light lavender, mildly flavored vinegar for gifts. Another storage method — recommended by horticulturist Frank Gouin as an excellent way of preserving chive flavor — is to layer alternately in a glass jar 1" of kosher salt with 1" of chives. Pack down each layer with a spoon. Use these chives in any dish, just as you would fresh chives; they're said to be especially good in soups. As an added bonus, the brine also can be used to flavor soups and other dishes.

Method	Taste
Frozen	Excellent
Dried	Fair–good
Fresh	Excellent

Selected Varieties

Chive: Common strain; mild onion flavor; great in salads, soups, potatoes, eggs, cheese, and fish dishes
Widely available

Curly Chive (*Allium senescens* var. "glaucum"): Ornamental; not edible; deep pink flowers
Companion Plants

Garlic Chive *(Allium tuberosum)*: Also known as Chinese leeks. Mild onion-garlic flavor; broader leaves than common strain; from Japan; pretty white flowers
Widely available

Grolau Chive: Similar to the common strain but developed for indoor growing; excellent in greenhouses; productive when cut continuously; good strong flavor
Nichols

Mauve Garlic Chive: Same as the garlic chive but with pretty mauve flowers
Richters

Yellow Garlic Chive: Same as the garlic chive but with light yellow flowers
Richters

Greenhouse

All chives grow well in a cool greenhouse or on a windowsill.

coriander/cilantro

Coriandrum sativum
Umbelliferae (Parsley Family)

CORIANDER is an annual cool-season herb, tender to frost and light freezes. In some warmer climates, coriander is self-seeding. It grows easily, although it does go to seed quickly when the weather turns hot. For a steady supply of the leaf, try sowing in succession every 1 to 2 weeks. If you're growing it for seed, stake the plant at the time of sowing or transplanting. All parts of the coriander are edible, including the root, which is similar in taste to the leaves but has an added nutty flavor. Unlike most herbs, fresh coriander leaf is usually either loved or hated. Make sure your guests can tolerate fresh coriander before including it in a dish. The coriander plant is said to attract useful insects, and coriander honey is famous for its flavor. An ancient herb with medicinal properties for aiding digestion and relieving rheumatism, coriander is recognized as the seed used in Indian and Mediterranean cuisine, whereas cilantro is the leaf used in Central and South American cuisines. Because it is used commonly in Chinese cuisine, cilantro is sometimes referred to as "Chinese parsley."

Site

Full sun to partial shade

Temperature

For germination: 50°F–70°F
For growth: Cool

Soil & Water Needs

pH: 6.0–7.0

Fertilizer: Light feeder; N reduces flavor

Side-dressing: Not necessary

Water: Average

Measurements

Planting Depth: ¼"–½"
Root Depth: 8"–18"

Height: 12"–21"
Breadth: 6"–12"

Space between Plants:
In beds: 6"–8"
In rows: 8"–12"
Space between Rows:
12"–15"

Propagation

By seed

Pests

Carrot rust fly

Diseases

None

Companions

Caraway, eggplant, fruit trees, potato, tomato

Incompatibles

None

Other

Coriander is said to enhance anise growth. It is also a putative companion to fruit trees because it attracts beneficial insects.

Harvest

Leaves, stems, roots, and seeds are used in food preparation. Snip stalks with small, immature leaves for fresh leaves with best flavor. Cut back the top growth up to 6" below the flower buds or ends. Do not wash or aromatic oils will be lost. For seeds, harvest when the seeds and leaves turn brown but before seeds drop; cut the whole plant.

First Seed-Starting Date

Germinate	+	Transplant	+	Days Before LFD	=	Count Back from LFD
12 days	+	30 days	+	7 to 21 days	=	49 to 63 days

Last Seed-Starting Date

Germinate	+	Transplant	+	Maturity	+	SD Factor	+	Frost Tender	=	Count Back from FFD
6 days	+	21 days	+	55 days	+	14 days	+	0 days	=	96 days

Storage Requirements

Coriander leaves store poorly unless preserved in something like salsa, but even then its flavor can fade in a day. Coriander seeds, on the other hand, keep well in airtight jars. To dry seeds, tie the plant upside down in a warm, dry, dark place for several weeks until the seeds turn brown. Place stalks in a paper bag and thresh until all seeds are removed from stems. Sift out seeds from chaff. Another method to finish the drying process is to remove seeds from the stems and dry them in a slow oven (100°F) until they turn a light brown. You can smell the difference between properly air-dried or roasted coriander seed and seed that is still green. For best flavor, grind the seeds just before use. Store seeds in an airtight jar in a cool, dark place.

Method	Taste
Fresh	Excellent; cuttings last well 3–7 days refrigerated in water that is refreshed daily, longer if covered with a plastic bag
Frozen	Some report excellent for leaf, some report poor
Dried	Excellent for seed, poor for leaf

Selected Varieties

Chinese Coriander: Especially good for fresh leaf production because it doesn't bolt as quickly
Richters

Coriander (standard): Fresh leaves are delicious in salsa and in Latin, Indian, Chinese, and many other dishes. The seeds are traditional in curry, chili, and some baked goods, and are a great addition to soups, stews, and Indian dishes. Good in all areas, including high and low desert; self-seeding in warmer climates
Widely available

Slow-bolt Coriander: Especially good for fresh leaf production because it doesn't go to seed as quickly
Shepherd's

Vietnamese Coriander *(Polygonium odoratum)*: Not a true coriander; almost identical to native smartweed or knotweed; a perennial in some climates; tastes remarkably like true coriander; can be grown indoors
Companion Plants, Richters

Greenhouse

Coriander grows well in the greenhouse.

Anethum graveolens
Umbelliferae (Parsley Family)

DILL is an annual or perennial warm-season herb, very sensitive to light freezes and frost. If it's not grown in a protected spot, stake it to keep the tall stocks upright. Because of its long taproot, dill does not transplant easily, so don't attempt to transplant it once it grows beyond the seedling stage. If dill is not planted early enough, seed may not develop until the beginning of the second year. It can be grown in the greenhouse if you provide a container large enough for its roots, at least 6 to 8 inches in diameter, and pot it in rich soil. In the garden, if allowed to go to seed without complete harvest, it will reseed itself and grow as a perennial. As a seed, it's used primarily for pickling. As dill weed, it's used to flavor sauces, fish, meats, soups, breads, and salads. For best flavor, snip the weed with scissors rather than mincing it with a knife.

Site

Full sun, sheltered from wind

Temperature

For germination: 50°F–70°F
For growth: Hot

Soil & Water Needs

pH: 5.5–6.5

Fertilizer: Light feeder; might need one application of compost or slow-release fertilizer

Side-dressing: Not necessary

Water: Average

Measurements

Planting Depth: ¼"–½"
Root Depth: Very long, hollow taproot

Height: 3'–4'
Breadth: 24"

Space between Plants:
For leaf: 8"–10"
For seed: 10"–12"
Space between Rows:
18"–24"

Propagation

By seed

Pests

Carrot rust fly, green fly, parsleyworm, tomato hornworm

Diseases

None

Companions

All brassicas, fruit trees

Incompatibles

Carrot, fennel (cross-pollinates with dill), tomato

Other

Some gardeners report that, in orchards, dill helps attract beneficial wasps, bees, and flies that pollinate blossoms and attack codling moths and tent caterpillars. Others report that it may help cabbage by repelling aphids, spider mites, and caterpillars. There is, however, no scientific study or proof of this.

Harvest

Cut the tender feathery leaves close to the stem. Herbs should be cut in the morning after the dew has dried. Do not wash or aromatic oils will be lost. The flavor of dill foliage is best before the flower head develops and when used the same day it's cut. If you want to harvest dill seed, let the plant flower and go to seed. Harvest when the lower seeds turn brown and before they

First Seed-Starting Date

Sow every 3 weeks for a continuous supply of leaves.

Germinate	+	Transplant	–	Days After LFD	=	Count Back from LFD
14 days	+	21 to 30 days	–	14 days	=	21 to 30 days

scatter. The lower seeds on a head will brown first; the upper ones can dry indoors. Finish drying by tying stems together and hanging them upside down in a cool, dark, dry place, or place in a paper bag with holes cut in the sides. Sift to remove the seed from chaff.

Storage Requirements

Fresh dill keeps for up to 3 days when stored in a jar, with stems submerged in water, covered with plastic. It will store for up to 3 months in the refrigerator when layered with pickling salt in a covered jar. To use, just brush off salt. To freeze, store on the stem in plastic bags. Cut off what you need and return the rest to the freezer. To dry the leaves, spread them over a nonmetallic screen in a dark, warm, dry place for several days. Then store in an airtight container. Crush or grind immediately before use.

Method	Taste
Frozen	Good for foliage
Dried	Good for foliage, excellent for seeds
Fresh	Excellent

Selected Varieties

Aroma Dill: Large yields and excellent aroma
Sandy Mush

Bouquet Dill: Best grown for dill seeds for pickles and potato salads; a compact and attractive plant but has larger seed heads and larger leaves than the common strain
Widely available

Dill: Old-fashioned dill; used for dill pickles and sauerkraut
Widely available

Dukat Dill: Bred for higher essential oil content; best for its fresh leaves; originally from Finland; slow bolting; flavor is mellow and aromatic; never bitter; excellent with shrimp, fish, and cucumbers
Widely available

Fernleaf Dill: AAS Winner; PVP variety (patented by Plant Variety Protection); early; compact; slow bolting; space-saving variety especially suitable for containers; 18" tall
Southern Exposure

Indian Dill *(Anethum sowa)*: Seeds used in Indian curries; leaves used in rice and soups; slightly more bitter than the common strain, smaller plant
Companion Plants, Sandy Mush

Greenhouse

Dill grows well in the greenhouse.

fennel

Foeniculum vulgare
Umbelliferae (Parsley Family)

FENNEL self-sows easily and is a perennial warm-season herb that is half-hardy to frost and light freezes. This herb is best sown in succession plantings in spring, as it tends to bolt in hot summers. For a fall crop, sow in July. If you're growing Florence fennel — renowned for its swollen and tender anise-flavored stalks — mound the soil around the base of the plant to promote more tender, blanched stalks. Fennel stems are a more common culinary item in Europe than in the United States. They're delicious raw in salads, steamed or sautéed with a little butter and wine, or added to soups, stews, and fish dishes. Fennel is reputed to have a broad range of medicinal uses that date back to the time of Hippocrates. Fennel leaves are thought by some to improve digestion, and fennel tea is believed to have antiflatulent and calming effects. With its mild aniseed flavor, fennel seed can be used whole or ground as a spice in breads, sausages, pizza and tomato sauce, stews, and fish and chicken dishes.

garlic

Allium sativum
Lillaceae (Lily Family)

GARLIC is an annual or perennial cool-season crop and is hardy to frost and light freezes. Plant cloves in a sunny location in rich, deep, moist, well-drained soil. You may have to add compost to the soil before planting. Although it can be started from seed, it's easiest to grow from individual cloves. Garlic bulbs mature in an average of 6 to 10 months. Cloves may be planted in either fall or spring, but fall plantings yield larger bulbs the next summer than do spring plantings harvested in fall. Remove flower buds if they develop. Early cultivars store poorly and have inferior quality. Garlic is renowned for its broad range of culinary and medicinal uses, from reportedly imparting strength to laborers who built the pyramids to more modern studies about its antibacterial and beneficial circulatory effects.

Anything not benefiting from the addition of chocolate will probably benefit from the addition of garlic.

— Culinary proverb

Site

Full sun

Temperature

For germination: 60°F–80°F
For growth: Cool

Soil & Water Needs

pH: 4.5–8.3

Fertilizer: Light feeder; use compost and liquid seaweed extract

Side-dressing: Not necessary

Water: Low; for perennial bulbs, withhold all water during summer, except in arid, dry areas

Measurements

Planting Depth: 1"–2" (pointed end up)
Root Depth: Bulbs: 2"–2'

Height: 1'–3'
Breadth: 6"–10"

Space between Plants:
In beds: 3"
In rows: 4"–6"
Space between Rows:
12"–15"

Propagation

Usually propagated by cloves (easiest method), though garlic can be started from seeds
Perennial Method: One Washington State gardener has a garlic patch that hasn't been planted or plowed for 20 years. His method? When the garlic is about 2' tall, pinch off the seed buds that develop to encourage garlic to form. Do not water at all during summer, except in arid, dry climates, where you should water regularly. Harvest the large plants only by simply pulling them; digging isn't necessary. The smaller plants die back and emerge again the

First Clove-Starting Date

Cloves can be planted about 6 weeks before the last frost.

Germinate	+	Transplant	+	Days Before LFD	=	Count Back from LFD
7 to 14 days	+	0 (direct)	+	14 to 21 days	=	21 to 35 days

Last Clove-Starting Date

Cloves can be planted in autumn for harvest in spring.

Germinate	+	Transplant	+	Maturity	+	SD Factor	+	Frost Tender	=	Count Back from FFD
7 days	+	0 (direct)	+	90 days	+	14 days	+	0 days	=	111 days

following year. After harvest, lightly till the surface with a hand cultivator and uproot weeds. Water well to germinate weed seeds, then cultivate again to get rid of young weeds. In October, spread a 3"–4" leaf mulch. In spring, garlic will grow through the mulch again.

Pests

Nematode

Diseases

Botrytis rot, white rot

Companions

Beet, brassicas, celery, chamomile, fruit trees, lettuce, raspberry, rose, savory, tomato

Incompatibles

All beans, peas

Other

Garlic spray has been found to have some antifungal properties in controlling bacterial blight of beans, bean anthracnose, brown rot of stone fruit, cucumber and bean rust, downy mildew, early tomato blight, leaf spot of cucumber, and soft-bodied insects such as aphids and leafroller larvae.

Harvest

Green garlic shoots, a gourmet treat in many locales, can be cut from the bulbs going to flower and used like scallions. Garlic bulbs are ready to harvest when the tops turn brown and die back. Do not knock the tops down to hasten harvest; some research indicates this practice shortens storage life. Withhold water and, in a few days, dig carefully to lift the plants. The tops usually are not strong enough to pull, with the exception of the perennial garlic described earlier (see page 265). Be careful not to bruise bulbs or they'll get moldy and attract insects when stored.

Storage Requirements

Cure bulbs in the sun for several days to 2 weeks to harden the skins and dry. To braid, keep the tops on. Otherwise, clip off dried leaves and the root bunches. Store in paper bags, net bags, or nylon stockings, tying a knot between each bulb in the stocking.

Method	Temperature	Humidity	Storage Life
Fresh	32°F	65%–70%	6–7 months

Selected Varieties

California White: Easy to grow; stores like onion; very mild flavor
Field's, Jung

Elephant Garlic *(Allium ampeloprasum):* Milder than regular garlic and about 6 times larger; hardy for fall planting; developed and named by Nichols; bake whole, roast, or use fresh in salads and other dishes where a mild garlic flavor is desired
Widely available

Italian Garlic or Rocambole or Serpent Garlic *(A. sativum ophioscordon):* Mild garlic flavor; easy to peel; aboveground bulblets and belowground bulbs are prized in Europe; the stems bearing the small bulblets are looped
Richters, Sandy Mush

Nichols Silverskin: Noted for its strong flavor, size, and easy peeling
Fox Hill, Nichols

Nichols Top Set: Pungent flavor; harvest medium bulbs at end of season; small bulblets at the top of each plant can be kept for early spring or fall planting
Nichols

Polish White (N.Y. White): Dependable and high yields; large cloves; good for Northern growers
Southern Exposure

Silver Rose (Silver Skin): Rose cloves in white skins; best for braiding; keeps up to 10 months
Park

Spanish Roja: One of the strongest garlics available; light purple skins; brown cloves; recommended for cooler areas
Park, Southern Exposure

Greenhouse

None; garlic doesn't do well in the greenhouse.

garlic

lavender

Lavendula angustifolia
Labiatae (Mint Family)

LAVENDER is a perennial warm-season herb, hardy to frost and light freezes in Zones 5 through 8. It likes full sun and light, well-drained soil. Lavender can be easily grown in containers in the greenhouse or on windowsills and trained to different forms. In addition to serving as a pretty border plant, dried lavender flowers are touted as an effective moth repellant, and lavender oil is used as an additive to soaps and sachets. For the best oil production, lavender needs hot, dry weather from May through August.

For centuries, oils have been used for a broad variety of purposes, from curing lice, repelling mosquitoes, and embalming to use in porcelain lacquers, varnishes, and paints. Lavender is also used to make a light perfume; simply add rose petals, lavender flowers, and jasmine flowers to distilled vinegar and store in airtight bottles. In the kitchen, lavender flowers and leaves can be used to flavor jellies and vinegars, herb mixes for salads, and even lavender ice cream.

Site

Full sun protected from wind; can be stony

Temperature

For germination: N/A
For growth: Hot

Soil & Water Needs

pH: 6.5–7.0

Fertilizer: Light feeder; lime may be needed to make sure soil is basic enough

Side-dressing: Not necessary

Water: Low

Measurements

Planting Depth: Cover roots to root cellar
Root Depth: Deep

Height: 14"–3'

Breadth: Up to 5'

Space between Plants:
In beds: 2'
In rows: 4'–6'
Space between Rows: 6'

Propagation

By cuttings or seed. Lavender is seldom started from seed because of its long germination time. In summer, take 2"–3" long cuttings from the side shoots; make sure there is some older wood in cutting. Place cuttings 3"–4" apart in moist, sandy soil in a shaded cold frame. When they're 1-year-old, plant in well-drained, dry soil that is protected from severe frost. The first year, plants should be pruned to keep them from flowering and to encourage branching.

Pests

Caterpillars

Diseases

Fungal diseases

Companions

None

Incompatibles

None

Other

Lavender leaves are reputed to repel insects.

Harvest

Don't cut until its second year as an outdoor plant. Pick the flowers when in blossom but just before full bloom, usually in August. On a dry and still day, cut early in the day after the dew has dried. Cut back the top growth up to 6" below the flower spikes. Do not wash or aromatic oils will be lost.

Storage Requirements

To dry, tie branches in small bunch and hang it upside down in a warm, dry, dark place. Then remove the flowers from the stems and keep whole in storage.

Method	Results
Dried	Excellent; remains aromatic for a long time

Selected Varieties

HARDY ENGLISH TYPES

English Lavender *(L.a.* or *L. vera):* 3'; the hardiest lavender; considered the best for culinary uses, perfumes, soaps, and oils; spreading; bushy growth; silvery gray and narrow leaves; light purple flowers; good for northern climates
Widely available

Grey Lady (*L.a.* "Grey Lady"): 18"; fast, compact growth; grey foliage; lavender-blue flowers; hardy; good for borders or pots
Companion Plants, Sandy Mush

Hidcote Lavender: Historical favorite of British royalty; flowers used in salads and sachets
Kitchen Garden

Jean Davis (*L.a.* "Jean Davis"): 18"; dainty compact mounds; low growth; bluish foliage; pink flowers; very hardy; excellent for borders and formal gardens
Dabney, Sandy Mush

Rosy (*L.a.* "Rosea"): 18"; attractive; compact plant; pink flowers; good for borders and formal gardens
Companion Plants, Richters

True Lavender: 3'; very hardy; good for clipped borders
Sandy Mush

TENDER TYPES

Good for frost-free areas; may also be dug up and potted as a winter houseplant.

French or Fringed Lavender *(L. dentata):* 3'; very green, toothed leaves; fragrant foliage; provides more but lower quality oil with camphor-rosemary scent
Companion Plants, Dabney, Richters, Sandy Mush

Spanish Lavender *(L. stoechas):* 30"; beautiful gray foliage with green overtone; toothed and narrow leaves; aroma is strong and resinous — reminds some of turpentine; flowers bloom nearly all year; light purple flowers; excellent year-round potted plant
Companion Plants, Sandy Mush

Spike *(L. latifolia):* Coarse, broad leaves; oil is used in soaps; high oil yields
Richters

Greenhouse

Tender varieties can be grown inside.

sweet marjoram

Origanum majorana
Labiatae (Mint Family)

MARJORAM is a perennial or annual warm-season herb, very tender to frost and light freezes. It is winter hardy only in the South, Zones 9 through 10, so most gardeners grow it outdoors as an annual or in pots that can be brought indoors through winter. Marjoram likes full sun and soil that is light, dry, well-drained, and very low acid. This herb has played varied roles over the centuries, from medicinal and culinary purposes, to aromatic uses in closets and sachets, to serving as a green wool dye. Don't confuse sweet marjoram with wild marjoram or the perennial oregano (*O. vulgare*). Sweet marjoram is used for cooking, whereas wild marjoram is used for medicinal purposes. Somewhat milder and sweeter than oregano but still perky with peppery overtones, marjoram can substitute for oregano in virtually any tomato-based dish, as well as in marinades, dressings, and stews. Marjoram harmonizes well with thyme, basil, garlic, onion, and bay and is particularly complementary to eggs, fish, savory meats, and green vegetables.

sweet marjoram

Site

Full sun

Temperature

For germination: 65°F–75°F
For growth: Hot

Soil & Water Needs

pH: 6.5–7.0

Fertilizer: Light feeder

Side-dressing: Not necessary

Water: Low

Measurements

Planting Depth: 0"; tiny seeds need light and should not be covered

Root Depth: 6"–12"

Height: 1'–2'
Breadth: 10"–18"

Space between Plants:
In beds: 6"–8"
In rows: 8"–10"
Space between Rows: 15"

Propagation

By seeds, cuttings, and root division. If you propagate by seed, either sow seeds directly outside after all danger of frost has passed, or start seeds indoors and then transplant in clumps of three, spaced 6"–8" apart. Germination takes 8–14 days. Pinch back plants before they bloom. After the first year, you can divide the plant roots and pot them

Storage Requirements

To dry, tie the cuttings in small bunches and hang them upside down in a warm, dry, dark place. When dried, remove the leaves from the stems and store whole. Crush or grind just before use. Store in airtight jars in a dark place.

Method	Taste
Frozen	Good
Dried	Excellent; retains much of its flavor
Fresh	Excellent

for your winter windowsill or greenhouse.

Pests

Aphid, spider mite

Diseases

Botrytis rot, damping off, rhizoctonia

Companions

Sage; marjoram is also said to generally improve the flavor of all vegetables

Incompatibles

Cucumber

Harvest

Pick marjoram all summer and, if you want, cut it down to 1" above the ground. In the North, don't cut it back severely in fall, as it will weaken the plants. Herbs should be cut in the morning after the dew has dried. Cut back the top growth up to 6" below the flower buds or ends. Do not wash or aromatic oils will be lost.

Selected Varieties

Golden Marjoram (*O. marjorana* "Aurea"): Crinkled; yellow-white leaves; attractive border plant; not strongly scented; not good for culinary uses
Companion Plants

Marcelka Marjoram: Similar to oregano with hint of pine; wonderful with mushrooms and tomato sauce; makes a soothing bath
Kitchen Garden

Pot Marjoram *(O. onites)*: 12"–15"; pretty plant; spreads slowly in all directions; attracts bees; profuse blooms in late summer; stronger flavor than true marjoram; hardier than true marjoram; larger flowers
Companion Plants, Fox Hill, Nichols

Sweet or Knot Marjoram: 12"–24"; best culinary variety, good in soups, salads, vinegars, and Italian, French, and Portuguese cuisines; an infusion is said to be good for coughs and sore throats; keeps well on a sunny windowsill through winter; small, oval, gray-green leaves
Widely available

Greenhouse

Any variety can be grown in a cool greenhouse in winter, but its flavor will be milder.

oregano

Origanum vulgare subs. *hirtum*
Labiatae (Mint Family)

OREGANO is a perennial warm-season herb, hardy to frost and light freezes through Zone 5. It likes full sun and well-drained, average soil. Flowering doubles the concentration of oil in oregano leaves, so for the strongest flavor don't harvest until the plants start flowering. Nonflowering varieties should be harvested in late spring, as the oil concentrations rise steadily in spring and then decline. But the late spring peak flavor of nonflowering varieties simply cannot compare with the flavor of autumn flowering varieties. For the strongest-flavored oregano, choose a variety that is autumn flowering. It's difficult to get a good-flavored oregano, so one strategy is to start from a cutting or plant that you have tasted or one that has an established track record. The best is *O. heracleoticum.* Oregano is a stronger, more peppery version of marjoram and is a staple in Mediterranean and Latino cuisine. It is used to enhance egg dishes, tomato-based dishes, savory meats, and vegetables.

Site

Full sun

Temperature

For germination: 65°F–75°F
For growth: Hot

Soil & Water Needs

pH: 6.5–7.5

Fertilizer: Light feeder

Side-dressing: Not necessary

Water: Low

Measurements

Planting Depth: 0"; tiny seeds should not be covered
Root Depth: Shallow

Height: 12"–24"
Breadth: 12"

Space between Plants: In beds: 18"
Space between Rows: 18"–20"

Propagation

By root division in spring, or by cuttings and seeds. The easiest method is to take cuttings in summer and root them in sandy compost. Seeds germinate in about 4 days.

Pests

Aphid, leafminer, spider mite

Diseases

Botrytis rot, fungal disease, rhizoctonia

Companions

All beans, cucumber, squash

Incompatibles

None

Harvest

One gardener claims that a first harvest when the plant is a mere 6" tall fosters bushy growth. When the plant is budding prolifically in June, she cuts the plant severely so that only the lower set of leaves remains. The plant reportedly leafs out again in a few weeks. She then cuts back the plant severely again in August. Generally, herbs should be cut in the morning after the dew has dried. Cut back the top growth up to 6" below the flower buds or ends. Do not wash the leaves or aromatic oils will be lost.

Storage Requirements

To dry, tie the cuttings in small bunches and hang them upside down in a warm, dry, dark place. When dried, remove the leaves from the stems, and keep leaves whole for storage. Store in airtight jars in a dark place. Crush or grind just before use. Fresh oregano has a milder flavor than does dried oregano.

Method	Taste
Frozen	Good
Dried	Good
Fresh	Excellent

Selected Varieties

Compact Oregano *(O. compacta Nana):* 2"–3"; excellent ground cover in sunny areas; strong flavor; does well in a cool window in winter
Sandy Mush

Dark Oregano: 2'; excellent seasoning; larger and darker leaves; more upright than standard
Fox Hill, Sandy Mush

Golden Creeping Oregano *(O. var. aureum):* 6"; decorative golden ground cover; needs winter mulch; very mild oregano flavor
Widely available

Greek Oregano *(O. var. hirtum/O. hirtum/ O. heracleoticum):* 18"; true wild oregano from Greece; excellent flavor but flavor intensity declines with age of plant; white flowers; bright green leaves; good for drying
Widely available

Italian Oregano: See Thyme, page 293; this has a strong oregano flavor but is actually a thyme

Mexican Oregano *(Lippia graveolens):* Not a true oregano but listed here for its oregano flavor; often sold in Mexico and the Southwest as true oregano; used in Southwest chili and Mexican dishes; very strong flavor resembling oregano; can be grown into a miniature tree in a greenhouse or bright window
Companion Plants, Dabney, Fox Hill, Richters

Mt. Pima Oregano *(Monarda austromontana):* Used as commercial oregano in Mexico; wild plant with an unusual flavor; for low and high desert
Native Seeds

Oregano Tytthantum (Khirgizstan Oregano): 18"; excellent flavor; bushy growth; glossy green leaves; pink flowers
Sandy Mush

Seedless Oregano *(O. viride):* 18"; excellent seasoning; as strong as Greek oregano but sweeter and less biting; leaves resemble sweet marjoram
Companion Plants

Greenhouse

Any variety will do well in the greenhouse.

parsley

Petroselinum crispum
Umbelliferae (Parsley Family)

PARSLEY is a cool-season biennial herb, hardy to frost and light freezes. It is notorious for its long germination time. Several methods can help speed things along: soak the seeds overnight or for 48 hours; refrigerate the seeds; freeze the seeds; or pour boiling water over the soil plug before covering it. If you soak the seeds longer than overnight, change the water twice. Be sure to discard the water, as it will contain some of the germination inhibitor, furanocoumarin. If you let some of the plants go to seed late in the season, they may produce seedlings that can be dug and grown on the windowsill for next year's crop. Parsley is hardy but will usually go to seed in its second year, so it's most often grown as an annual. While some say it's difficult to transplant, you can direct sow or transplant parsley from pots. An excellent source of vitamin C, iron, and minerals, parsley is more than a pretty garnish. Its refreshing flavor is a great addition to soups, salads, sauces, and many other dishes.

Site

Full sun to partial shade

Temperature

For germination: 50°F–85°F
For growth: 60°F–65°F

Soil & Water Needs

pH: 6.0–7.0

Fertilizer: Heavy feeder

Side-dressing: Two to three times during the season, apply compost or spray with liquid seaweed.

Water: Low

Measurements

Planting Depth: ¼"
Root Depth: Shallow to 4'

Height: 12"–18"
Breadth: 6"–9"

Space between Plants:
In beds: 4"
In rows: 6"–12"
Space between Rows: 12"–36"

Propagation

By seed

Pests

Cabbage looper, carrot rust fly, carrot weevil, nematode, parsleyworm, spider mite

Diseases

Crown rot, septoria leaf spot

Allies

Uncertain: Black salsify, coriander, pennyroyal

Companions

Asparagus, corn, pepper, tomato

Incompatibles

None

Harvest

Cut as needed. To keep it productive, frequently cut back the full length of the outside stems and remove all flower stalks. Herbs should be cut in the morning after the dew has dried. Cut back the top growth up to 6" below the flower buds or ends. Do not wash the leaves or aromatic oils will be lost.

First Seed-Starting Date

Germinate	+	Transplant	+	Days Before LFD	=	Count Back from LFD
11 to 42 days	+	7 to 14 days	+	10 days	=	28 to 66 days

Last Seed-Starting Date

Germinate	+	Transplant	+	Maturity	+	SD Factor	+	Frost Tender	=	Count Back from FFD
11 to 42 days	+	7 to 14 days	+	63 to 76 days	+	14 days	+	14 days	=	109–160 days

Selected Varieties

PLAINLEAF GROUP
(P. crispum neapolitanum)

Best for drying, or for use in soups and stews.

Giante D'Italia or Giant Italian: 2'–3'; sweet, intensely flavored leaves; favorite for Italian foods; thick stalks; great in soups, stews, and salads
Cooks, Nichols, Kitchen Garden

Italian Broad Leaf or Single Italian: Stronger flavor than the curled; best for drying; flat dark green leaves
Companion Plants, Dabney, Richters

Plain or Single Leaf: Earlier maturing than others; deeply cut, bright green leaves; excellent flavor
Nichols, Richters, Shepherd's

CURLED GROUP
(P. crispum)

Best for garnishing

Decora: Best for warm climates; hot weather doesn't slow growth; deep-green curled thick leaves; good aroma and flavor
Richters

Exotica or Forest Green: Grows vigorously even in cool weather; deep-green curled leaves; good aroma and flavor
Nichols, Richters

Mosscurled: Standard curly variety; bright green; good aroma and flavor; best for freezing
Widely available

Petra: "Triple curled" and mosslike from Holland; ideal for pots and kitchen gardens; grows back after cutting
Kitchen Garden

HAMBURG GROUP
(P. crispum ruberosum)

Parsnip Rooted or Hamburg: The parsniplike root of this variety can be added to soups and stews; leaf can be used but its flavor isn't rich
Burpee, Nichols

Greenhouse

Any variety thrives in a cool greenhouse.

Storage Requirements

To dry, tie the cuttings in small bunches and hang them upside down in a warm, dry, dark place. When dried, remove the leaves from the stems and keep whole for storage. (If you prefer, the drying job can be hastened in an oven or microwave.) Store in airtight jars in a dark place. Crush or grind just before use.

Method	Taste
Frozen	Good for curly parsley
Dried	Good for broad leaf Italian
Fresh	Excellent (stored in jar in refrigerator, with stem submerged in water, covered with plastic bag)

rosemary

Rosmarinus officinalis
Labiatae (Mint Family)

ROSEMARY is a perennial warm-season herb, very tender to light frost and freezes. It can tolerate a wide variety of soil conditions, ranging from 4.5 to 8.7 pH, and anywhere from 12 to 107 inches of water per year. Rosemary favors a sunny location and does not transplant well. If grown from seed, do not harvest for 3 years. It benefits from frequent pruning at any time of year; don't hesitate to cut it back severely. "Dr. Rosemary," nurseryman Thomas De Baggio in northern Virginia, recommends that north of Zone 8 rosemary be grown in pots year-round and brought indoors for the winter, as most varieties are hardy to only 15°F to 20°F. "Arp" is the only variety that is hardy to –10°F and can be grown outdoors north of Zone 8. If you grow "Arp" outside, support its limbs and shield it with burlap in winter — not polyethylene, which can capture heat underneath — leaving the top open to minimize fungal disease. Overwatering causes tips to turn brown, followed by all leaves browning and dropping off.

Site

Full sun; good drainage essential

Temperature

For germination: 65°F–75°F
For growth: Hot

Soil & Water Needs

pH: 6.0–7.0

Fertilizer: Light feeder

Side-dressing: Two to three applications per season of liquid seaweed

Water: Low

Measurements

Planting Depth: ¼"
Root Depth: 12"–24"

Height: 2'–6'
Breadth: 12"–24"

Space between Plants: 12"–24"

Propagation

Stem-tip cuttings taken in spring or fall root easily in sandy compost. Propagating from seed is not recommended because the seeds are unreliable and rapidly lose viability. If you do start from seeds, however, use seeds that are less than 2 weeks old. Rosemary seeds take a long time to germinate — up to 3 weeks — so be patient. Make sure the seed pots have excellent drainage and keep soil moist.

Storage Requirements

To dry, tie cuttings in small bunch and hang them upside down in a warm, dry, dark place. When dried, remove the leaves from the stems and keep whole for storage. Store in airtight jars in a dark place. Crush or grind just before use.

Method	Taste
Frozen	Unknown
Dried	Good
Fresh	Excellent

Pests

Greenhouse: Spider mite and whitefly
Outdoors: Mealybugs and scale

Diseases

In humid climates, fungal botrytis rot, rhizoctonia

Companions

All brassicas, beans, carrot, sage

Incompatibles

Cucumber

Other

Rosemary is said to deter bean beetles, cabbage moth, and carrot rust fly.

Harvest

When the plant matures you can harvest it year-round. Cut 4" branch tips, but do not remove more than 20 percent of the plant's growth. Herbs should be cut in the morning after the dew has dried. Do not wash or aromatic oils will be lost.

Selected Varieties

CULINARY

Common or Upright Rosemary *(R. officinalis)*: 2'–6'; hardy to 15°F; the only rosemary good for cooking
Widely available

ORNAMENTAL

Most of these can be used for culinary purposes but are better ornamentals.

Arp (*R.o.* "Arp"): 5'; hardiest cultivar to –10°F; gray-green leaves; shrubby; very fragrant, light blue flowers; ornamental
Dabney, Fox Hill

Benendon Blue or Pine-Scented: 2'–5'; hardy to 20°F; very strong pinelike aroma; dark blue sparse flowers; tall; narrow growth habit; used in potpourri; the only nonculinary variety
Companion Plants, Fox Hill, Richters, Sandy Mush

Blue Boy: 2'; hardy to 20°F; smallest habit and leaves of any rosemary; especially suited for indoor growing; good on windowsills and for bonsai; prostrate type; mild fresh fragrance; light blue flowers abundant in summer
Companion Plants, Dabney, Fox Hill

Creeping or Prostrate Rosemary *(R.o. prostratus)*: 6"–12"; hardy to 20°F; excellent for bonsai; espalier; hanging baskets and living wreaths; its long branches twist and curl; short and narrow leaves; mild fragrance; pale blue flowers; blooms almost year-round
Companion Plants, Dabney, Sandy Mush

Joyce deBaggio or Golden Rain: 5'; hardy to 20°F; gold-edged leaves; strong fragrance like common rosemary; dark blue sparse flowers; compact; bush; self-branching; very pretty landscape specimen; looks gold from a distance
Companion Plants, Dabney

Greenhouse

Any variety; rosemary grows beautifully in the greenhouse.

rosemary

sage

Salvia officinalis
Labiatae (Mint Family)

SAGE is a perennial shrub and is hardy to –30°F, if covered. In the North, cover with a loose mulch or hay or evergreen boughs. A June-bloomer, sage likes well-drained and moderately rich soil, though it will tolerate poor soil and drought. Sage seeds store and germinate poorly. When started from seed, it takes about 2 years to grow to a mature size. Most gardeners start sage from cuttings or division, using the outer, newer growth. Some gardeners recommend replacing sage plants every several years when they become woody and less productive. Sage has putative antibacterial activity and has been used as a natural preservative for meats, poultry, and fish. Distilled sage extracts have been made into antioxidants that are used to increase the shelf life of foods. Some research also indicates that sage lowers blood sugar in diabetics and may have estrogenic properties, which might account for the popular folk belief that sage dries up milk. Fresh sage has a lemony and slightly bitter flavor, while dried sage has a mustier flavor. It adds depth to egg dishes, breads, and a variety of meats.

Why should a man die if he has sage in his garden?

— Dabney Herbs catalog

Site

Full sun; good drainage essential

Temperature

For germination: 60°–70°F
For growth: Warm

Soil & Water Needs

pH: 6.0–7.0

Fertilizer: Light feeder

Side-dressing: An occasional spray of liquid seaweed will benefit the plant.

Water: Average

Measurements

Planting Depth: ¼"
Root Depth: Shallow

Height: 12"–40"
Breadth: 15"–24"

Space between Plants:
In beds: 18"–20"
Space between Rows: 3'

Propagation

By layering, stem cuttings, seed. Crowns of old sage plants can rarely be divided successfully, but dividing may work on younger plants. Start seeds indoors 1–2 months before the last frost or sow directly outdoors 1–2 weeks before the last frost; they germinate in about 21 days. Transplant outdoors about 1 week before the last frost. In fall, take a 4" cutting and root it for planting the following spring.

Pests

Slug, spider mite, spittlebug

Diseases

Powdery mildew, rhizoctonia, verticillium wilt

Companions

All brassicas, carrot, marjoram, rosemary, strawberry, tomato

Incompatibles

Cucumber, onion family

Other

Sage can be used to deter cabbage moths and carrot flies.

Harvest

If you want to keep these plants through winter, harvest lightly in the first year and no later than September. Herbs should be cut in the morning after the dew has dried. Cut back the top growth up to 6" below the flower buds or ends. Don't wash or aromatic oils will be lost.

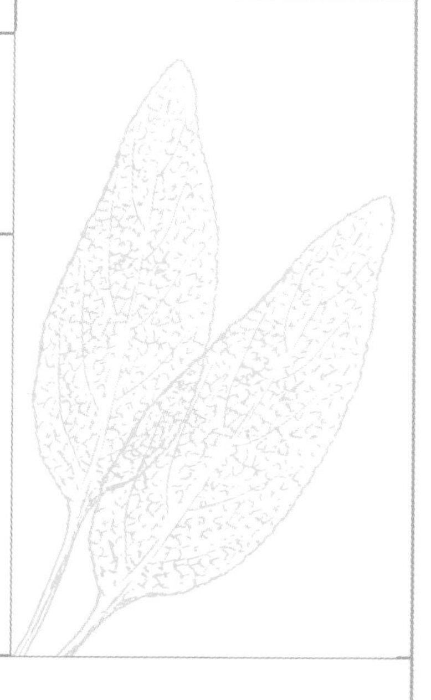

Selected Varieties

Blue or Cleveland Sage *(S. clevelandii):* 3'; royal blue flowers; pleasant aroma; good for culinary use and potpourri
Companion Plants, Dabney, Fox Hill, Richters

Broad Leaf Sage: 14"–30"; flavoring for meats, dressings, and sauces
Southern Exposure

Clary or Muscatel Sage *(S. scleria):* 2'–4'; fragrant; very ornamental; masses of 18" pale purple, white, or pink flowers; large, hairy, toothed leaves; sometimes added to Muscatel wine; said to have medicinal value; grows well in heavy soils
Widely available

Common or Garden Sage: 2'–3'; gray-green pebbly leaves; pale blue flowers; good culinary herb for meat, stuffing, sausage, omelettes, cheese, and bean dishes; said to have medicinal value
Companion Plants, Dabney, Richters, Sandy Mush

Golden Sage (*S. off.* "Aurea"): 18"; hardy to 20°F; chartreuse-yellow on edges of dark green leaves; good culinary and border plant; compact growth
Companion Plants, Richters, Sandy Mush

Pineapple Sage *(Salvia elegans):* 2'–4'; brilliant red flowers with pineapple scent; attracts hummingbirds; good culinary sage for drinks, chicken, jams, and jellies; good indoor plant; needs good light indoors
Companion Plants, Dabney, Richters, Sandy Mush

Purple Sage (*S. off.* 'Purpurea'): 18"; compact; aromatic red-purple foliage; good culinary sage for stuffings, sausage, eggs, soups, and stews; preferred variety for tea; needs winter mulch; good in or outdoors in full sun
Companion Plants, Fox Hill, Richters, Sandy Mush

Tri-color Sage (*S. off.* 'Tricolor'): 2'–3'; hardy to 20°F; variegated white, purple, and green leaves; true sage flavor; very showy; excellent culinary and decorative border plant where hardy; needs winter mulch
Widely available

Storage Requirements

To dry, tie cuttings in small bunches and hang them upside down in a warm, dry, dark place. When dried, remove the leaves from the stems, and keep whole for storage. Store in airtight jars in a cool, dry, dark place. This herb can also be frozen in airtight containers.

Method	Taste
Frozen	Good
Dried	Fair
Fresh	Excellent

Greenhouse

Any variety; sage grows well in the greenhouse.

tarragon

Artemisia dracunculus
Compositae (Sunflower Family)

TARRAGON is an aromatic, perennial herb, hardy to frost and light freezes through Zone 4. It likes full sun to partial shade and rich, sandy, well-drained loam. The plants should be divided every 2 or 3 years for flavor and vigor. All flower stems should be removed to kccp the plant productive. Tarragon most often fails due to soil that is too wet or too acidic. It can be grown in containers with good drainage in the greenhouse or on a windowsill. The only tarragon worth growing for culinary use is French tarragon, but it is somewhat harder to find than Russian tarragon. Tarragon is an excellent flavor enhancer in vinegars, salad dressings, and chicken, cheese, and egg dishes. It is important in French cuisine, particularly in béarnaise and hollandaise sauces, sauce tartare, and herb mixtures such as herbs de provence. Tarragon reportedly aids digestion and when made as an infusion is said to soothe rheumatism, arthritis, and toothaches.

tarragon

Site

Full sun to partial shade

Temperature

For germination: 75°F
For growth: Warm

Soil & Water Needs

pH: 6.0–7.0

Fertilizer: Light feeder

Side-dressing: Not necessary

Water: Average

Measurements

Planting Depth: Cover roots
Root Depth: 6"–12"

Height: 12"–36"
Breadth: 24"

Space between Plants:
In beds: 18"
In rows: 2'
Space between Rows: 30"

Propagation

By cuttings, divisions, or seed. French tarragon cannot be grown from seed, so if you grow tarragon seeds it is the less aromatic, more common Russian tarragon. To propagate French tarragon, take cuttings or divisions in early spring and transplant 2' apart.

Pests

None

Diseases

Downy mildew, powdery mildew, rhizoctonia (root rot)

Companions

Tarragon is alleged to enhance the growth of most vegetables

Incompatibles

None

Harvest

Begin harvest 6–8 weeks after transplanting outside. The leaves bruise easily, so handle gently. Herbs should be cut in the morning after the dew has dried. Cut back the top growth up to 6" below the flower buds or ends. Don't wash or the aromatic oils will be lost.

Selected Varieties

French Tarragon *(A. dracunculus* var. *sativa):* Classical French Tarragon; cannot be grown from seed; best if shaded during the hottest part of the day; divide every 4–5 years to maintain flavor; narrow pointed leaves
Widely available

Russian Tarragon *(A. dracunculus):* Not recommended; almost flavorless; tarragon seeds are always this variety
Widely available

Tarahumara (Mt. Pima Avis, aka: Mexican Mint Marigold) *(Tagetes lucida):* A substitute for tarragon; perennial native to Mexico; used as a tea; wonderful aroma; for low and high desert
Native Seeds

Greenhouse

Tarragon does well with 16 hours of light over a 6-week period at 40°F.

Storage Requirements

To dry, tie cuttings in small bunches and hang them upside down in a warm, dry, dark place. When dried, remove the leaves from the stems and keep whole for storage. The leaves will brown slightly during the drying process. Crush or grind just before use. Store in airtight jars in a dark place. Fresh tarragon can also be preserved in white vinegar (which preserves flavor better than drying) and by freezing the leaves in airtight plastic bags.

Method	Taste
Fresh	Excellent
Frozen	Good (better than dried)
Dried	Fair

thyme

Thymus vulgaris
Labiatae (Mint Family)

THYME is a perennial herb and is hardy to frost and light freezes in Zones 5 through 9. It likes full sun and does well in light, dry to stony, poor soils. Good drainage is essential or the plant will be susceptible to fungal diseases. Keep it sheltered from cold winds. It may not survive severe winters unless covered or heavily mulched. The plant may become woody and straggly in 2 to 3 years. Either replace it or try cutting back three-fourths of the new growth during the growing season to rejuvenate it and keep it bushy. French thyme is difficult to propagate with cuttings but the following method is suggested: Prune the plant severely in mid-June, and take small ½- to 1-inch softwood cuttings in mid-July; root the cuttings in a sand bed, and cover the bed with a milky white plastic "tent" 6 to 8 inches above the top of the cuttings; mist the cuttings once a day at midday. Important in French and Greek cuisine, thyme is wonderfully versatile, enhancing everything from breads, vinegars, and butters to egg, fish, poultry, other meats, and vegetables.

Site

Full sun to partial shade

Temperature

For germination: 60°F–70°F
For growth: Warm

Soil & Water Needs

pH: 5.5–7.0

Fertilizer: Light feeder

Side-dressing: Not necessary

Water: Average

Measurements

Planting Depth: 0"–¼"
Root Depth: 6"–10"

Height: 3"–12"
Breadth: 18"–3'

Space between Plants: 8"–12"

Propagation

By cuttings, divisions, layering, seed. Start seeds indoors 2–3 weeks before the last frost, keeping seeds dry and uncovered. For layering, divisions, and cuttings, snip 3" from fresh new green growth, place in wet sand, keep moist for 2 weeks, and transplant when rooted. The best time for divisions or cuttings is spring; also early summer.

Pests

Aphid, spider mite

Storage Requirements

To dry, tie cuttings in small bunches and hang them upside down in a warm, dry, dark place. When dried, remove the leaves from the stems and keep whole for storage. Crush or grind leaves just before use. Store in airtight jars in a dark place or freeze in airtight containers.

Method	Taste
Fresh	Excellent
Frozen	Fair
Dried	Excellent

Diseases

Botrytis rot, fungal diseases, rhizoctonia (root rot)

Companions

All brassicas, eggplant, potato, strawberry, tomato

Other

May repel cabbageworm and whitefly

Incompatibles

Cucumber

Harvest

Cut as needed before the plant blossoms in midsummer. Alternatively, harvest the entire plant by cutting it down to 2" above the ground; the plant will grow back before the season ends, but this method renders the plant less hardy for the winter. Herbs should be cut in the morning after the dew has dried. Cut back the top growth up to 6" below the flower buds or ends. Do not wash the leaves or aromatic oils will be lost.

Selected Varieties

Caraway *(T. herba-barona):* 4"; rapid spreader, good ground cover and in rock gardens; rose flowers; used for meats, soups, and vegetables
Companion Plants, Dabney, Nichols, Richters

Common, Garden, or English *(T. vulgaris):* 14"; small upright shrub; source of antiseptic oil, Thymol; common, English, and French variations are most often used in cooking and also are excellent bee plants
Widely available

Creeping *(T. glabrescens):* 6"; main use as a good ground cover; a dense mat with purple flowers; fast spreader; early bloomer
Companion Plants

English, German or Winter (*T. vulgaris* var.): 8"; broad dark green leaves; robust growth habit
Companion Plants, Fox Hill, Richters, Sandy Mush

French or Summer (*T. vulgaris* var.): 12"; grayer and sweeter than the English; needs some winter protection; pink flowers; trim upright plants
Widely available

Italian Oregano Thyme: 10"; strong oregano taste and aroma; tender perennial
Companion Plants, Nichols, Sandy Mush

Lemon and Golden Lemon *(T. citriodoratus):* 12"; low-growing with strong lemon scent; delicious in teas and in fish and chicken dishes; not as hardy as other thymes; may need winter protection; the Golden Lemon variation has sharply defined yellow edges on leaves and is a beautiful plant
Widely available

Mother-of-Thyme or Creeping *(T. praecox* subsp. *articus):* 2"; beautiful ground cover; good bee plant; rose-purple flowers; dark green leathery leaves; used in formal herb gardens, between stepping stones, and on earth benches; releases fragrance when stepped or sat on; comes in red-flowered and lemon variations
Dabney, Richters, Sandy Mush

Silver *(T. vulgaris "Argenteus"):* 10"; somewhat sprawling, striking plant; pretty white margins on green leaves; pale blue flowers; flavor like common or garden thyme; good for edging and hanging baskets
Companion Plants, Dabney, Fox Hill, Richters, Sandy Mush

Greenhouse

Any variety does well, but keep foliage dry to prevent rot.

2-year rotation). If possible, routinely rotate on a 3- to 8-year basis. *Same soil site* is defined as a radius of 10 feet from where the vegetable has been planted. So, rotation can occur within the same growing bed as long as the vegetable is planted at least 10 feet away from where it was the previous year.

Intercrop. Try to avoid *monocrops,* single crops of, say, just corn or beans. Space and time permitting, if you want to grow a large amount of one vegetable, try planting it in several different places or breaking up a large patch with plantings of another vegetable. Intercropping discourages some pests from feasting conveniently on their favorite food. In fact, studies have shown that random mixing of plant cultivars in a stand greatly reduces disease severity. Also, certain vegetables may grow better in one spot than in another, so experimentation with placement may prove helpful.

Test different varieties of the same crop. Each variety of a particular vegetable has slightly different growing habits and different disease and insect resistance. One variety may thrive where another dies. The more varieties you try, the higher intrinsic resistance your garden has to widespread disease and insect damage. Over time, you will identify the varieties that fare best in your particular microclimate.

Encourage good air circulation by giving each plant adequate space to grow to maturity. Prune trees and canes annually to prevent overcrowding of limbs and branches. In some cases, if the area becomes too crowded, you may need to remove entire trees.

Select varieties that are resistant to or tolerant of the problems specific to your climate. Your Extension agent or local organic association should be able to tell you what the predominant pest problems, whether fungal, bacterial, viral, or insect, are in your area. For some diseases, selecting resistant varieties is the only prevention and remedy.

Buy certified disease-free seed or plants whenever possible. For certain diseases predominant in the eastern United States, buying seed grown west of the Rockies is an important preventive measure.

SECOND LINE OF DEFENSE

prevention strategies

MOST OF THE DEFENSIVE ACTIONS DISCUSSED IN THIS SECTION also happen to be organic remedies for a host of diseases and animal pests. If you adopt these practices at the outset, you will need to consult the specific remedies offered in the rest of the chapter much less frequently.

Some steps may be free, some inexpensive, others more expensive. Do what you can, when you can, choosing options that are consistent with your goals. A purple martin house, for example, may seem like a pricey investment initially, but it will last twenty years, and purple martins will greatly reduce the need for insecticidal spray, chemical or organic. Bluebird and bat houses, by contrast, can be built easily from scrap lumber or bought at modest cost. In sufficient numbers, bluebirds and bats can accomplish the same goal.

Plant Maintenance

These steps are critical for prevention. As a group, they are also the most common and important remedies for disease and insect damage. If you experience disease or insect problems, review this list. If there is a step that you haven't already taken, implement it as a remedy.

Avoid deep planting where fungal root rot (rhizoctonia) is a problem. Shallow planting encourages the early emergence of seedlings and gives them a better chance of survival.

Water from below using some form of irrigation. A cheap method is to use gallon-sized plastic jugs whose bottoms have been cut out. Simply insert the necks of the jugs into the soil at intervals appropriate to the plants' needs, and fill the jugs with water. More expensive methods include soaker hoses and emitter tubes. Watering from below helps to prevent and remedy fungal disease by discouraging its growth and minimizing its spread through water droplets. It is also more efficient than regular watering and reduces evaporative loss.

Water before noon. This allows the plants to dry off in the middle of the day, before nightfall, which also discourages fungal growth.

Don't work around or touch wet plants. Diseases are frequently transmitted by human hands and infected tools, so this preventive measure is standard practice.

Don't touch healthy plants after working with diseased plants. Again, human hands and infected tools are common methods of disease transmission, so this should be avoided.

Avoid high-nitrogen fertilizers. Fertilizers can harm soil organisms and also can stimulate the plant to put its energy into leafy growth rather than fruits.

Feed the soil with compost. For some diseases, a major remedy is to spread 1 inch or more of compost through the garden. (See pages 6–9 for more information.)

Mulch in spring and fall. Mulch is another occasional remedy for fungal, bacterial, and certain insect pest problems. Apply several inches of mulch in spring and 4 to 6 inches in fall, after harvest. (See "Undertake fall cultivation" below; see also page 9 for more information on mulch.)

Control weeds in and around the garden. Weeds can harbor insect and disease pests. If you weed regularly in the early summer months, you should not experience a severe weed problem during the rest of the season. Removing perennial weeds and thistles within 100 yards of the garden is considered a remedy for several diseases.

Remove rotting or dead leaves, stalks, weeds, and plants. These also can be a breeding ground for pests. Rid the garden of fall leaves, too, even if you aren't growing any fall crops.

Move piles of wood and garden debris to a spot away from the garden. These also can be a breeding ground for pests.

Don't allow large, stagnant pools of water. Small birdbaths and running water are not a problem.

Mow grass in orchards. Orchard grass is another potential breeding ground for pests. One "natural" orchard plan involves planting companion grasses or plants that attract beneficial insects and allowing them to naturalize. Most orchards, however, are surrounded by orchard grass, which should be kept mowed to discourage pests.

Disinfect tools periodically. Even if there's no sign of disease, make an effort to disinfect tools regularly. And when working with diseased plants, *always* disinfect tools between pruning cuts.

Shade plants in extremely hot weather if they wilt continuously. Use a shading material such as cheesecloth. If plants wilt one day, they won't necessarily die the next, but don't be afraid to shade your plants. Recent studies by NASA indicate that plants integrate light over time, and that long, sunny summer days provide more light than plants require. If plants are wilting from heat stress, make sure they are receiving adequate water.

Protect plants from freezes. Whenever a freeze is anticipated, protect plants with a light material. Even newspaper works.

Promptly harvest fruits and vegetables when ripe. Allowing fruits and vegetables to stay on the stem too long promotes certain diseases.

Remove plants immediately after harvest. Don't allow harvested plants to just sit in the garden. If they're not diseased, remove them and add them to the compost pile, or, where appropriate, work them back into the soil.

Always remove and destroy infected or diseased leaves, canes, or the entire plant when necessary. For some disease and insect pests, removing the infected leaves may be sufficient. For other pests, which may be systemic, removing the affected leaves may not be sufficient. If all other measures discussed in this section fail to control the problem, you may need to resort to various sprays — whether simple home remedies or insecticidal soaps, or harsher remedies such as the copper-based fungicides and botanical sprays.

Undertake fall cultivation. After plants have been harvested and the garden cleaned of plant debris, leave the soil bare for a few days. For normal purposes, cultivate the soil no more than 3 inches. For certain insect pest problems, cultivate the soil to a depth of 6 to 8 inches to expose eggs and larvae to birds and other predators. Two weeks later, lightly cultivate the soil a second time with a rake, now to a depth of just 2 inches. Leave soil bare for a few more days, then plant a cover crop or apply a layer of winter mulch 4 to 6 inches deep.

Helpful Supports and Devices

These devices may seem unnecessary, but they can make the difference between the presence or absence of a problem. In some cases, they are recommended as remedies to specific disease or animal pests.

Borders. Borders of certain plants around vegetable beds have been found by some research to be helpful in maintaining garden health. Plant in small ratios relative to the size of the garden bed. Try planting borders of dead nettle, valerian, hyssop, lemon balm, and yarrow.

Trellises, stakes, A-frames, tepees. These and other forms of plant support accomplish multiple goals. Not only do they make the most efficient use of growing space, but they promote better light penetration and air circulation, both of which help minimize disease. Supports also make it much easier to prune and harvest. Use supports for vegetables such as tomatoes, pole beans, peas, cucumbers, melons, and squash. Melons and large squash can be supported on trellises by tying them with panty hose or similar materials. If grown on the ground, melons should be raised off the soil with steel cans or plastic containers to minimize rot. Cane fruit, such as blackberries, raspberries, and grapes, also grow effectively when trellised.

Row covers. These are helpful throughout the growing season. In spring and fall, they keep your crops warm and protected from frost, thereby prolonging the season. In spring and summer, they are an easy and effective protection against insect attack. Before applying row covers, however, make sure that harmful insects are not trapped underneath.

Row covers are usually made of a lightweight woven material, such as polyester or polyvinyl alcohol, that allows in air, light, and water and requires no supports. To enhance warmth under row covers, fill gallon jugs with water and place them every few feet along the rows. The water in the jugs absorbs heat during the day and radiates the stored warmth through the night.

Wire cages, stakes, and water-based white paint. These are critical aids to saplings. Wire cages provide an effective barrier against animals. When planting a sapling, install a ½- to ¼-inch hardware cloth around the trunk, so that it extends 4″ below the ground to prevent rodents from burrowing under and 18 to 24 inches above the ground to prevent rabbit damage, particularly in winter months. Make sure the wire is spaced a couple of inches away from the trunk to allow for growth.

For trees on dwarf rootstock characterized by poor anchorage, place a 4- to 6-foot stake close to the trunk before filling the planting hole with soil, and secure it with special ties that permit growth and flexing. Tree trunks can be painted anytime, but if done while young, it will help prevent sunscald or winter injury, a condition promoted by sunny winter days and resulting in cracks or cankers.

Electrical fences or other barriers. In rural areas, barriers against large animals such as deer are almost essential. Many a gardener has gone to bed, content with a prospering garden, only to awaken to find devastation. Some gardeners initially feel the problem doesn't warrant the expense, but their minds are changed pretty quickly when their entire garden, flowers and vegetables, are destroyed in short order. (For suggestions on different types of barriers, see the remedies for deer in the animal pest chart on page 365.)

Attract Beneficial Animals and Insects

Birds, bats, bees, predatory and parasitic wasps, ladybugs — these are just a few of the beneficial animals and insects that are critical to a self-sustaining garden. Beneficials are one of the easiest ways to prevent disease and to deter animal pests that spread disease. They save you both money and time — and time *is* money, after all.

In the beginning, your garden's ecosystem may not contain all of the beneficials that you want or need. You may choose to import beneficial insects to fight a specific problem, for example. Such a practice is part of integrated pest management (IPM), a common approach in commercial crop management. IPM relies extensively, but not exclusively, on precisely timed releases of beneficial insects to fight pest problems. Backyard gardeners who don't have the opportunity or inclination to investigate the proper timing or method of releasing purchased beneficial insects probably shouldn't use this remedy. If you can follow through on the timed releases, however, importation of beneficials can be an effective remedy. Beneficial insects are listed, where applicable, in the animal pest chart on pages 348–417; suppliers can be found in appendix D. Always follow the supplier's instructions for quantity and timing to ensure success.

Some purists believe that the importation of beneficial insects disrupts the development of a balanced local ecosystem. This argument has some merit. But be aware that some beneficials are self-limiting: when their food source dies, they die.

Ideally, a self-sustaining garden should not need imported help, but you may decide to resort to imported beneficials when faced with acute infestations. The decision depends on your individual goals. If you've implemented the first and second lines of defense, the probability of extensive damage to any single crop or crop variety is low. To attract beneficial insects, plant herbs, flowers, and clovers around the borders of your garden. (For more information on beneficials, see pages 6–11 and also page 297.)

THIRD LINE OF DEFENSE identify the problem

IDENTIFICATION MAY SEEM AN IMPROBABLE DEFENSIVE ACTION, but it is absolutely essential in the garden. If you saw water spots on your ceiling, you wouldn't run right out to replace the roof; first you seek the cause. The spots might simply result from condensation due to inadequate ventilation of the roof rafters. Likewise, when plants get sick, make every effort to diagnose the problem accurately before attempting remedial action. What appears to be a fungus at first glance may simply be a lack of phosphorus.

The most common garden problems are frequently the simplest ones to fix: overwatering or underwatering, inadequate nutrients, poor drainage, or lack of ventilation. Take your time, and examine any problems closely before deciding which remedy or combination of remedies is appropriate.

The balance of this chapter is intended to help you pinpoint the possible causes of your plant problems and to help you conduct informed discussions with professionals. If you want to experiment with your garden — and you are not committed to saving every plant, maximizing yields, or obtaining picture-perfect produce — the guidelines and remedies that follow can be for you a source of fun, discovery, and learning. If you are a market gardener with an investment in yields, both quality and quantity, obtain an accurate diagnosis from a professional before implementing a remedy. Don't forget: advice and soil tests from Extension agents are free.

Regular Monitoring

Monitor the garden daily, if you have time, or at least weekly. Act quickly to prevent a pest or disease from becoming a problem, starting with the most benign methods of control, which are also usually the easiest and least expensive. The simplest methods include handpicking and trapping, or spraying plants with strong jets of water to clear off bugs. Work your way up from these to stronger methods, such as biological predators. If the problem becomes particularly serious, you might try stronger controls like botanical insecticides, but use them sparingly and only as a last resort. (See the warning under Botanical Controls on page 317.) The key is to find the control for your garden that is both the most effective and the most benign.

If you hear hoofbeats, look for horses not zebras.

Systematic Examination of the Environment

Before attempting diagnosis, you must first learn how to observe. Begin by taking a general accounting of the plant and determining the parts that are affected. Following is a partial list of things you might look for.

- **Leaves:** Inspect edges (margins), veins, top side, and bottom side. Notice wilting, general coloration and distortion (curling, crinkling), holes, spots, eggs, and insects.
- **Blossoms and fruits:** Check petals, blossom end of the fruit, and skin. Look for spots, discolorations, decay, cuts, holes, premature drop, lack of fruit setting, eggs, and insects.
- **Stems or bark:** Examine from the soil level, or even slightly below the soil level, to the top. Notice cuts, cracks, or splits in the tissue, blisters, growths (cankers or galls), discoloration, wilting, stunting, twisting, spindly growth, sticky coating, gummy exudate, spots, eggs, and insects.
- **Roots and surrounding soil:** If the plant is sufficiently diseased, you may want to dig it up. Check the length and breadth of main roots and root hairs. Notice nodules, knots, decay, underdevelopment, twisting or distortions, eggs, cocoons, and insects.

Start with the Basics

After examining the plant and its environment, consider the basics, not the unusual. Start with the most common garden problems — water, nutrients, drainage, air circulation — even if insects or disease are clearly present. Many pests don't attack a plant unless it is already weakened and stressed. Before reaching for the fungicide, check to see whether the roots are too wet and compacted to get oxygen. Global garden destruction can be due to animals, such as deer and rodents. And global garden sickness can be caused by acid rain, pollution, or smog, modern afflictions we often forget to consider during diagnosis.

Identifying and Treating Common Problems

For more detailed information on remedies, see the Glossary of Organic Remedies on pages 315–325.

Acid Soil

Nitrogen, phosphorus, and other nutrient deficiencies appear because an acid medium makes them less biologically available. The plant performs poorly and exhibits multiple symptoms.

Symptoms: pH below 6.0; excess aluminum and manganese, because an acid medium makes these more biologically available.
Where It Occurs: Common in areas with acid rain, especially in the Northeast.
Diagnostic Tools: Soil and pH tests
Organic Remedies

- Earthworms
- Compost
- Ground dolomitic limestone (slow release, contains magnesium and calcium)
- Ground calcitic limestone (slow release, no magnesium and calcium)
- Wood ashes (fast release, caustic)
- Organic material, because lime speeds decomposition of organic matter

Alkaline Soil

Symptoms: pH above 7.0; micronutrient deficiencies; poor plant performance.
Diagnostic Tools: Soil and pH tests
Organic Remedies

- Earthworms
- Compost and other organic materials
- Gypsum (calcium sulfate)
- Aluminum sulfate or powdered sulfur

Water

Too Much: The soil around the stem is soaked. Mold, moss, or fungus may be growing on top of the soil. Other symptoms include wilting, yellowing, and dead leaf margins.
Too Little: Plants wilt due to loss of tissue turgor, which is maintained by water pressure. Leaves may eventually brown and die. With prolonged water shortage, growth is stunted.
Diagnostic Tools: Why guess? Buy a simple moisture meter to find out whether the root medium is too wet or dry. Buy a rain gauge and mount it where it can be easily checked.
Organic Remedies

- Earthworms
- Compost

Do what's logical: If it's too wet, let the root medium dry out; if too dry, provide more water.

Nitrogen (N)

Nitrogen is essential for all phases of growth. Because it can be rapidly depleted, garden soil needs a slow constant supply such as compost provides.

Too Little: Slow growth. Lighter green leaf color is followed by yellowing tips, usually starting at the bottom of the plant. Leaf undersides may become blue-purple. Plant eventually becomes spindly, and older leaves drop. Fruits are small and pale

before ripening and overly colored when ripe.
Where It Occurs: Widespread.
Too Much: Lush, green foliage. Little or no fruit, because the plant is putting all of its energy into growth.
Where It Occurs: Usually where there is an excess of fast-acting fertilizer.
Diagnostic Tools: Soil test kit
Organic Remedies

- Earthworms
- Compost (slow release)
- Foliar spray of diluted fish emulsion or other liquid fertilizer
- Fish meal (slow release)
- Composted manure (slow release)
- Hoof and horn meal (slow release)
- Cottonseed meal (medium release)
- Blood meal (fast release)

Avoid: Urea, the various nitrates, ammonium phosphate

Phosphorus (P)

Phosphorus is essential for proper fruiting, flowering, seed formation, and root branching. It also increases the rate of crop maturation, builds plant resistance to disease, and strengthens stems.

Too Little: Leaf undersides have blue-red spots that expand to entire leaf. Darkened leaves develop a blue-green tinge. Green crops have reddish purple color in stems and leaf veins. Fruiting crops are leafy, with fruits setting and maturing late, if at all.
Where It Occurs: Widespread, except in the Northwest. Most severe in the Southeast from the Gulf of Mexico to North Carolina.
Diagnostic Tools: Soil test kit
Organic Remedies

- Earthworms
- Compost
- Bone meal (slow release)
- Soft phosphate (even slower release)
- Phosphate rock (very slow release)

Avoid: Phosphoric acid, superphosphates, highly soluble compounds

Potassium (K)

Potassium is essential for regulating water movement in plants and helps with production of sugars, starches, proteins, and certain enzyme reactions. It also increases cold-hardiness, especially in root crops.

Too Little: Leaves at the plant base turn grayish green. Leaf edges yellow, brown, or blacken, and curl downward. Black spots appear along leaf veins. New leaves curl and crinkle. Flowers and fruit are small and inferior. Stems are hard and woody. Plant and roots are stunted. Leaves may turn bronze, yellow-brown, or mildly bluish.
Where It Occurs: Most common east of the Mississippi and in coastal fog areas; most severe in parts of Texas through Florida, and north to Virginia.
Diagnostic Tools: Soil test kit
Organic Remedies

- Earthworms
- Compost
- Greensand (very slow release)
- Crushed granite (very slow release)
- Rock potash (slow release)
- Kelp meal (medium release)

- Wood ash (fast release, caustic)
- Feldspar dust

Avoid: Potassium chloride

Magnesium (Mg)

Magnesium is important for chlorophyll production and respiration.

Too Little: Lower, older leaves yellow between the leaf veins and eventually turn dark brown. Leaves get brittle and curl upward. Fruit matures late. Symptoms usually appear in late season.
Where It Occurs: Acidic soils, leached, sandy soils, or soils high in potassium and calcium.
Diagnostic Tools: Soil and pH tests
Organic Remedies
- Earthworms
- Compost
- Seaweed meal
- Liquid seaweed (foliar spray)

Calcium (Ca)

Calcium is necessary for water uptake and proper cell development and division.

Too Little: Newer, upper leaves turn dark green, sometimes curl upward, and leaf edges yellow. Weak stems, poor growth, early fruit drop, fruit cracking. Fruits develop water-soaked decaying spots at the blossom end. Examples are blossom end rot in tomato, tip burn in lettuce, black heart in celery, and Baldwin spot in apple and pear flesh (brown spots under the skin and bitter flesh).
Where It Occurs: Acidic, leached, very dry soils, or soils high in potassium.
Diagnostic Tools: Soil and pH tests
Organic Remedies
- Earthworms
- Compost
- Ground limestone (slow release), applied once in fall
- Ground oyster shells (slow release)
- Crumbled eggshells (slow release)
- Avoid fertilizers high in nitrogen and potassium
- Wood ashes, applied once in spring

Avoid: Quick lime, slake lime, hydrated lime

Boron (B)

Boron is important for cell wall formation and carbohydrate transport.

Too Little: Bushy growth from lower stems. Newer shoots curl inward, turn dark, and die. Young leaves turn purple-black. Leaf ribs get brittle. Fruit develops cracks or dry spots.
Where It Occurs: Eastern United States, especially in alkaline soils.
Diagnostic Tools: Soil and pH tests
Organic Remedies
- Granite dust (slow release)
- Rock phosphate (slow release)
- Liquid kelp foliar spray (fast acting)
- Boric acid spray on newly opened fruit blossoms (0.02 pound to 1 gallon water)

Iron (Fe)

Iron is important for chlorophyll formation.

Too Little: Similar to nitrogen deficiency, but leaf yellowing is between the veins and starts first on upper, not lower, leaves.
Where It Occurs: Often in soils with pH above 6.8.
Diagnostic Tools: Soil and pH tests
Organic Remedies
- Compost
- Lower soil pH to 6.8 or less
- Add peat moss manure (lowers pH)
- Glauconite (source of iron)
- Greensand (source of iron)

Drainage

Too Much: Sandy soils do not sufficiently retain water.
Too Little: Long-standing puddles. Slowly melting snow.
Diagnostic Tools: Dig a gallon-sized hole and fill it with water. Let it drain, fill it again with water, and time how long it takes the hole to drain; more than 8 hours indicates poor drainage.
Organic Remedies *(for both conditions)*
- Earthworms
- Compost or other organic matter (helps soil retain and drain water)

Poor Ventilation

Plants wilt due to heat stress. Dew dries very slowly. Mold, moss, mildew, and other symptoms of too much moisture are present.

Organic Remedies
- Stake plants, or use some support structure to keep plant erect
- Prune trees and plants; remove suckers between stem crotches
- If the garden is too crowded, remove weakest, smallest, or least-desired plants

Ozone (O_3)

This is a major photochemical irritant in smog. Plants become more brittle and generally attract more insects. One theory suggests ozone alters amino acid and sugar levels, making the plant more attractive to pests.

Organic Remedies: There is no known remedy for smog pollution. If gardening in a smog-polluted area, do everything possible to maintain soil and plant health, and don't assume the plant is diseased.

Walnut Wilt

Roots of black walnut and butternut trees secrete an acid (juglone) that is toxic to tomato, potato, pea, cabbage, pear, apple, sour cherry, and others. Plants can suddenly wilt and die. Their vascular systems brown, which may cause a false diagnosis of fusarium or verticillium wilts.

Diagnostic Tools: If affected plants are growing near a black walnut or butternut tree, the acid from the tree roots is most likely the problem.
Organic Remedies: Remove the tree, or plant the garden a distance from the tree that exceeds the height of the tree.

Micropests and Macropests

Microdestructive agents receive the designation *micro* because they cannot be seen by the naked eye. They include fungi, bacteria, and viruses. By contrast, macrodestructive agents, ranging from whiteflies to deer, are those that you can see and identify.

Identification of macropests is fairly simple and normally possible without the aid of a professional. Micropests are another matter, however. A definitive diagnosis of most microscopic organisms cannot be made without specialized training, and sometimes, without laboratory tests. The difficulty is compounded by the fact that microscopic organisms share numerous symptoms. Several broad differences do exist. A fungus infection frequently results in mold growth, whereas bacteria may cause bad-smelling plant parts. Mottled coloring in the leaves is often a symptom of a virus. These differences are not reliable for diagnosis, however, because they don't consistently accompany the disease. When a definitive diagnosis is necessary, consult a trained professional.

The Blue-Cheese Syndrome. A garden is much like blue cheese, which contains about thirteen critical microorganisms that produce its distinctive taste and texture. In blue cheese, about half of these microorganisms are fungal and half bacterial. Similarly, every garden contains a broad variety of microorganisms and macroorganisms, all acting and interacting to produce health or disease. In the advanced stages of disease, therefore, there is probably no such thing as a "pure" infection by a single organism or even by a single family of organisms. Disease is usually promoted by a host of factors, from nutrients to microbial activity. Once a plant is weakened, it can be attacked by several different organisms simultaneously.

The good news is that some organic remedies are broad-spectrum; they help attenuate a broad range of diseases. Many of these remedies are the same preventive practices mentioned previously, such as ensuring good soil drainage or watering from below to prevent further contamination through water droplets. One of the most basic broad-spectrum remedies is to remove and destroy diseased leaves, fruit, branches, or plants. When the basics have been ruled out — and disease has been ruled in — a thorough and accurate diagnosis of the major pathogens responsible for the disease may not be necessary if the broad-spectrum remedies work.

Differential diagnosis is important, however, when insect vectors play a role, for the insects then can be controlled. A differential diagnosis also is especially critical any time the gardener wants to use remedies specific to a particular disease, such as a botanical, mineral, or other chemical spray.

What follows is an outline of microagents and macroagents, symptoms, and promoting factors and remedies. The list is neither definitive nor a substitute for professional advice or textbooks. A single plant may exhibit only one or multiple symptoms. And, occasionally, there are exceptions to the promoting factors and remedies listed. Remember, many of the remedies are also good preventive garden practices.

Overview of Symptoms and Remedies

Fungal Diseases

Symptoms

Fungal-specific symptoms: Mold on any plant part; may be fuzzy, flat, or colored. This group of diseases also includes the blights, which are characterized by rapid withering or tissue decay with no apparent rotting. The blights are perhaps the most difficult to identify because they can mimic each other and often deviate from specific symptoms.

Spots (on leaves, stems, or bark, or fruit and flowers): Watery, soft, sunken, dry, shriveled, colored.

Leaves: Yellowed, wilted, fallen, defoliation, curled, wrinkled.

Fruit: Shriveled, misshapen.

Flowers: Spots, discolored petals.

Stems: Rot (decay), sunken areas, girdling, watery blisters, and other types of cankers, cracks, and dark swellings.

Roots: Rot (decay), discoloration, knots, cankers.

Promoting Factors

Usually fostered and spread by prolonged periods of rain, moisture, dew, humidity. Spread by tools, gardeners, wind, seeds (sometimes), and insects. Can be soilborne and harbored in plant debris. Some fungal pathogens can live up to 15 years in the soil.

Immediate Remedies

- Remove and destroy diseased plants or parts of plants.
- Water from below.
- Don't touch plants when wet, to avoid spreading disease.
- Disinfect hands after working with diseased plants.
- Disinfect tools between cuts.
- Improve ventilation.
- Keep orchard grass mowed.
- Spray fruit trees with liquid seaweed every 2 to 3 weeks to prevent spread of fungus; this coats fruits with a protective filament.
- For some fungal problems, you can spray with copper-based fungicides, Bordeaux mixture, sulfur, or lime-sulfur. Never use sulfur on apricots or high doses on cucurbits, which are sensitive to sulfur and will be injured. Copper can kill earthworms.
- There are an increasing number of antifungal commercial sprays and wettable powders available for the organic grower. Check in catalogs and Web sites for sprays that might apply to your specific fungal problem.
- For certain fruit tree fungi, follow a spray program starting in spring when the tree is dormant, such as that recommended by Necessary Trading Company (see appendix D).
- Find out which spray is appropriate for the disease.

Long-Range Remedies

- Import earthworms.
- Compost.
- Use long crop rotations of 3 to 5 years for all fungal wilts.

- Plant in a well-drained area.
- Select resistant varieties.
- Solarize garden soil (this is especially effective for controlling verticillium wilt).
- Where recommended, use hot water seed treatment as follows:

  For brassicas: Place seeds in 122°F water for 25 minutes.

  For celery: Place seeds in 118°F water for 30 minutes.

  For solanaceae: Place seeds in 122°F water for 30 minutes.
- For fruit trees, paint trunk and lower limbs to reduce cold injury, because disease spreads through injury.
- Don't prune trees until buds swell, because disease may spread through dead buds and bark cuts.
- Prune limbs or canes 4″ below the infection (cankers, galls), and disinfect tools between cuts.
- Research plots in California's San Joaquin Valley showed that composted yard trimmings rid peaches of brown rot.
- Increase organic matter in soil to discourage wilts.
- Store harvested produce in well-ventilated, cool, dry areas.
- Applying soil taken from underneath birch trees or surrounding birch roots to soil where diseased plants have been removed is thought to have beneficial effects.
- A biologically beneficial strain of the fungus *Gliocladium virens* is commercially available (Soil Guard) and reportedly provides long protection against pathogenic fungi that live in the soil.

Bacterial Diseases

Symptoms

Bacteria-specific symptoms: Bad smells associated with fruit, roots, stems, or leaves.

Circular and angular spots (on leaves, stems, and bark): Sunken, raised, and water-soaked tissue may drop out of spots, leaving holes.

Leaves: Yellowed, curled, wilted, stunted.

Fruit: Slimy, spots.

Blossoms: Withered, dead.

Stems: Lesions, wilted, blackened, dead, wartlike growth, oozing.

Roots: Soft, slimy.

Promoting Factors

Usually spread by rain and some beetles (e.g., cucumber beetles, flea beetles).

Immediate Remedies

- Apply insecticidal soap spray at specific site of infection.
- Remove and destroy infected plants.
- Disinfect tools.
- Don't touch plants when wet.
- Clean cultivation.
- For trees, prune out infected parts, disinfect tools between cuts, and destroy removed parts.
- Where severe, remove and destroy tree. If feasible, solarize soil.
- Where recommended (usually for fruit trees), spray with a copper-sulfur blend. (This can kill earthworms.)
- Where recommended (usually for vegetables), apply micronized sulfur.

Long-Range Remedies

- Import earthworms.
- Compost.
- Plant resistant varieties.
- Crop rotation.
- Many peppers, tomatoes, and brassicas crops have bacterial diseases inside the seed. Experimental results suggest that seeds be given hot-water treatments at 122°F for 20–25 minutes, followed by cool water. Seeds should be wrapped loosely in cloth and weighted to submerge them in the hot water bath. Carrots require 122°F for 10 minutes. *Solanaceae* plants require 122°F for 25 minutes.
- In the Eastern United States, use seeds grown west of the Rockies that are free of bacterial blight.
- Improve soil drainage.

Viral Diseases

Symptoms

Virus-specific symptoms: Mottled coloring in leaves.
Leaves: Mottled coloring; misshapen growth; yellowing; curling downward; crinkling; unusual narrow, pointed, or fern-like leaves; veins may disappear.
Fruits: Misshapen, premature ripening.
Blossoms or flowers: Misshapen, underdeveloped, few.
Stems: Stunted, twisted, misshapen.

Promoting Factors

Usually spread by insect or human vectors (e.g., aphids, cucumber beetles, grasshoppers, leafhoppers, thrips, smokers [tobacco mosaic], human hands, and tools).

Immediate Remedies

- Control insect vector.
- Disinfect tools and hands before working with plants.
- Remove and destroy infected plants, or parts of plants.
- Clean cultivation.

Long-Range Remedies

- Import earthworms.
- Compost.
- Plant resistant varieties.
- Remove weeds that harbor insect vectors.
- Plant in areas protected from wind to minimize aphid contamination.
- Use closer spacing where leafhoppers are a problem.

Insect Pests

Symptoms

Visible presence of insect (but those that suck plant juices are more difficult to spot).
Leaves: Chewed or ragged holes, defoliation, tunnels, blotches, skeletonization, wilting, webbing, leaf drop, eggs on upper or lower leaf sides.
Fruit: Small, round or other-shaped holes; premature drop.
Flowers: Malformed.
Stems and bark: Holes or sunken area; severed, weak, stunted or distorted, wilted.
Roots: Malformed, poorly developed, eaten; decay in storage.

Immediate Remedies

- Handpick bugs off plants.
- Remove and destroy eggs.

- Use traps; most are insect specific.
- Mulch, where appropriate.
- Remove and destroy heavily infested plants.
- Clean cultivation.
- Where appropriate, apply sprays of insecticidal soaps, oils, etc.
- Spray fruit trees with liquid seaweed every 2 to 3 weeks, which masks fruit scent and protects the trees against insect damage.

Long-Range Remedies

- Import earthworms.
- Compost.
- Plant resistant varieties.
- Use row covers before pest emerges.
- Use fall cultivation.
- Encourage insect-eating birds and other beneficial animals.
- Solarize soil, if feasible.

Mycoplasmas

Symptoms

Because mycoplasmas are life-forms intermediate between viruses and bacteria, plant symptoms may resemble those of viruses and bacteria. Mycoplasmas are the smallest organisms lacking cell walls capable of self-replication and are known to cause diseases in humans, animals, and plants.

Promoting Factors

Usually spread by insect vectors such as leafhoppers.

Remedies

See remedies for fungal, bacterial, and viral diseases.

Glossary of Organic Remedies

Following is a detailed description of the common remedies for micropests and macropests that are given in the charts in abbreviated form. Suppliers of these remedies are listed in appendix D.

When selecting a remedy, the rule of thumb for the noncommercial gardener is to start with the least invasive and least toxic remedies. If these are ineffective, then consider removing the pest-damaged plants. Use stronger remedies (botanical sprays) only if your need to save the pest-damaged plants outweighs the harmful effects the sprays may have on the garden ecosystem. Apply all remedies in the smallest effective quantities and in the most limited areas necessary for control.

Bacillus thuringiensis (Bt). Something akin to the wonder drug of organic growing, Bt is a bacterium that was discovered in 1901. It comes in a powder form that is biodegradable and alleged to be so safe that it can be sprayed or dusted right up to the day of harvest. Bt targets and kills leaf-eating caterpillars, leaving unharmed other insects, animals, and birds. It works by paralyzing the insect's gut. New strains of Bt are now being isolated that target specific larvae, such as the new M-One that attacks Colorado potato beetle larvae.

WARNING: Legitimate concern exists about the possibility that exclusive or immoderate use of Bt will foster Bt-resistant pests in future generations. Such resistance has already been demonstrated in at least one moth pest. Many specialists believe that as such problems arise, new strains of Bt can be developed to conquer the newly resistant pests. This argument is identical to one used by pharmaceutical companies to justify new strains of penicillin to combat penicillin-resistant bacteria. Although new forms of penicillin have been developed successfully, the toxicity to humans of the newer penicillins exceeds the original by ten to twelve times. Severe allergic response to the original penicillin was less than 1 percent; severe allergic response to "improved penicillins" are as high as 10 to 20 percent. Therefore, noncommerical growers may decide, as I did, not to use Bt or, if necessary, to use it only as a last resort.

Beneficial insects and plants. (See also Integrated Pest Management on page 321.) Beneficial insects are attracted to flowering *Umbelliferae,* such as dill and carrots in their second year, yarrow, sweet cicely, and fennel. (For more information on beneficials, see pages 6–11, 297, and 304.) The Henry Doubleday Research Association in Coventry, England, has shown that the following beneficial insects are attracted to the noted plants:

- **Hover fly**: Pot marigolds, *Nemophilia*, bush morning glory, poached-egg plant (*Limnanthes douglasii*)
- **Parasitic wasps and flies**: Yarrow (*Achillea*), flowering fennel and carrots, angelica
- **Lacewings and parasitic wasps**: Mustard

Beneficial nematodes. Unlike the various nematodes that cause damage, beneficial nematodes such as *Neoplectana carpocapsae* are a fast and totally safe control for various pests. A number of species and varieties of beneficial nematodes exist. As juvenile-stage microscopic organisms, these parasites inject the insect with bacteria that kill the host within 24 hours. They don't harm humans, plants, pets, birds, earthworms, honeybees, or beneficial insects. A self-limiting control, beneficial nematodes seek out soil grubs (young phase of certain insects), feed, reproduce, and, when their food supply is exhausted, die. Only effective in the juvenile stage, these nematodes can be stored in the refrigerator for up to 2 months.

Beneficial nematodes must be applied at sufficient rates to be effective, about 50,000 per foot of standard row, or per square foot in raised beds. Numerous destructive pests are susceptible to these nematodes, including borers, weevils, cutworms, cucumber beetles, and even gypsy moths. One quart is sufficient for about 50 square feet — just sprinkle the mixture where the pests are a problem. One source suggests using nematodes in compost heaps to eradicate harmful larvae.

Botanical controls. Derived from naturally occurring sources, these insecticides are considered "organic" by many people because of their alleged lack of persistence in the environment. Botanicals are also generally thought to be relatively safe for birds, pets, humans, and other wildlife. Some of the new synthesized chemicals, however, which are not considered organic, seem to be gentler on beneficial insects in and above the soil than some of the older "organic" sprays. Strictly speaking, minerals, such as copper, are not botanicals, but they are included under this rubric for easy reference.

WARNING: Insecticides and botanicals should *not* be used unless crops have suffered significant damage. No matter how safe an insecticide is reputed to be, the introduction of botanical poisons, minerals, and other substances alters a garden's natural system of checks and balances. Rotenone and pyrethrins can kill beneficial organisms in and above the soil just as easily as the pests you're trying to eradicate. Ryania and sabadilla are somewhat less treacherous to beneficial insects. The definition of *significant damage* is a matter of personal interpretation, but bear in mind that most plants can lose up to 20 percent of their foliage and still produce yields equal to those of plants with no foliage loss.

Even when significant damage occurs, you might consider the size of the area you will need to spray and the value of your crop, compared to the potential damage to the ecology of the rest of your garden.

Insecticides and fungicides should *only* be used as an absolute last resort, after all other methods have failed. Apply the botanical control just to the leaves, plants, or areas that are affected and at the minimal concentrations necessary to do the job. (For more information, see the information that follows about copper, neem, pyrethrum, rotenone, ryania, and sabadilla.)

Bug Juice. Considered an old home remedy, bug juice may or may not be effective. It is thought that, on death, some insects release a pheromone to warn their kind to stay away. But applying bug juice might also help spread disease to other plants. If you

What Goes Up Must Come Down

Substances that may damage insects above the soil also have the potential to damage beneficial organisms in the soil. Such remedies may treat the plant at the expense of the ecosystem. When possible, the potential effects these remedies may have on beneficial organisms above and in the soil have been noted.

Some perspective may be helpful here. Most garden vegetable plants last for a year, whereas the soil takes years to build into a healthy medium. If one year you drive out beneficial organisms in order to get rid of a pest problem and harm the soil in the process, the following year there is a distinct possibility that your garden will have even more problems and require even more remedial measures. This type of vicious cycle leads to open-field hydroponics, a condition discussed on page 3.

Although there may be times you want to use these remedies to save some plants, do so with consideration for the soil. Apply the remedy in the smallest effective quantities and in the smallest area necessary for control. Massive preventive spraying is usually neither necessary nor recommended for the backyard gardener. Commercial growers and orchardists may have little choice at the moment, but the backyard gardener does.

want to try it experimentally, capture ½ cup of the pest, crush it, and liquefy it with 2 cups of water. Do *not* use a household blender to accomplish this; instead, use an old blender jar or a mortar and pestle specifically intended for garden use. Strain to remove particles before spraying. Spray both sides of the leaves and stems.

Clean Cultivation. (See Plant Maintenance on page 301.) Generally, *clean cultivation* means you should remove and destroy all infected plants or parts of plants, and clean out weeds, dead and rotting vegetation, piles of junk, and plants that are finished bearing. Destroy diseased plants by burning (where permitted), burial, or some other appropriate means. Remove stagnant pools of water and piles of brush, lumber, or stones near the garden. Don't work around or touch wet plants, particularly when they're diseased. Don't add diseased plant material to compost piles.

Colored Mulch. Yellow mulch may serve as an attractant to plant pests and can serve as a way of "trapping" plant pests in a concentrated zone. If you are going to spray, vacuum, or do something else, applying yellow mulch may make it easier for you to accomplish your goal while having less impact on the larger garden area.

Cooking Oil Spray Mix. The U.S. Department of Agriculture recommends the following home oil spray for controlling aphids, white flies, and spider mites: (1) mix 1 tablespoon of plain dishwashing liquid with 1 cup of cooking oil; (2) add 1 to 2½ teaspoons of the oil/detergent solution to 1 cup of water; (3) spray directly on plants every 10 days.

The USDA finds that this spray works well on carrots, celery, cucumber, eggplant, lettuce, pepper, and watermelon but cautions that it may burn squash, cauliflower, and red cabbage.

Copper. (See the warning included under Botanical Controls.) Copper is usually considered a fungicide and can be dusted, sprayed, or combined with other substances, such as rotenone. It is an effective control of fungal and bacterial diseases, such as anthracnose, all blights, brown rot, downy and powdery mildews, leaf spot, rust, and scab. The main problem with using copper in an organic garden is that it has been shown to kill earthworms through drips and fallen leaves. (For a more detailed discussion of the effects of copper and other substances on earthworms, see Jerry Minnich, *The Earthworm Book* [Emmaus, Penn.: Rodale, 1977].) The National Organic Standard allows garden use of copper (fixed copper in oxides, and copper sulfate) provided that copper-based materials are not used as herbicides and are used in a way that minimizes accumulation in the soil.

Crop Rotation. At a minimum, plant the same vegetable in the same soil site only once every other year, rotating on a 2-year basis. Some diseases require a longer minimum rotation. *Same soil site* is defined as being within a 10-foot radius of where the vegetable was planted previously. So rotation can occur within the same growing bed as long as the new site is at least 10 feet away from where the crop was grown the previous year or from where there is refuse from the planting. If possible, routinely rotate on a 3- to 5-year basis.

Diatomaceous Earth (DE). This is an insect remedy with very low toxicity. Diatomaceous earth is a hydrophilic (water-loving) form of silicon dioxide (sand) that rapidly takes up water. Harvested from riverbeds, it is made from petrified skeletons of water-dwelling microorganisms known as *diatoms*. It works by desiccating insects.

Apply a dusting of diatomaceous earth after a light rain, dew, or after lightly spraying plants with water. Dust all plant surfaces, starting from the base and moving upward. You can also spray diatomaceous earth in a weak insecticidal soap solution (¼ pound diatomaceous earth to 5 gallons water to 1 teaspoon insecticidal soap), which helps the spray adhere to plant surfaces. Fruit trees can be painted with the same solution. Whether dusted or sprayed, reapply after rainfall. Diatomaceous earth is effective against most soft-bodied insects (aphids, mites, slugs, etc.) but also may work against some beetles and weevils.

Fall Cultivation. Leave the soil completely bare for a few days. Cultivate soil to a depth of 6 to 8 inches to expose eggs and larvae to birds and other predators. Two weeks later, lightly cultivate the soil a second time with a rake, this time to a depth of only 2 inches. Leave the soil bare for a few more days, then plant a cover crop or apply a layer of winter mulch 4 to 6 inches deep.

Garlic Spray. This falls into the category of old home remedies, which means it may or may not work for you. Garlic has been found to have some antifungal properties (see Other listed for garlic on page 266), but the methods of extraction and the quantities necessary haven't been widely tested. This spray should be considered experimental. (See Hot Pepper Spray for ideas on how to prepare garlic spray.)

WARNING: Earthworms will not eat plants in the onion family, so garlic spray could drive them away from the area sprayed.

Handpick. Handpicking is often the most effective control against insect pests. Wear

Bad Times for Applying Pesticides

Pesticides such as botanicals and homemade sprays, soaps, and oils should not be applied under any of the following conditions:

- In the rain
- In windy conditions
- In intense sunlight
- In the middle of hot, dry days (pesticide may volatilize before reaching insect or leaf)
- When mixed with noncompatible materials
- On open blossoms, when bees are present
- When plant is moisture stressed (too much or too little)
- When leaf is wet with dew (unless such moisture is specifically recommended to assist adherence)

gloves at all times to prevent allergic reactions. To kill bugs, drop them into soapy water, boiling water, or water that has some insecticidal soap in it. Small amounts of kerosene or oil also work but are more difficult to dispose of in an ecologically safe way.

Horticultural Oil. The two horticultural grades of oil work by suffocating mature insects, larvae, and eggs. The older and heavier grade of horticultural oil, sometimes called *dormant oil,* is applied to fruit trees in fall or early spring when the trees are still dormant. If applied to large areas of a plant, foliage oil can clog leaf pores, thereby preventing respiration and severely injuring or killing the plant. The newer and lighter grades of horticultural oil, sometimes called *superior horticultural oil,* are less injurious to some plants because they evaporate more quickly. They can be used up to 1 month before harvest, as long as soil moisture is adequate, relative humidity favors evaporation, and the plant is relatively healthy. Woody ornamentals are most likely to tolerate this oil in their foliage state. To determine whether a plant will tolerate the superior oil, spray it on a few leaves. Leaf tips and margins will yellow after several days if the plant is damaged by the oil. Horticultural oils are considered relatively safe for humans, other warm-blooded animals, and beneficial insects.

Hot Pepper Spray. Red peppers have been shown to contain an active ingredient called *capsaicin,* which repels onion fly maggots when as little as 1 milligram is sprinkled around the plant base. It is believed to be effective against other insects as well. It apparently does not work well in monocultures, but it does work well in gardens having a variety of plants and offering insects multiple feeding sites. Capsaicin kills nerve fibers in mammals, so it may act against insect pests by affecting their nervous systems. Capsaicin can be purchased commercially or you can make your own. One commercial product combines hot pepper extract with paraffin wax and claims to repel insects for up to 30 days.

WARNING: The effects of capsaicin on beneficial insects and earthworms are not well known.

To make your own hot pepper spray, try one of these recipes.

Recipe 1: Chop, grind, or blend hot peppers. Mix ½ cup of ground hot peppers with 2 cups of water. Strain to remove particles.

Recipe 2: Mix together 2 tablespoons garlic powder, 1 to 2 tablespoons Tabasco sauce or Louisiana Hot Sauce, a dash of dishwashing liquid, and 2 cups of water. The soap helps the spray adhere better to leaves and bugs. The spray works well against many soft-bodied insects, such as aphids, white flies, mealy bugs, and most larvae (except that of gypsy moth). *(Recipe courtesy horticulturist Carl Totemeier, Ph.D.)*

Apply hot pepper spray twice, allowing a 2- to 3-day interval between each spraying. The spray must make physical contact with the pest to be effective. Generally, pepper spray usually repels, but does not kill, the bugs.

Hot Water Seed Treatment. Place seeds in a cheesecloth bag; place bag in water at the designated temperature and for the specified number of minutes. Stir continuously, adding hot water periodically to keep the water temperature consistent. Don't pour hot water directly on the seeds. When finished, remove the seeds and immerse them in cold water. Drain and dry.

Insecticidal Soap Spray. This is not the same thing as homemade soap, and it is not effective for household cleaning. Insecticidal soap is derived from fatty acid salts that are able to penetrate the cell membranes of insects. They work best on soft-bodied insects such as aphids, mealybugs, mites, and whiteflies. Insecticidal soap spray biodegrades in 2 to 14 days, and the residue on the plant, once dry, is considered ineffective.

Insecticidal soaps are specifically formulated to attack only harmful pests and, unless specified, are not harmful to humans, pets, and most beneficial insects. They have been used successfully in conjunction with beneficials such as *Encarsia formosa* and ladybug larvae. Insecticidal soaps are generally better for plants than are household soap sprays because they don't usually burn plant leaves, they lack the added ingredient of household soap, and they contain known quantities of active ingredients (fatty acid salts). Any phytotoxic symptoms appear within the first 48 hours of spraying, so if you're unsure whether the soap will burn or injure a plant, test it on just a few leaves and wait for two days to see if wilting, yellowing, or other symptoms of burn appear.

Insecticidal soap works on contact only, so spray it directly on the pest. And spray only those areas affected by the pest, not the entire patch of plants and not necessarily the entire plant. Coat both sides of any infested surface. This is called *spot spraying.*

Generally, insecticidal soap is not very effective against hard-bodied insects, but adding isopropyl alcohol increases its effectiveness. The alcohol carries the soap through the pest's protective coating to its body. Simply add ½ cup isopropyl alcohol to every 4 cups of insecticidal soap. Some commercial insecticidal soaps are premixed with alcohol, though they may not advertise this fact.

It's best to spray in early morning. Don't spray insecticidal soap at temperatures above 90°F or in full sun. And don't spray newly rooted cuttings, new transplants, or blooming fruit and nut trees. If a plant begins to wilt in the first few hours after spraying, rinse all soap residues from the plant.

Integrated Pest Management (IPM). In the latter twentieth century, significant advances were made in understanding how pests can be controlled through the carefully timed and controlled release of beneficial insects that eat pests and/or their larvae, interrupt the pest cycle, or otherwise reduce or prevent crop damage by pests. Once considered by many to be too experimental or too difficult, IPM has become a mainstream practice for both large- and small-scale growers due to increased education and improved technologies. IPM is one of the many effective tools that can be used by the organic gardener.

Neem. (See the warning included under Botanical Controls.) Available under various labels such as Margosan-O, Bioneem, Neemisis, neem is a relatively new botanical insecticide derived from the neem tree of arid, tropical zones in Asia and Africa. The active ingredient, azadirachtin, is unlike most other botanical poisons, which typically act on the stomach or nerves. Neem acts as a feeding deterrent and interrupts normal hormonal activity so that insects cannot molt properly; they die in their own skin. Neem does not persist in the soil and is not known to be toxic to mammals, birds, or beneficial insects, including honeybees. It has been shown to be effective against numerous pests, including the more difficult pests: Colorado potato beetle, gypsy moth, grasshopper, Japanese beetle, and Mexican bean beetle.

Pyrethrum. (See the warning included under Botanical Controls.) This botanical poison is derived from the pyrethrum flower, a member of the chrysanthemum family. Pyrethrum is considered a "knock-out" contact insecticide and stomach poison; it works by paralyzing the insect. If the insect receives less than a lethal dose, it may revive.

Pyrethrum is considered safe for warm-blooded animals and relatively safe for honeybees and ladybug larvae, but only because it degrades within 6 hours in temperatures above 55°F. It is often used in combination with rotenone and is effective against most greenhouse pests, especially flying insects such as whitefly. If possible, use ryania or sabadilla before using pyrethrum. Use pyrethrum with extra caution. Spray at dusk when honeybees are less active, and don't spray at all if a heavy dew is expected. The foliar spray is considered to be less harmful to bees than is the dust application.

Rotenone. (See the warning included under Botanical Controls.) This botanical poison refined from the root of a tropical plant also can be found at a 5 percent concentration in the roots of a weed native to the eastern and southern United States, Devil's shoestring (*Tephrasis virginiana*), which Native Americans have used as a fish poison. It is used at two concentrations, 1 percent for the more easily killed insects and 5 percent for the more difficult bugs. Rotenone kills beneficial insects just as easily as harmful insects and is extremely toxic to fish. If possible, use ryania or sabadilla before using rotenone. Rotenone degrades within 3 to 7 days in the presence of light and oxygen.

Row Covers. Row covers are an extremely effective barrier against egg-laying insects. They are made of lightweight material that lets in air, light, and water. Cover transplants immediately or as specified for pest emergence times. Seal the edges to the ground, and leave the row covers on all season, unless otherwise noted. Allow extra material to accommodate the growth of the tallest plant. For plants that need pollination, such as most cucumbers, lift the edges of the row covers during bloom time in the early morning for 2 hours, twice a week. This is all the time bees and other pollinators need to do their good work. When the blooming time is past, secure the row cover edges again until harvest time.

Do *not* use row covers if the pest was seen in that same spot the previous season or if it overwinters or lays eggs in the soil.

Row covers usually raise the temperature underneath, which is helpful for extending the growing season in spring and fall. If you plan to keep the covers on through the warm season, however, especially if you live in one of the southern growing zones, the temperature rise may not be beneficial to plants that don't tolerate heat well.

Tufbel is reputed to be one of the best row covers. The suppliers listed in appendix D offer a variety of row covers.

Ryania. (See the warning included under Botanical Controls.) This botanical poison is refined from the South American ryania shrub. As a contact and stomach poison, it controls pests by making them extremely sick. It is considered to be safe for humans and other warm-blooded animals and plants. It is also considered to be less toxic to beneficial insects, including honeybees, than are other botanical controls. The foliar spray is less harmful to bees than is the dust application. Ryania degrades quickly, so it can be used close to harvest.

Sabadilla. (See the warning included under Botanical Controls.) Derived from the seeds of a South American lily, this botanical poison is a very old insecticide that dates back to the sixteenth century. Sabadilla is a powerful contact and stomach poison. It is considered somewhat less harmful to beneficial insects than are other botanicals, although honeybees are vulnerable. If applied as a dust, humans and pets may experience irritated mucous membranes and sneezing fits.

Soap Spray (homemade). Homemade soap spray is not the same as insecticidal soap spray. Tests of various types of household soap sprays show that they differ considerably in their effectiveness against bugs and in the degree to which they burn plant leaves. Begin by using 2 to 3 tablespoons of soap per gallon of water; never use more than 4 tablespoons of soap, for it will burn the plant leaves. Test different soaps to determine the brand and concentration that works best without burning the leaves. You might also choose to rinse the plant well 1 to 2 hours after spraying it, which prevents harm to the plant. Avoid using soaps that are dyed or perfumed. Soap flakes, although more difficult to prepare, seem to be the least damaging to plants. Some dishwashing liquids are mild enough to be left on the plant without rinsing. Soap spray is a contact insecticide (*see* Insecticidal Soap Spray).

WARNING: Unless you are absolutely certain that the soap you use is biodegradable, you may do more harm than good if you spray it widely. Spot spraying, however, should not be a problem. Commercial manufacturers may label a kitchen soap "biodegradable," but that begs the questions: Under what conditions and in how many years? The label may be appropriate for a septic tank, but always check with the manufacturer to be sure the soap will biodegrade rapidly. If in doubt, use an insecticidal soap that is designed to biodegrade rapidly and to target specific pests.

Soap and Lime Spray. Mix agricultural lime at the rate of ¼ to ½ cup per gallon of water. Add 2 to 3 tablespoons of insecticidal soap per gallon. The soap helps the lime adhere to insect bodies. The lime can dry out small insects, kill some small insects and mites, and irritate adult insects. Test the spray first on a small area of the plant, because some plants may react adversely to these materials. Wait several days before judging the results and deciding whether to spray the remainder of the plant. As with other soap sprays, spot spray to avoid harming beneficial insects.

Stem Collar. Stem collars provide physical barriers against crawling pests. To make stem collars, use waxed paper cups, which hold up well in the rain, or some other weather-resistant material, such as cardboard or tar paper. Cut a hole in the bottom of each cup to accommodate the stem, then cut slits radiating out from the hole to allow for stem growth. Place the inverted cup around the base of the stem, sinking the edges into the soil at least 1 inch. If using something other than a cup, simply cut a 3- to 4-inch collar out of the material, place it around the stem, push it 1 inch into the soil, and fasten the sides together securely. Cardboard toilet paper tubes work well and eventually disintegrate.

Sticky Balls. These traps may function best as early-warning devices to detect the presence of pests in sufficient numbers to require further action. Usually red, the balls are hung in trees as both warning and control devices. As a monitoring device, count the pests trapped every 2 to 3 days and recoat the ball every 2 weeks. As a pest control, use one to four balls per tree, depending on tree size. Clean and recoat balls as needed, and remove all balls after 4 weeks to avoid catching too many beneficial insects.

Sticky Bands. Sticky bands are wrapped around tree trunks to trap larvae, egg-laying females, or other pests. Use cotton batting, burlap, or heavy paper that's at least 6 inches wide. Place bands around the tree trunk at about chest height. Cover the batting with a 6- to 12-inch piece of tar paper coated with Tanglefoot or with a mixture of either pine tar and molasses or resin and oil. Wrap tightly around the trunk and secure with wire or a tie. Renew the sticky substance periodically. If you're trapping larvae, remove the bands and destroy the larvae once weekly in warm weather or biweekly in cool weather. If you're trapping a crawling insect that is not bearing or laying eggs (e.g., aphids), clean as necessary and renew the sticky substance.

Sticky Traps. These traps may function best as early-warning devices to detect the presence of pests in sufficient numbers to require further action. For monitoring, count the pests trapped every 2 to 3 days; recoat the traps every 2 weeks. These traps are usually yellow or white boards, about 8×10 inches, coated with a sticky substance such as Tangle Trap or petroleum jelly. Do not use motor oil (recommended by some), because it will change the board's color. Place the traps adjacent to susceptible plants; do not place them above the plants. Clean and recoat traps as needed. Try not to use these traps for more than 3 to 4 weeks at a time to avoid capturing too many beneficial insects. In a greenhouse, place the traps at the height of the plants' canopy.

Synthetic Pesticides. Designed for quick results and long-lasting activity, synthetics are totally man-made poisons, often acting on the nervous systems and digestive tracts of insects. Some synthetics are considered milder and less toxic than the more "natural" botanical poisons. They are considered less safe for organic growing, however, because their breakdown products often persist in the soil longer and are more toxic than their original formulations. Some synthetic pesticides persist so long, in fact, that they can be found in soils more than 20 years after their application. Botanicals are favored in organic growing because they don't persist in the environment and their breakdown products are safer.

WARNING: Studies have shown that when used in greenhouses the residues of some pesticides may linger longer than specified. Therefore, safe reentry time may be longer than expected. If you are a greenhouse grower, contact the manufacturer to determine whether the pesticide has been tested under conditions similar to those in your greenhouse.

Traps. Traps are an easy, nontoxic way to monitor and control many different insects. They usually use pheromones and colors to attract the targeted bug. Monitoring traps is useful to determine when the insect emerges, where it comes from, and how many insects are present, before resorting to chemical remedies. Generally, traps are a safe and cost-effective tool for a self-sustaining garden. Traps are available for such insects as the slug, Japanese beetle, codling moth, gypsy moth, oriental fruit moth, corn earworm, cherry fruit fly and husk fly, apple maggot (see Sticky Balls), and some scales, among others.

Some Caveats on Traps

Traps that are not pheromone-specific may capture beneficial insects as well as harmful ones; consequently, traps shouldn't be left out longer than recommended. Pheromone traps that work over long distances may actually attract more of the undesirable targeted insect to your garden. Some pheromone traps are best suited to large commercial operations. Before investing in traps, consult the supplier about your particular situation and determine whether traps are best placed in the garden or away from it.

Identifying and Remedying Specific Micropests and Macropests

The two sections that follow outline in detail how to identify and remedy a range of diseases and how to control the insect and animal pests that may affect your plants. Keep in mind that the charts are meant to assist, not to discourage. If you rotate crops, destroy diseased material from garden areas, use resistant varieties, and start with pathogen-free transplants, most diseases shouldn't be a problem. As for insects, most gardens experience very few, usually no more than three to five major pests. Proper controls adopted in a timely fashion should minimize any major threats to your crops. And remember that plants can lose up to 20 percent of their foliage and still produce yields equal to those of plants with no foliage loss.

Only when you have made a positive identification of the microscopic destructive agent should you apply remedies specific to that disease. Specific disease remedies, such as botanical sprays, are generally more injurious to an ecosystem than are the broad-spectrum, preventive remedies discussed on pages 301–304. An application of a special remedy that doesn't target the appropriate disease is not only a waste of money and time but it also can be potentially harmful to the plant and ecosystem. An antibacterial spray when used inappropriately, for example, could destroy beneficial bacteria both above and in the soil. The balances above and in the soil take time to build and should not be sacrificed lightly. If you are ready to spray or use some other pest-specific remedy, take time to consult a textbook, an Extension agent, or another professional to ensure that you have correctly identified the pest.

Prevention is by far the best approach to disease and insect control, but it is most critical with diseases. Postinfection remedies for specific macropests are often more varied and usually include some easy, nontoxic methods that don't harm beneficials above or in the soil.

In the sections that follow, I have tried to avoid, or at least minimize recommendations of, controls that might be difficult to administer, could potentially harm earthworms and soil microorganisms, or could cause ecological imbalances in future years. Not everyone agrees as to the appropriate use of certain substances. When in doubt, every effort has been made to err on the side of safety.

A Word about Weeds. It is often said that a weed is any plant that grows where it is not desired. Some people keep meticulous gardens, weeding every last whisper of grass or dandelion green throughout the season. Research in north central California has shown that such meticulous care may not be necessary and may even be harmful. Hoeing and other forms of weeding were shown to reduce yields in some cases. The research concluded that cucumbers, lettuce, and cauliflower require only 2 weeks of weeding to be well established. A balanced approach might be to weed meticulously during the first few weeks of plant establishment and then to follow your own aesthetic. Here are some suggestions for dealing with weeds.

- **Solarize the soil,** which kills some weed seeds.
- **Add radish juice** to the growing medium to completely inhibit germination of some important weed species. Research studies have found radish juice to be an effective herbicide.
- **Consider lamb's quarters.** The common weed lamb's quarters (*Chenopodium album*) has been found to strongly inhibit germination and growth of some plants, according to researchers in Oklahoma. Both water extracts and pulverized residues of fresh lamb's quarters can significantly reduce germination and growth of some plants. Pulverized lamb's quarters gradually loses its herbicidal properties over time.
- **Use hot water.** Hot water melts the waxy coating on weed leaves or breaks down the plant's cellular structures. Treated plants are unable to retain moisture and dehydrate within hours or a few days. Hot water kills new as well as mature plants.
- **Consider corn-gluten meal.** Corn-gluten meal is believed by some to be an effective pre-emergent herbicide; it is available commercially.
- **Consider a handheld commercial propane device.** These are said to "kill weeds safely by heating — not burning." It is lightweight and portable.
- **Till in the dark.** Tilling in the dark may cut some weed problems in half, according to research from the U.S. Department of Agriculture research in Rosemont, Minnesota. Seeds of certain weed species need a flash of light to break dormancy and tilling in the dark prevents this. This approach is recommended only for those with good night vision.
- **Spray white vinegar.** Studies have shown that vinegar (5% acetic acid) can be effective against grass and some weeds if sprayed at least three times.

Controlling Diseases

Selection Criteria: Diseases are included in this chart *only* if they meet the following criteria: (1) the remedies for the disease go beyond the remedies mentioned in the earlier parts of this chapter (to avoid unnecessary repetition); (2) insects are known to spread the disease organisms (the disease may be attenuated by insect control); and (3) the disease organism is especially virulent and the affected plants need to be removed immediately.

If you do not find a disease or insect pest listed in the charts that follow, refer to pages 312–315 for generic remedies that may apply.

Pathogens: Diseases in different plants may share a common name (e.g., bacterial wilt) but are often caused by different pathogens. Consequently, the "same" disease may have distinct symptoms in different plants. To make matters more difficult, the same pathogen occasionally causes different symptoms in different species. To simplify matters in the chart, diseases have been grouped by their common names and, when possible, the different symptoms are described.

Remedies: In addition to implementing the specific remedies listed in this chart, remember *never* to work with wet plants and to practice clean cultivation as described on page 318, both helpful remedies for many diseases.

WARNING: Diseases are more difficult than insect pests to diagnose accurately. This chart demonstrates how challenging disease identification can be, given the similar symptoms shared by a variety of diseases. The information presented here is intended to help you conduct a more informed discussion with professionals; it is not meant to be a definitive diagnostic tool.

Before using botanical or other sprays or instituting large-scale or costly remedies, make every effort to obtain an accurate diagnosis. If attempting a diagnosis on your own, carefully read each description of possible diseases for the particular plant; don't stop at the first one, even if it seems to match the symptoms of your plant. Because this section is not exhaustive, consult other texts as well, including those listed in the bibliography. If in doubt, take a plant sample to a local professional to obtain a diagnosis.

Anthracnose

Type: Fungus

Plants Affected: Numerous plants

Where It Occurs: Widespread. Overwinters in seed and residues of infected plants. It can be spread by wind, rain, animals, clothing, and tools. It is most common in cool wet conditions.

Symptoms

- Numerous strains of this fungus affect different crops and ornamental plants.
- Characterized by spots that become brown, grayish, or black; the spots are often sunken and may ooze.
- May cause lesions on stems and vines, and may infect the roots.
- Fruits can develop circular cankers of varying size and color, depending on the crop; in moist conditions these spots can contain salmon gelatinous spores.

Organic Remedies

- See Overview of Symptoms and Remedies for fungal diseases on page 312.
- Remove diseased refuse, use a 3-year rotation, and ensure well-drained soil and good air circulation.
- Select varieties that are resistant to this mildew.
- Remedy, a commercial fungicide containing baking soda, can be used on various ornamentals and fruit trees and is said to be effective against black spot, powdery mildew, leaf spots, anthracnose, phoma, phytophthora, scab, and botrytis.

Armillaria Root Rot

also known as Honey Mushroom, Mushroom Root Rot, Oak Root Fungus, Shoestring Fungus

Type: Fungus

Plants Affected: Fruit trees, nut trees, ornamentals

Where It Occurs: Western states, Atlantic Coast, Florida, and Gulf Coast. Worst in heavy, poorly drained soils.

How It Is Spread: Fungus spreads underground 1 foot at a time.

Symptoms

- White, fan-shaped fungus appears between the bark and wood.
- Lower trunk decays.
- Crown is girdled by fungus.
- Mushrooms appear in autumn around the plant base.
- Roots rot and die slowly. If stressed by other factors, the tree will die suddenly.

Organic Remedies

In mild cases: Remove the soil around rotted trunk areas; cut out dead tissue; let trunk dry out during summer and replace soil when a freeze approaches.
In severe cases: Remove the plant, stump, and, if possible, the roots.

Asparagus Rust

Type: Fungus

Plants Affected: Asparagus

Where It Occurs: Widespread. Worst in moist seasons.

How It Is Spread: Wind and spores.

Symptoms

- Leaves and stems develop orange-red spots or blisters that, in time, burst open with orange-red spores.
- Tops yellow and die prematurely.

Organic Remedies

- Cut, remove, and destroy affected tops.
- Do not start new plantings next to old plantings.
- Destroy wild asparagus.
- If necessary, apply a sulfur spray every 7–10 days until 1 month before harvest.
- Use heavy mulch and drop (not overhead) irrigation to minimize spreading disease through splashing.
- Select resistant varieties.

Aster Yellows

see Yellows

Bacterial Canker

also known as Bacterial Blast, Bacterial Gummosis

Type: Bacteria

Plants Affected: Almond, apricot, blueberry, cherry, peach, tomato

Where It Occurs: Widespread, especially in cool, windy, moist weather.

How It Is Spread: Transmitted by wind, rain, infected seeds, and debris. It enters through skin wounds.

Symptoms

In almond, apricot, peach: Small purple spots develop on leaves, black spots on fruits, and cankers on twigs.
In blueberry: Stems develop reddish-brown to black cankers and nearby buds die. Plants eventually die.
In cherry: Leaves wilt and die. Cracks and stems may ooze in spring and fall. Limbs may die.

In tomato: Oldest leaves turn downward first; leaflets curl and shrivel. Only one side may be affected. A stem cut lengthwise may reveal creamy white to reddish-brown discoloration. Young infected fruits are stunted and distorted. Fruits may develop small, white round spots.

Organic Remedies

For tomato: Hot water seed treatment at 122°F for 25 minutes. Plant resistant varieties or certified seed. Rotate crops.
For trees: Prune immediately. Between cuts, disinfect tools and hands. If uncontrollable by pruning, destroy tree.

Bacterial Wilt

also known as Stewart's Disease *in corn*

Type: Bacteria

Plants Affected: Beans, corn, cucurbits, eggplant, tomato

Where It Occurs

Corn: Central, South, and East.
Cucurbits: Northeast and North Central States. Some wilts are fostered by moist soil and soil temperatures above 75°F.
Solanaceae: South.

How It Is Spread: Transmitted in beans by seeds; in cucurbits by both types of cucumber beetle; in corn by the corn flea beetle. The pathogen affecting beans and tomatoes overwinters in debris.

Symptoms

In beans: Seedlings usually die before reaching 3 inches tall. Mature vines wilt, especially midday, and die.
In cucurbits: Plants wilt rapidly and, even while green, can die. Cut stems produce oozy strings. This bacteria blocks the plant's vascular system.
In corn: Plants may wilt and leaves can develop long water-soaked or pale-yellow streaks. A cut in the lower stem oozes yellowish droplets that can be drawn out into fine, small threads.
In tomato: Plants wilt and die rapidly, starting with young leaves first. Lower foliage may yellow slightly. A lengthwise stem cut reveals an oozy gray-brown core.

Organic Remedies

- Immediately remove and destroy infected plants.
- Plant resistant varieties.
- For corn and cucurbits: Control the insect vector (see Symptoms).
- For beans and tomato: Plant only certified wilt-free seeds. Rotate crops on a 4–5 year basis. Fumigate soil.

Bean Rust

Type: Fungus

Plants Affected: Bean

Where It Occurs: Occurs along the Eastern seaboard and in irrigated areas in the West. Fostered by relative high humidity for 8–10 days.

How It Is Spread: Spores are spread by wind and water.

Symptoms: Leaf undersides develop small, red-orange to brown blisters full of spores. Leaves yellow, dry, and drop prematurely.

Organic Remedies

- Don't reuse vine stakes.
- Use long crop rotations.
- If necessary, apply a sulfur spray every 7–10 days until 1 month before harvest.
- Use heavy mulch and drop (not overhead) irrigation to minimize spreading disease through splashing.
- Select resistant varieties.

Black Heart

Type: Environmental

Plants Affected: Celery, potato

Where It Occurs: Widespread, but rare in the Northwest and North Central States.

Symptoms

In celery: A low calcium-potassium ratio causes spreading brown, water-soaked areas on leaves.
In potato: Low oxygen levels at the tuber center cause purple, black, or gray areas.

Organic Remedies

For celery: Make sure the soil contains adequate calcium.
For potato: Improve soil drainage to reduce chances of it recurring. Don't leave potatoes in or on very hot soil (over 90°F).

Black Knot

Type: Fungus

Plants Affected: Apricot, cherry, plum

Where It Occurs: East of the Mississippi.

How It Is Spread: Spread by wind and rain. Harbored in the tree knots.

Symptoms

- Coal black, hard swellings appear on twigs and limbs.
- Olive green growths develop in late summer, then blacken. They can be 2–4 times the thickness of the branch.
- Limbs weaken. Trees die slowly.
- The disease is detected the year after infection.

Organic Remedies

- Cut off all twigs and branches at least 4 inches below swellings. Destroy cuttings. Cover wounds with paint or wax.
- Remove all infected wild cherry and plum.
- Where significant damage is expected, spray with a lime-sulfur or Bordeaux mix at the first bud stage.
- Plant resistant varieties.

Blotch Disease

also known as Sooty Blotch

Type: Fungus

Plants Affected: Apple, citrus, pear

Where It Occurs: In the eastern, central, and southern states to the Gulf of Mexico.

Symptoms

- Only skins are affected. Fruits are still edible.
- Mottled, irregular shaped spots (up to ¼") appear.
- "Cloudy fruit" has spots that run together.

Organic Remedies

- Scrub citrus fruit skins.
- Peel skins off fruits other than citrus before eating.
- Prune and space trees for better air circulation.
- Prune and destroy infected twigs.

Botrytis

also known as Gray Mold

Type: Fungus

Plants Affected: Numerous plants

Where It Occurs: Widespread. This fungus can overwinter in infected plant debris and can be carried into storage. It favors cold wet weather.

Symptoms

- Several different strains of this fungus affect different crops and ornamental plants.
- On leaves, light tan or gray or whitish spots develop; leaves are then covered by a darker colored fungus. The leaf withers and the fungus progresses into the stem. In crops like lettuce, the leaves become a slimy mass.
- Infected fruits are water soaked and soft.
- Infected roots may have light brown and water-soaked lesions anywhere, but they occur most commonly on the crown.
- Grayish-brown mold develops on surfaces and, in storage, can spread into "nests" of mold.

Organic Remedies

- See Overview of Symptoms and Remedies for fungal diseases, page 312.
- Remedy, a commercial fungicide containing baking soda, can be used on various ornamentals and fruit trees and is said to be effective against black spot and powdery mildew, as well as leaf spots, anthracnose, phoma, phytophthora, scab, and botrytis.
- Research in Israel suggests that biological control methods might be enhanced if more than one biocontrol agent is used simultaneously; significant increases in efficacy were achieved by combining a beneficial yeast and a beneficial bacterium to suppress gray mold (*Botrytis cinerea*) on strawberry foliage.
- USDA scientists at the Appalachian Fruit Research Station in West Virginia evaluated the juices and essential oils from various plants to determine their antifungal activity, particularly, their activity against gray mold due to *Botrytis cineria.*

Juices completely inhibiting germination for 48 hours: Garlic creeper (*Adenocalyma alleaceaum*); elephant garlic (*Allium ameloprasum*); fragrant-flower garlic (*ramosum*); serpent garlic (*sativum*); society garlic (*Tulbaghia violacea*); several sweet and hot pepper cultivars, including habanero and tabasco.

Extracts completely inhibiting germination for 40 hours (only the most commonly available plants are listed here): Chamomile mixta (*Anthemis mixta*), 1.56% and higher; citronella (*Cymbopogon nardus*), 6.25% and higher; clove buds (*Eugenia cayophyllata*), 1.56% and higher; geranium (*Pelargonium roseum*), 25% and higher; lavandin (*Lavadula fragrans*), 50% and higher concentrations; lavender, 100%;

lemongrass, 12.5% and higher; marjoram, 50% and higher; oregano (*Coridothymus capital*), 6.25% and higher; peppermint, 50% and higher; rosemary, 100%; sage lavanulifорlia (*Salvia officinalis*), 50% and higher; savory (*Satureja montana*), 12.5% and higher; spike (*Lavandula angustifolia*), 25% and higher; lemon thyme (*Thymus hiemualis*), 100%; red thyme (*Thymus zygis*), 0.78%.

- Israeli researchers are finding that foliar fertilizer therapy can induce disease resistance. Leaves are sprayed with solutions that contain minor trace plant nutrients at very low concentrations, which apparently induce systemic resistance to powdery mildew in cucumbers and common rust in corn.
- Mycostop, a biological control, is the living formulation of the *Streptomyces* bacterium found in sphagnum peat. Originally developed in Europe, it is now available in the United States. The wettable powder can be applied as a drench or soil spray or as a seed treatment to control fusarium, alternaria, phomopsis, pythium, phytophthora root rots and botrytis gray molds, with negligible toxicity to humans and animals.

Canker Dieback

Type: Fungus

Plants Affected: Apple, pecan

Where It Occurs: Widespread. For apples, especially a problem in moist springs and early falls. For pecans, may be caused by inadequate water in winter or heavy soils with poor drainage.

Symptoms

In apple: Watery blisters on bark. Oval dead patches (1"–1') may become sunken. Branches wilt and die from the tip down. Eventually leaves yellow.

In pecan: Branches wilt and die from the tip down.

Organic Remedies

- Avoid wounding tree trunks and stems.
- Cut off infected branches well below infection.
- For small cankers, gouge out with sharp knife and treat with tree paint or Bordeaux paste. Disinfect tools between cuts.
- Destroy all pruned matter.
- **For pecan:** Ensure adequate drainage in hardpan soils and deep watering in sandy soils.

Cedar Apple Rust

Type: Fungus

Plants Affected: Apple, pear

Where It Occurs: In the eastern and central states and in Arkansas.

How It Is Spread: Transmitted in wind and rain by spores from red cedars and junipers. Has a 2-year life cycle.

Symptoms

- Light yellow spots on leaves become bright orange spots.
- Fruits may develop spots.

Organic Remedies

- Remove all red cedars, junipers, wild apple, and ornamental apples within 300 yards of infected trees. Alternatively,

plant a windbreak between the disease hosts and apple trees.
- If necessary, apply a sulfur spray every 7–10 days until 1 month before harvest.
- Use heavy mulch and drop (not overhead) irrigation to minimize spreading disease through splashing.
- Select resistant varieties.
- Copper may be used, but can harm earthworms.

Celery Mosaic

Type: Virus

Plants Affected: Celery

Where It Occurs: Widespread

How It Is Spread: Spread by aphids. Continuous celery cultivation and certain weeds (wild celery, poison hemlock, wild parsnip, and mock bishop's weed) can cause virus to persist.

Symptoms
- Leaves turn yellow and mottled green.
- Stalks are stunted, twisted, and narrow.

Organic Remedies
- Control aphids.
- Destroy infected plants.

Cherry Leaf Spot & Plum Leaf Spot

also known as Yellow Leaf *in cherry*

Type: Fungus

Plants Affected: Cherry, plum

Where It Occurs: East of Rockies; particularly bad in warm, wet conditions.

How It Is Spread: Harbored in debris and spread by wind.

Symptoms
- Purple spots on leaves, followed by bright yellow foliage.
- Spots often drop out of leaves, followed by defoliation.
- Heavy fruit drop in plums.
- Very damaging.

Organic Remedies
- Destroy fallen leaves and fruit; pinch off infected leaves.
- A spray of wettable sulfur from petal fall to harvest is virtually the only control.
- Fall cultivation of soil — no more than 2 inches — reduces spore spread.

Common Mosaic

Type: Virus

Plants Affected: Beans

Where It Occurs: Widespread

How It Is Spread: Spread by aphids and gardeners. Overwinters in perennial weeds like Canadian thistle.

Symptoms
- Severely stunted plants with few pods.
- Mottled green elongated leaves crinkle and curl downward at the edges.
- Infection occurs near bloom time.
- Plants eventually die.

Organic Remedies
- Control aphids.
- Destroy infected plants, no matter how mildly affected.

- Remove all perennial weeds within 150 feet of the garden.
- Plant resistant varieties.

Cottony Rot

also known as Pink Rot *and* Watery Soft Rot *in cabbage;* White Mold *in beans*

Type: Fungus

Plants Affected: Beans, cabbage, celery, lettuce

Where It Occurs: Widespread. Most common in cool, moist conditions.

How It Is Spread: Spread by small black bodies, which can survive in soil for up to 10 years.

Symptoms

- White mold develops and small, hard black bodies form on or within the mold.
- **In beans:** Stems develop water-soaked spots; branches and leaves follow. White mold develops in these spots.
- **In cabbage and lettuce:** Stem and leaves near ground become water-soaked, leaves wilt, and plant collapses. White mold grows on the head.
- **In celery:** White, cottony growth appears at stalk base. Stalks rot and taste bitter.

Organic Remedies

- Remove and destroy infected plants, if possible before black bodies form.
- Plant in well-drained soil; raised beds help greatly.
- Rotate with immune or resistant crops such as beets, onion, spinach, peanuts, corn, cereals, and grasses. Avoid successive plantings of beans, celery, lettuce, or cabbage.
- **For beans and celery:** Where flood irrigation is feasible, usually in muck or sandy soil, flood the growing area for 4–8 weeks, or alternate flooding and drying, which kills the black bodies.

Crown Gall

Type: Bacteria

Plants Affected: Almond, apple, apricot, blackberry, blueberry, cherry, filbert, grape, peach, pear, pecan, plum, raspberry, walnut

Where It Occurs: Widespread

How It Is Spread: Transmitted through wounds in the roots, crowns, and stems by tools and soil water.

Symptoms

- Fruit trees and brambles weaken and produce small, poor fruit.
- *Galls* are swellings that circle roots and crowns; they are sometimes several inches in diameter and can be spongy or hard.
- Plants can survive for many years but are very susceptible to other stresses.

Organic Remedies

- **For brambles:** Destroy canes and plants with symptoms.
- **For trees:** Prune galls and treat with tree surgeon's paint or a bactericidal paint. Cover with soil after painting. Destroy infected portions. Disinfect hands and tools between cuts.

- Remove badly infected trees, trunks, and roots and destroy. Don't plant another susceptible tree in infected location for 3–5 years.
- Propagate fruit trees only by budding.
- Select resistant rootstocks.
- Use Gall-trol or Norbac 84-C at planting (competitive bacteria).
- Ensure good soil drainage.

Crown Rot

see Southern Blight

Crown Rot

Type: Fungus

Plants Affected: Almond, apple, cherry, pear

Where It Occurs: Widespread where trunks are wet at the soil line.

How It Is Spread: Like collar rots and damping off, this fungus attacks at or below the soil surface.

Symptoms

- Late in the season, one or more branches turn reddish.
- Leaves turn yellow or brown and wilt.
- Dead bark tissue appears at the soil line, with sunken and sometimes girdling cankers.

Organic Remedies

- Rake soil away from tree crown and, if necessary, expose upper roots. This improves air circulation and may correct the problem.
- Avoid deep planting of new young trees.
- Avoid standing water or continuously wet conditions around the trunk.
- Allow soil to dry out thoroughly between waterings.

Cucumber Mosaic

also known as CMV, Mosaic; Yellows *in spinach*

Type: Virus

Plants Affected: Cucurbits, lettuce, pepper, potato, raspberry, spinach, tomato

Where It Occurs: Widespread

How It Is Spread: Spread by aphids (in cucurbits, lettuce, pepper, tomato), striped or spotted cucumber beetles (in cucurbits), and gardeners. It overwinters in many perennial weeds.

Symptoms

General: Yellow-green mottling and curled foliage. Plants are weak, stunted, may have few blossoms, poor fruit, and may die. Fruits are misshapen and mottled. Distorted leaves are common.

In cucurbits: Leaf margins can curl downward, and cucumbers and summer squash fruits become mottled yellow-green.

In pepper and tomato: Older leaves look like oak leaves and develop large yellowish ring spots. Affected leaves can drop prematurely. Fruits develop concentric rings and solid circular spots, first yellow then brown. Fruits flatten and roots are stunted.

In potato: Tubers may develop brown spots, and the plant yellows and dies.

In raspberry: Dry, seedless, crumbly fruit. Canes may droop, blacken, and die.

Organic Remedies

- Control appropriate insect vector — aphids or cucumber beetles (see Symptoms). Apply row covers until blossom time to prevent aphids and cucumber beetles.
- Remove and destroy infected plants and, if severe, surrounding plants.
- Remove all perennial weeds within 150 feet of the garden.
- Plant resistant varieties.

Curly Dwarf

Type: Virus

Plants Affected: Artichoke

Where It Occurs: Pacific Coast and southern coast of Texas.

Symptoms

- Stunted plants.
- Aphids and leafhoppers.

Organic Remedies

- Control aphids and leafhoppers.
- Remove and destroy all infected plants.
- Remove all milk thistle and other nearby weeds.

Curly Top

also known as CTV; Western Yellow Blight *in tomato*

Type: Virus

Plants Affected: Beans, beet, cucurbits, pepper, spinach, tomato

Where It Occurs: Widespread

How It Is Spread: Beet leafhoppers

Symptoms

General: Stunted plants have numerous small leaves that pucker, crinkle, curl, and yellow. Fruits are few, dwarfed, or may ripen prematurely.

In beans: Leaves are darker green. Young plants may die, but older plants usually survive. Plant produces few or dwarfed pods and looks bushy.

In cucurbits: Leaves may be mottled.

In tomato: Branches are very erect; leaflet veins turn purple; plant turns a dull yellow.

Organic Remedies

- Control beet leafhoppers.
- Apply row covers until blossom time to reduce leafhoppers.
- Prune and destroy infected parts of plants immediately. If serious, remove and destroy all infected plants and all nearby plants.
- Grow tomatoes away from beets, spinach, melons, or other leafhopper hosts. Also, space them closely to discourage leafhoppers.
- Select resistant varieties.

Downy Mildew

Type: Fungus

Plants Affected: Numerous plants

Where It Occurs: Widespread. Fungal spores are spread by the wind and can be carried by rain, wet clothing, and tools. Unlike some other mildews, this fungus may overwinter in some areas in seed, in diseased roots, and as tiny oospores (sexual fruiting bodies).

Symptoms

- Numerous strains of this fungus attack different crops and ornamental plants. Many may be confused with mosaic.
- **In corn:** Leaves develop a characteristic chlorosis (yellowing) at the base of the oldest leaf, with a distinct margin between this and the green upper portion of the leaf.
- **In cucurbits:** Leaves develop pale-green areas that change to yellow angular spots along leaf veins. If moisture is present, the leaf underside may develop a faintly purplish hue or a range of colors from white to black. Older leaves tend to be affected first, then it moves to younger leaves. Fruit may be dwarfed and have poor flavor. In moist conditions, plants may be overcome by the fungus in days.

Organic Remedies

- See Overview of Symptoms and Remedies for fungal diseases on page 312.
- Select resistant varieties.

Enation Mosaic

also known as Pea Virus I, Leaf Enation

Type: Virus

Plants Affected: Pea

Where It Occurs: Widespread, but particularly a problem in the Northwest.

How It Is Spread: Spread primarily by aphids; overwinters in various clovers, vetch, and alfalfa.

Symptoms

- Young leaves become mottled.
- Leaf undersides develop small outgrowths known as enations.
- Vine tips become misshapen and internodal distance shortens.
- Stunted plants have few, if any, pods. Pods may have yellow seeds.
- Extremely damaging.

Organic Remedies

- Control aphids.
- Plant early to avoid high aphid populations.
- Select resistant varieties.

Fire Blight

Type: Bacteria

Plants Affected: Apple, pear

Where It Occurs: Widespread, particularly in areas of high humidity, dew, and rain. Bacteria remain dormant through the winter inside cankers.

How It Is Spread: Aphids, psylla, bees, and rain. Enters through blossoms or new growth.

Symptoms

- Infected shoots turn brown and black, as though scorched by fire.
- Lesions may ooze orange-brown liquid.
- Blossoms wither and die. Reddish, water-soaked lesions develop on bark.

Organic Remedies

- Control aphids and psylla.
- Remove and destroy all suckers and infected branches. Control is most effective

when infected areas are removed as soon as seen, no matter the season. Cut at least 12 inches below point of visible wilt. After each cut, disinfect tool in bleach solution (1:4 dilution). In winter, repeat and treat cuts with asphalt-based dressing.

- Spray the tree regularly while in bloom with a solution of 4% Clorox (4 ounces Clorox in 3 gallons water). This sterilizes the blossoms so bees do not spread the blight.
- Avoid heavy pruning or high nitrogen fertilizer. Both stimulate rapid twig growth.
- Check soil acidity. The more acid the soil, the more prone to fire blight.
- If significant damage occurs, spray with a copper-sulfur blend labeled for fire blight.
- Plant resistant varieties.

Fusarium Wilt

also known as Fusarium Yellows or Yellows

Type: Fungus

Plants Affected: Asparagus, brassicas, celery, lettuce, melon, pea, potato, spinach, sweet potato, tomato, turnip, watermelon

Where It Occurs: Widespread, but occurs mainly east of the Rockies. Worst in the South; in light, sandy soil; and in dry weather with temperatures of 60°–90°F. Temperatures above 90°F retard the disease.

How It Is Spread: Water, tools, and seeds. The disease can live in soil for 20 years. Thrives in warm, dry weather.

Symptoms

General: Yellowing, stunting, and wilting (often rapid). Lower leaves may wilt first. Plants usually die, but seedlings can die quickly. In some plants, a sliced lower stem may reveal discoloration originating from the roots.

In celery: Ribs also redden.

In muskmelon: One side of the vine develops a water-soaked yellow streak near the soil line that darkens to brown.

In potato: In storage, blue or white swellings may develop on brown decayed areas.

Organic Remedies

- Remove and destroy infected plants.
- Ensure good soil drainage.
- Don't plant susceptible crops for 8 years in places where fusarium was last seen; rotate crops on a regular basis as a preventive measure.
- Solarize garden soil where possible; sterilize potting soil.
- Select resistant varieties, one of the best controls.
- For asparagus: Use seed labeled "treated with a Clorox solution" (available in New Jersey).
- Mycostop, a biological control, is the living formulation of the Streptomyces bacterium found in sphagnum peat. Originally developed in Europe, it is now available in the United States. The wettable powder can be applied as a drench or soil spray or as a seed treatment to control fusarium, alternaria, phomopsis, pythium, phytophthora root rots and botrytis gray molds with negligible toxicity to humans and animals.

Gray Mold

see Botrytis

Leaf Blight

also known as Bacterial Blight

Type: Bacteria

Plants Affected: Carrot

Where It Occurs: Arizona, Indiana, Iowa, Michigan, New York, Oregon, and Wisconsin.

Symptoms: Spots on seedling leaves start out yellow-white. In time they turn brown and look water-soaked.

Organic Remedies
- Hot water seed treatment: 126°F for 25 minutes.
- Plant disease-free seed.
- At harvest, remove and destroy carrot tops.
- Rotate on a 2- to 3-year basis.

Oak Wilt

Type: Fungus

Plants Affected: Chestnut, oak

Where It Occurs: Occurs in states bordering the Mississippi; west to Oklahoma, Kansas, Nebraska; east to Pennsylvania; and south to all states in the Appalachian Mountains.

How It Is Spread: White oak is a major reservoir for the fungus.

Symptoms
- A very serious disease in Chinese chestnut; better tolerated by white oak.
- Water-soaked spots form on leaves, usually along the tip and margin. Leaves turn brown and fall off.
- Symptoms start at the top of the tree, then move to the trunk, which develops short, bulging, vertical splits in the bark.

Organic Remedies
- Once this disease is identified, immediately cut and destroy the tree. Remove the stump and major roots.
- The fungus is thought to enter through wounds that penetrate the tree bark, so you might try dressing wounds with a fungicide or tree paint as soon as they are spotted.

Orange Rust

Type: Fungus

Plants Affected: Blackberry, raspberry

Where It Occurs: Widespread

How It Is Spread: Wind. Overwinters in plant stems and roots.

Symptoms
- Yellow spots appear on both sides of leaves in early spring.
- Several weeks later, the leaf underside ruptures with masses of orange-red powdery spores.

Organic Remedies
- Remove/destroy infected canes, roots, suckers, and wild berries within 500 yards.
- Mulch heavily with straw and leaf mold.
- Apply lots of compost in autumn with extra P and K. (Low P and K encourages rust.)
- Select resistant varieties.

Pear Decline

Type: Mycoplasma

Plants Affected: Pear

Where It Occurs: Habitat areas of the pear psylla, which are principally east of the Mississippi and northwestern states

How It Is Spread: Transmitted by the pear psylla. Especially affects those on oriental rootstocks.

Symptoms: Tree weakens and slowly dies.

Organic Remedies

- Control pear psylla.
- Use only *P. communis* rootstock.

Pecan Bunch

Type: Mycoplasma

Plants Affected: Pecan

Where It Occurs: Arkansas, Georgia, Kansas, Louisiana, Mississippi, Missouri, New Mexico, Oklahoma, and Texas.

Symptoms: Similar to Walnut Bunch.

Organic Remedies: See Walnut Bunch.

Phytophthera cinnamomi

Type: Fungus

Plants Affected: Blueberry (Highbush)

Where It Occurs: Emerges in hot, moist conditions, especially in Florida, Georgia, Arkansas.

How It Is Spread: Lives indefinitely in all soil types, except perhaps sand. Most prevalent in wet soils with poor drainage.

Symptoms

- Leaves yellow; leaf margins turn brown, wilt, and fall off.
- Stunted growth.
- A "flag" branch (dead branch with dried leaves still attached) may be seen on an otherwise healthy-looking bush.

Organic Remedies

- Best prevention is to grow plants in raised beds to ensure good drainage.
- Avoid stressing the plant with under- or over-fertilization or over-watering.
- Cut down to the ground the infected part of the bush; remove and destroy.
- Remedy, a commercial fungicide containing baking soda, can be used on various ornamentals and fruit trees, and is said to be effective against black spot and powdery mildew, as well as leaf spots, anthracnose, phoma, phytophthora, scab, and botrytis.

Powdery Mildew

Type: Fungus

Plants Affected: Numerous plants.

Where It Occurs: Widespread. Fungal spores survive in frost-free areas and are blown long distances. Insects also help spread the fungus in more local areas. High humidity and moisture on leaves is not necessary for this fungus to take hold. Leaves are most susceptible within 2 to 3½ weeks of unfolding.

Symptoms

- White powdery growth like talcum powder, frequently starting on the leaf underside for curcurbits and on the upper leaf side for lettuce.
- Growth can spread to cover most of the leaf.
- Numerous strains of powdery mildew affect different crops and ornamental plants.
- In crucifers: Talcum-like growth appears in spots or in larger areas on upper side of leaf and stems. Leaves may pale, then turn yeallow or tan.
- In cucurbits: Curcurbit leaves wither and die, and dry and become brittle. Vines may also wither and become whitish. Fruit can sunburn more easily, ripen prematurely, and have poor taste and quality.
- In lettuce: Lettuce leaves tend to curl before turning yellow, then brown.

Organic Remedies

- See Overview of Symptoms and Remedies for fungal diseases, page 312.
- Select varieties that are resistant to this mildew.
- E-RASE and Surround, the jojoba oil products, are registered for powdery mildew on grapes and ornamentals indoors and out.
- Swedish researchers report that spraying cucumber plants with a 5% emulsion of garlic extract resulted in substantial protection against powdery mildew infection (Erysiphe cichoracearum).
- Researchers in Germany treated 'Bacchus' grape plants with sodium bicarbonate (baking soda) and found that 1.0% solutions worked best against powdery mildew (Uncinula necator).
- Remedy, a commercial fungicide containing baking soda, can be used on various ornamentals and fruit trees, and is said to be effective against black spot and powdery mildew, as well as leaf spots, anthracnose, phoma, phytophthora, scab, and botrytis.
- Plant pathologists in India report reductions in symptoms of powdery mildew after applying copper, molybdenum, or 50 parts per million of gibberellic acid.
- For cucurbits, Israeli researchers report that foliar sprays containing potassium nitrate (readily available at garden centers as a fertilizer) can protect against powdery mildew. Spraying every 7 or 14 days with a 25 millimolar solution of potassium nitrate (plus surfactant) did not harm cucumber plants and did a good job of preventing infestations of powdery mildew due to *Sphaerotheca fuliginea*.

Psyllid Yellows

Type: Phytotoxemia, caused by a toxic substance thought to be a virus.

Plants Affected: Potato, tomato

Where It Occurs: Habitat of psyllids, also known as "jumping plant lice," which resemble tiny cicadas. Pysllids overwinter in Texas and Mexico, and are abundant after cool, mild winters. They occur in February and March in southern Texas, then migrate to areas north, including California and Colorado.

How It Is Spread: Spread by a toxic substance released by the tomato psyllid.

Symptoms

- In potato: Young leaves yellow or redden, brown, and die. Sprouts emerge on young tubers and form new tubers, creating eventually a chain of deformed tubers.
- In tomato: Older leaves get thick and curl upward. Young leaves turn yellow with purple veins. Plant is dwarfed and spindly. Developed fruit is soft. If the plant is attacked while young, no fruit will appear.

Organic Remedies

- Clear away Chinese lantern and ground cherries, both of which are hosts.
- Control the tomato psyllid; garlic spray may work.

Scab

Type: Fungus (many different strains)

Plants Affected: Almond, apple, apricot, beet, cucumber, melon, peach, pecan, potato, pumpkin, squash, watermelon

Where It Occurs: Widespread, particularly in the humid Southeast. Fostered by humidity.

How It Is Spread: Spread among fruit and nut trees primarily by wind. Overwinters in fallen leaves and dead twigs. Dry soil favors this fungus, so keep soil moist.

Symptoms

- Symptoms vary in different plants.
- The disease develops rapidly at 70°F.
- In beet and potato: Ugly corky, wartlike lesions on the outside of roots. Tubers are still edible if damaged areas are removed.
- In cucurbits: Leaves develop water-soaked spots and can wilt. Stems can develop small cankers. Immature fruit develop gray concave spots that darken, become deeper, and develop a velvety green mold.
- In fruit trees: When fruit is half-grown, small, greenish, dark spots eventually turn brown. Branches and twigs may develop yellow-brown spots. Fruit can crack.
- In pecan: In spring when leaves unfold, irregular, olive-brown to black spots appear, usually on leaf undersides. Concave spots appear on nuts. Nuts and leaves drop prematurely.

Organic Remedies

- Remove and destroy all diseased leaves, plants, and fruit. Mow under trees. Research shows chopping or burning apple leaves soon after they have dropped in the fall can significantly lower scab spore survival.
- For beet and potato: Lower soil pH to below 5.5.
- For fruit trees: Spray or dust with sulfur 3 weeks after petal drop, and repeat 2 weeks later. The University of Tennessee has shown that dormant soybean oil, degummed ("slightly refined"), is highly effective against scale when sprayed on apple and peach trees (5% emulsion in water on a volume basis) in early February. Also, this might be a good miticide/insecticide during the growing season at a lower concentration (1–2%).

• Remedy, a commercial fungicide containing baking soda, can be used on various ornamentals and fruit trees, and is said to be effective against black spot and powdery mildew, as well as leaf spots, anthracnose, phoma, phytophthora, scab, and botrytis.
• For pecans: Remove all leaves, shucks, and dead leaves. Burn, if legal. Hot composting methods will destroy the organism.
• Ensure good drainage.
• For all vegetable crops: Rotate on a 3-year basis at minimum.
• For melons: Ensure full sun by planting in a sunny location.
• Plant soybeans in infested soil and turn under.
• Select resistant varieties.
• Grow chives near the affected plant roots.
• Apply composted pigeon droppings to plant soil.
• Spray trees in spring and early summer with Equisetum (horsetail) tea.

Smut

Type: Fungus (different strains)

Plants Affected: Corn, onion

Where It Occurs
• Most likely to occur in climates with hot dry season and dry spring followed by a wet spell. Spores survive for years.
• Onion: Prevalent in northern states.

Symptoms
In corn: Stunted or misshapen stalks. Smut develops anywhere on the leaves, stalks, ears, or tassels. Ugly white or gray galls are covered with a shiny milky membrane that ruptures and releases more black spores.
In onions: Black spots on leaves or bulbs. Cracks containing black powder appear on the sides of spots. Onion seedlings may die in a month. If not infected by the time its first true leaves appear, the seedling will usually escape this disease.

Organic Remedies
• For corn: Remove and destroy smut balls before they break. If severe, remove and destroy infected plants. Remove all stalks in fall. To reduce injury, control corn borers when tassels first appear.
• Avoid manure fertilizer, which may contain spores.
• Rotate crops on a 3+ year basis.
• Select disease-resistant corn and disease-free onion sets.
• If you can produce it reliably, you might sell young corn smut to specialty restaurants, a new gourmet item.

Southern Blight *(Sclerotium rolfsii)*

also known as Crown Rot, *Sclerotium* Root Rot, Southern Wilt

Type: Fungus

Plants Affected: Artichoke, bean, okra, peanut, pepper, soybean, tomato, watermelon

Where It Occurs: Prevalent in southern states below 38° latitude, coast to coast. This fungus prefers warm soil (80°F or higher), moisture, and sandy soils low in nitrogen.

How It Is Spread: This fungus can spread over the soil to other plants. It overwinters 2 to 3 inches below the soil line in the mustard seed–like bodies. It is known to infect 200 plant species.

Symptoms

- Leaves yellow, wilt, and drop, starting at the bottom, followed by vine wilt and death.
- White mold grows on the stem at or near the soil line. The white mold may harden and crust over.
- Round, yellow, tan, or dark brown bodies the size of mustard seeds appear on lower stems and on soil, and may develop white mold growth.
- Some fruits and roots develop round lesions.
- Storage rot may occur in cabbage, squash, potato, and sweet potato.

Organic Remedies

- In warm climates, immediately dig up infected plants and destroy. In cool cli mates (above 38° latitude), some suggest this disease may not need controls.
- Black plastic mulch is reported by Auburn University in Alabama to help control this disease and is an easier control than soil solarization.
- After harvest, plow in the plant stubbles deeply.
- Rotate crops and don't plant susceptible vegetables near each other.
- Solarize soil.
- Use wide spacing.
- Plant early where there is a history of southern blight.

Southern Wilt

see Southern Blight

Spotted Wilt

Type: Virus

Plants Affected: Tomato

Where It Occurs: Widespread

How It Is Spread: Transmitted almost entirely by thrips.

Symptoms

- Leaves develop small orange spots.
- Older leaves brown and die.
- Plant is stunted.
- Green fruits can develop yellow spots that develop concentric zones of brown, green, pink, or red shading.

Organic Remedies

- Control thrips. Use a reflective mulch such as aluminum foil or black plastic sprayed with aluminum.
- Grow two plants per pot and set out together. Only a maximum number of plants per given area get infected, so yields can be kept high.

Stewart's Disease

see Bacterial Wilt

Sunscald

also known as Winter Injury

Type: Environmental

Plants Affected: Apple, cherry, onion, pecan, pepper, tomato

Where It Occurs: Widespread. Dark tree bark is the most susceptible. Worst in drought years with cold, sunny, winter days.

How It Is Caused: The southwestern side of the plant warms during the day, then cells rupture during cold nights.

Symptoms

- Vegetables develop white or yellow wrinkled areas.
- Onions get bleached and slippery tissue during curing.
- Fruit tree bark darkens and splits open in long cracks or cankers.

Organic Remedies

For fruit trees: Shade bark by covering with burlap or apply a white interior latex paint.
For vegetables: Keep as much foliage as possible to shade fruit; don't prune suckers at the plant base. If necessary, prune above the first group of leaves above fruit.
Onions: Don't cure in direct sun.

Tobacco Mosaic (TMV)

Type: Virus

Plants Affected: Eggplant, pepper, tomato

Where It Occurs: Widespread

How It Is Spread: Spread by tools and hands (especially those of smokers). Virus is known to live in cured tobacco for up to 25 years.

Symptoms

- Misshapen leaves in young plants.
- Dark green mottled leaves tend to be pointed or fernlike; leaves may curl and wrinkle and have a grayish coloring.
- In the final disease stage, leaves drop, branches die, and fruits yellow and wrinkle.
- Very difficult to control.

Organic Remedies

- Never smoke near plants; have smokers scrub hands before touching plants.
- Destroy infected and nearby plants. Clear nearby perennial weeds.
- Disinfect all tools.
- Spray infected seedlings with skim milk or reconstituted powdered-milk solution. Spray until seedlings are dripping.
- Select resistant varieties.

Walnut Bunch

also known as Witches'-Broom, Brooming Disease

Type: Uncertain; some suspect mycoplasma or virus

Plants Affected: Strawberry, hickory, pecan; particularly serious in butternut; walnut

Where It Occurs: Northeast and Midwest

How It Is Spread: Uncertain; insect vectors are suspected.

Symptoms

In strawberry: Symptoms are on shoots only. On swollen stems, deformed shoots are bushlike and broomlike. Occurs on woodland borders and where balsam fir is present.
In walnut: Lateral buds don't remain dor-

mant, but produce bushy, densely packed shoots and undersized leaves, often 2 weeks earlier than healthy branches. Few nuts are produced. Nuts are shriveled, soft-shelled, poorly developed, and have dark kernels. Diseased shoots enter dormancy in fall, very late.

Organic Remedies

For strawberry: Remove all infected berries and plants. Eliminate nearby balsam firs.

For walnut: Removing all diseased branches can be an effective control. Make cuts well back from the infected area. Disinfect tools. Propagate only from disease-free trees.

Western Yellow Blight

see Curly Top

Winter Injury

see Sunscald

Yellows

Type: Fungus; *see* Fusarium Wilt

Yellows

Type: Virus

Plants Affected: Beet, celery, spinach

Where It Occurs: California, Oregon, Washington, Utah, Colorado, Michigan, Nebraska, Ohio, Maryland, and Virginia.

How It Is Spread: Spread by aphids.

Symptoms

- Leaves yellow, starting at tips and margins.
- Outer and middle leaves can get thick and brittle.
- Stunted plants and roots.

Organic Remedies

- Control aphids, using a reflective mulch such as black plastic sprayed with aluminum.
- Don't plant winter spinach near beets.
- Plant vegetables in protected areas to minimize aphids spreading by wind.

Yellows

also known as Aster Yellows

Type: Mycoplasma

Plants Affected: Broccoli, carrot, celery, lettuce, onion, strawberry, tomato

Where It Occurs: Widespread, but particularly destructive in the West, where it affects more than 200 plant species.

How It Is Spread: Leafhoppers, particularly the six-spotted leafhopper.

Symptoms

- Young leaves yellow.
- Top growth is yellow and bushy like witches'-broom.
- Older leaves may become distorted.
- Carrot tops turn reddish-brown in mid- to late season.
- Flowers may be absent, green, underdeveloped, misshapen, or fail to produce seed or fruit.
- Immature leaves are narrow and dwarfed.

• Stunted plants.
• Sterile seeds.
• Vegetables ripen prematurely.
• Yields and quality are severely affected.

Organic Remedies

• Control leafhoppers; eradicate weeds to destroy leafhopper eggs, especially early in the season.

• Immediately remove and destroy infected plants and plant residues.
• Plant tolerant varieties.

Controlling Insects and Animals

About this chart:

Clean cultivation. Clean cultivation and sanitation practices are assumed (see pages 301 and 318). Both are important to the prevention and control of many diseases and insects.

Damage not requiring control. When assaulted by chewing pests, plants can tolerate up to 20 percent defoliation without significant loss in yield or quality. The exception is plants whose leaves are the product, such as spinach or cabbage.

Remedies. The use of aluminum foil mulch for prolonged periods may result in aluminum leaching into the soil. If you're planning to leave the mulch on for long periods, consider a substitute, such as black plastic sprayed with reflective aluminum paint.

Experimental Remedies. These remedies may or may not work; efficacy may depend on the severity of the pest problem as well as method of application. Many of these are homemade; for recipes see Glossary of Organic Remedies beginning on page 315.

Allies. Ally listings in this chart should be checked against the chart on allies and companions (pages 420–437), because an alleged ally may help only one of the crops listed. Also, remember that while an ally may effectively deter one insect pest it may, at the same time, attract others.

Aphid

many species

Description

• Tiny (1⁄16"–1⁄8") soft-bodied insects are pear-shaped and can be brown, black, pink, white, or green. They have long antennae, two tube-like projections from the rear, and may have wings.
• Aphids transmit many viral diseases.
• The phylloxera aphid attacks grapes and nearly wiped out the French wine industry in the 1800s.

Plants Affected: Many herbs, most fruits, most vegetables

Where It Occurs: Widespread; produces 20+ generations per year. The phylloxera aphid is primarily a problem in the West (e.g., California and Arizona) where European vines are grown.

Signs

- Foliage curls, puckers, or yellows.
- Plants can be stunted or distorted.
- Cottony masses may appear on twigs of trees and shrubs.
- Presence of sticky "honeydew," which attracts ants and supports black, sooty mold.
- Aphids of different species suck leaves, fruit, stems, bark, and roots.

Organic Remedies

- Symptoms may result from too much nitrogen or pruning. Check for the presence of aphids.
- Use row covers.
- Control aphids and ants, which carry aphid eggs, with sticky bands, sticky yellow traps, or yellow pans filled with soapy water.
- Keep grass mowed around garden.
- A reflective mulch like aluminum foil repels them.
- Application of fermented extract of stinging nettle is thought to repel black aphids. Use experimentally; see page 348.
- Homemade sprays: Use experimentally; see page 348.
  – Forceful water jets 2–3 times per day.
  – Bug juice (see page 317).
  – Garlic, onion, or pepper (see page 319).
  – Oxalic acid from leaves of rhubarb or spinach (boil 1 pound of leaves for 30 minutes in 1 quart of water; strain; cool; and add a touch of soap, not detergent).
  – Tomato or potato leaf juice.
- Other sprays:
  – Light or superior horticultural oil (3% solution), applied in the plant's dormant or active phase.
  – Insecticidal soap.
  – Strong lime solution spray.
  – Cooking oil spray mix, as recommended by USDA (see page 318).
  – Make sure all sprays contact the aphids.
- Dust with diatomaceous earth to dry up aphids — or with calcium to control ants.
- Fall cultivation.
- Rotate crops (more effective with root-feeding species).
- Japanese researchers report that the organosilicone non-ionic surfactant "Silwet L-77" is toxic to green peach aphids, especially at a high relative humidity.
- Biological controls: Larvae and adults of the green lacewing and ladybug. Syrphid fly larvae, minute pirate bug, damsel bugs, big-eyed bugs. In the greenhouse, PFR (*Paecilomyces fumosoroseus*) is a beneficial microorganism available in granule form that attacks whiteflies, mites, aphids, thrips, mealybugs, and certain other greenhouse pests and is claimed to be safe for humans and beneficials.
- Cinnamon oil sprays may be highly effective for controlling mites, aphids, and even fungi that attack greenhouse-grown plants with minimal concerns about health and safety. The Mycogen Corporation recently registered their product called Cinnamite.
- **Botanical controls**: Sabadilla dust, rotenone (1% solution), pyrethrin, neem.
- **Allies**: Anise, broccoli, chives, clover, coriander, cover crops of rye and vetch,

fennel, French beans, garlic, lamb's quarters, nasturtium, onion family, tansy. (See chart on pages 420–437.)

Apple Maggot

Description

- Small (¼") white-yellowish worms hatch in mid- to late summer. They create winding brown tunnels as they feed inside fruit. Once the fruit falls, larvae emerge to pupate in the soil.
- The adult is a small black fly with yellow legs, a striped abdomen, and zigzag black markings on its wings. Females lay eggs in puncture wounds in fruit.

Plants Affected: Apple,* blueberry, cherry, pear, plum

Where It Occurs: Northeast, west to the Dakotas, and south to Arkansas and Georgia. Produces one to two generations per year.

To Monitor: Prepare traps by mixing 2 teaspoons ammonia, ¼ teaspoon soap flakes, and 1 quart water. Hang traps in jars on the sunny side of trees, shoulder high. Ten jars per orchard should give a good picture of the maggots' presence. Count the flies every 2–3 days.

Signs

- Slight cavities and holes indicate eggs are present.
- Brown streaks on fruit skins.
- Premature fruit drop.
- After fruit falls or is picked, flesh becomes brown pulpy mess.
- Very damaging.

Organic Remedies

- Red sticky balls. Hang these on perimeter trees at shoulder height, using about 4 balls per tree.
- Make a control trap of 1 part molasses/9 parts water; add yeast; and pour into wide-mouthed jars. When fermentation bubbling subsides, hang the jars in trees. Use experimentally; see page 348.
- Remove and destroy badly infested fruit or feed it to animals. Drop mildly damaged fruit into water to kill maggots; these are okay for cider.
- Fall cultivation.
- **Biological controls:** Beneficial nematodes, applied in late summer to early fall.
- **Botanical controls:** Rotenone (5% solution).

Asparagus Beetle

Description

- The small (¼") metallic blue-black beetle, with three yellow-orange squares on each side of its back, lays dark shiny eggs the size of specks on leaves and spears.
- Eggs develop into orange larvae with black heads and legs, then green-gray grubs with dark heads.
- Beetles overwinter in garden trash.

Plants Affected: Asparagus

Where It Occurs: Widespread, but rare in the Pacific coastal areas, the Southwest, Florida, and Texas. Produces 2 generations per year in cold areas, 3–5 in warm areas.

* *Plant most frequently attacked by the insect.*

Signs

- Defoliation.
- Misshapen spears.
- Adult beetles eat leaves, fruit, and spears.

Organic Remedies

- Use row covers.
- Harvest regularly in spring.
- Vacuum adults off ferns. Immediately empty the bag and destroy the bugs or they can crawl back out.
- Fall cultivation to destroy overwintering pests; also in fall, let chickens in the garden to eat beetles.
- **Biological controls:** Encourage birds, and import ladybugs, chalcid and trichogramma wasps, and Encarsia formosa.
- **Botanical controls:** Ryania, rotenone (1% solution).
- **Allies:** Basil, marigold, nasturtium, parsley, tomato. Also, beetle allegedly dislikes bone meal. (See chart on pages 420–437.)

Bean Jassid

see Potato Leafhopper

Bean Leaf Beetle

Description: The adult is a small, ¼" reddish-tan beetle with 3–4 black spots on each side of its back. It lays eggs in the soil at the base of seedlings. Larvae, slender and white, feed on roots. Produces 1–2 generations per year.

Plants Affected: Bean,* pea

Where It Occurs: Primarily in the Southeast.

Signs

- Holes in leaves, particularly young seedlings.
- Adult beetles feed on leaf undersides and on seedling stems.
- Larvae attack roots but usually don't affect the plant.
- Not a frequent pest.

Organic Remedies

- Use same controls as for Mexican Bean Beetle; see page 384.
- **Biological controls:** Ladybugs and lacewings eat the eggs. Beneficial nematodes, mixed into seed furrow and in mulch, may kill larvae and emerging adults.
- **Botanical controls:** Rotenone (1% solution).

Beet Leafhopper

also known as Whitefly *in the West*

Description: See Leafhopper. Produces up to 3 generations per year.

Plants Affected: Beet,* bean,* cucumber, flowers, spinach, squash, tomato

Where It Occurs: West of Missouri and Illinois, except not in coastal fog areas.

Signs: The beet leafhopper spreads curly top virus, which is characterized by raised leaf veins; stunted plants; small, wartlike bumps on the leaf undersides; and curled, brittle leaves.

Organic Remedies: See Leafhopper controls on page 381.

** Plant most frequently attacked by the insect.*

Birds

Description: Can be major pests of all fruits and berries, and corn and peas.

Plants Affected: Fruits, nuts, many vegetables

Where It Occurs: Widespread

Organic Remedies

- **For trees:** Netting must be held out away from the tree by supports. Otherwise birds peck through the netting. Secure the netting tightly around the bottom. For more complete protection you can use cheesecloth, which is harder to peck through.
- **For growing beds:** Use row covers over corn and peas.
- **For corn seed:** Spread lime down seed rows — use experimentally; see page 348. Plant corn seeds deeper than usual, or mulch heavily. Erect several taut strands of black thread or fishing line over each seed row. Or, start seeds inside and transplant once established.
- **For mature corn:** When silks brown, tie small rubber bands over each ear.
- Construct growing boxes that are screened on all sides.
- For berries and other large plants, erect around the plant a tepee structure out of lumber or poles, and drape netting over it.
- **For strawberries:** Paint large nuts red and place in the strawberry patch before fruit has ripened. Some gardeners report that birds tire of pecking at these fake berries before the real ones ripen.Use experimentally; see page 348.
- Scare-eye balloons have been reported to be effective. Use experimentally; see page 348.
- Mulberry and elderberry trees offer berries in fall that birds eat first — use experimentally; see page 348. Mulberry grows rapidly to a breadth of 50'–60'.
- Use noise-making devices to scare birds away. One gardener reported that talk shows with human voices worked best. Use experimentally; see page 348.
- BirdXpeller Pro produces specific predator sounds, which are said to alarm and deter pigeons, sparrows, starlings, gulls, woodpeckers, crows, blackbirds, grackles, and geese. It is programmable with different predator sounds.

Blackberry Sawfly

Description

- Adults are small wasplike flies with two pairs of transparent wings hooked to each other. They lay white eggs on leaf undersides in May.
- Blue-green larvae (¾") roll leaves closed with webs and feed inside the webs through July. They then drop to the ground to pupate in the soil.

Plants Affected: Blackberry

Where It Occurs: Unable to determine; likely widespread

Signs: Rolled and webbed leaves.

Organic Remedies

- Use sticky traps to catch adults.
- Handpick eggs and larvae in spring.
- **Botanical controls:** Rotenone (1% solution); pyrethrin.

Blister Beetle

see Striped Blister Beetle

Boxelder Bug

Description: Medium-sized (½") bugs suck plant juices. They look like squash bugs but are brown with red markings. Nymphs are bright red.

Plants Affected: Almond, ash, boxelder,* maple*

Where It Occurs: Present wherever boxelders grow.

Signs: Damaged foliage and twigs.

Organic Remedies

- Remove any nearby female boxelder trees.
- Insecticidal soap spray, though this isn't a very effective control of hard-bodied insects.

Brown Almond Mite

see Mite

Cabbage Butterfly

see Imported Cabbageworm

Cabbage Looper

Description

- This large (1½") pale green worm with light stripes down its back doubles up, or "loops," as it crawls. It hides under leaves in hot, dry weather. The worm overwinters as a green or brown pupa in a thin cocoon attached to a plant leaf.
- The adult moth (1½") is night-flying, brownish, and has a silver spot in the middle of each forewing. It lays greenish-white round eggs, singly, on leaves.
- Eggs hatch in 2 weeks and the looper feeds for 3–4 weeks.
- Produces 4 generations or more per year.

Plants Affected: Bean, brassicas, celery, lettuce, parsley, pea, potato, radish, spinach, tomato

Where It Occurs: Widespread

To Monitor: Use pheromone traps to monitor population levels of adult moths, starting soon after planting.

Signs

- Ragged holes in leaves.
- Seedlings can be destroyed.
- Worms bore into the heads of all cabbage family plants.

Organic Remedies

- Use row covers all season.
- Handpick loopers.
- Plant spring crops to avoid the cabbage looper peak.
- Stagger planting dates to avoid entire crop susceptibility.
- Hot pepper spray. Use experimentally; see page 348.
- Spray with soap and lime, or dust wet plant with lime.
- In the deep South, practice a thorough fall cultivation and also rotate crops on a 3- to 5-year basis. This insect doesn't overwinter in the North, so these steps aren't necessary.

* *Plant most frequently attacked by the insect.*

• Make viral insecticide from loopers infected with nuclear polyhedrosis virus (NPV). Loopers are chalky white, sluggish or half-dead, may be on top of leaves or hanging from their undersides. They turn black and liquefy within days. Capture three and make bug juice. Spray from three bugs covers ¾ acre. Loopers take 3–6 days to die, but one spraying can last the entire season. Note: Spray only when it is cool and damp, and when you see that loopers are not hiding under leaves.
• Encourage predators such as toads, bluebirds, chickadees, robins, and sparrows.
• **Biological controls:** Bt, applied every 2 weeks until heads form (for broccoli and cabbage); lacewings and trichogramma wasp.
• **Botanical controls:** Sabadilla, rotenone (1% solution), pyrethrin, neem.
• **Allies:** Dill, garlic, hyssop, mint, nasturtium, onion family, sage, and thyme. (See chart on pages 420–437.)

Cabbage Maggot

Description

• The adult resembles a housefly but is half the size and has bristly hairs. It emerges from an underground cocoon in spring at cherry blossom time, in early summer, or in autumn. It lays eggs on plant stems near the soil line or in cracks in the soil.
• The small (⅓"), white, legless worm with a blunt end attacks stems below the soil line.
• To overwinter, maggots pupate 1"–5" deep in the soil and emerge on the first warm spring day.
• The maggot transmits both bacterial soft spot and black leg.
• In season, each life cycle takes about 6–7 weeks.
• Produces numerous generations per year.

Plants Affected: Brassicas, pea, radish (late spring), turnip

Where It Occurs: Widespread, particularly in western and northern United States. It thrives in cool, moist weather.

To Monitor: Yellow sticky traps are a good early-warning device for adults. Another method is to take a scoop of soil from the plant base, place it in water, and count the eggs that float to the top to determine the extent of the problem.

Signs

• Seedlings wilt and die.
• Stems are riddled with brown, slimy tunnels.
• Stunted, off-color plants.

Organic Remedies

• Use row covers in early season.
• Maggots don't like alkaline environment. Circle plants with a mixture of lime and wood ashes (moistened to prevent blowing) or diatomaceous earth — use experimentally; see page 348. Replenish after rain. Mix wood ashes into the surrounding soil. Do not replenish more than one or two times; continued use of wood ashes raises the pH excessively by its addition of potassium.
• Use a 12" square of tar or black paper to prevent larvae from entering the soil.

• Plant in very early spring or in fall to avoid maggot peak in May and early June.
• **Biological controls:** Beneficial nematodes are effective when applied before planting.
• **Allies:** Clover, garlic, onion family, radish, sage, and wormwood. (See chart on pages 420–437.)

Cabbage Moth

see Imported Cabbageworm

Cabbageworm

also known as Cross-striped Cabbageworm

Description: New worms are gray with large round heads. Mature worms (⅔") are green to blue-gray with long dark hairs and at least three distinct black bands across each segment. The adult moth is small, pale yellow with mottled brown on its forewings.

Plants Affected: Cabbage

Where It Occurs: Widespread

Signs: See Imported Cabbageworm

Organic Remedies: See Imported Cabbageworm

Caneborer

also known as Rednecked Cane Borer

Description

• Blue-black beetles (⅓") with coppery red thorax appear in May and lay eggs in June in cane bark near a ragged or eaten leaf.
• Flat-headed larvae bore into canes and cause swellings or galls in late July and August. Cut open galls to find creamy white grubs.

Plants Affected: Blackberry, raspberry

Where It Occurs: Present in the Northeast; other types are prevalent in other parts of the United States.

Signs

• Large cigar-shaped swellings on canes.
• Where cane joints swell, the cane may break off and die.

Organic Remedies: Cut out all canes with swellings and burn or destroy.

Cankerworm, Fall and Spring

also known as Inchworm, Measuring Worm

Description

• Small worms (1") are green, brown, or black, have a yellow-brown stripe down their back, and drop from trees on silky threads when ready to pupate. They pupate 1"–4" deep in the ground in cocoons near the trees.
• Male adults are grayish moths. Females are wingless. In spring, they lay brownish-gray eggs in masses on tree trunks or branches; in fall they lay brown-purple egg masses beneath the bark. Eggs hatch in 4–6 months.

Plants Affected: Apple,* apricot, cherry, elm, maple, oak, plum

Where It Occurs: Widespread. The fall variety is worst in early spring in California, Colorado, Utah, and northern United

** Plant most frequently attacked by the insect.*

States. The spring variety is worst in the East, Colorado, and California.

Signs

- Skeletonized leaves.
- Few mature fruit.
- Trees may look scorched when damaged by the spring variety.
- Defoliation may occur 2–3 years in a row.

Organic Remedies

- Use sticky bands from October to December, and again in February, to catch wingless females crawling up trees to lay eggs. Renew sticky substance periodically.
- Encourage predators such as bluebirds, chickadees, and nuthatches.
- Spray horticultural oil before leaves bud in spring.
- **Biological controls:** Trichogramma wasps work on spring species. Chalcid wasps. Bt, applied every 2 weeks from the end of blossom time to 1 month later.
- **Botanical controls:** Sabadilla.

Carrot Rust Fly

Description: Small (⅓"), yellow-white larvae burrow into roots and make rusty-colored tunnels. The adult fly is slender, shiny green, and lays eggs at the plant crown in late spring. Produces several generations per year.

Plants Affected: Caraway, carrot, celery, coriander, dill, fennel, parsley, parsnip

Where It Occurs: Widespread, but mostly a problem in northern and Pacific Northwest states.

Signs

- Stunted plants.
- Leaves wilt and turn yellow.
- Soft rot bacteria in the carrot.

Organic Remedies

- Use row covers all season.
- Fall cultivation and early-spring cultivation disrupt overwintering maggots.
- Rotate crops.
- Avoid early planting. Plant all target crops after maggot peak.
- Yellow sticky traps. British horticulturists have shown these traps are most effective for this pest when placed at a 45-degree angle.
- Sow seeds with used tea leaves — use experimentally; see page 348.
- Spread wood ashes (moistened to prevent blowing), pulverized wormwood, or rock phosphate around the plant crown to repel egg-laying — use experimentally; see page 348.
- **Allies:** Black salsify (oyster plant), coriander, lettuce, onion family, pennyroyal, rosemary, and sage. (See chart on pages 420–437.)

Carrot Weevil

Description: Small (⅓"), pale, legless, brown-headed worms tunnel into roots and celery hearts. The tiny (⅕") dark brown, hard-shelled adult is snout-nosed and overwinters in garden litter and hedgerows. Produces 2–3 generations per year.

Plants Affected: Beet, carrot,* celery, parsley, parsnip

** Plant most frequently attacked by the insect.*

Where It Occurs: East of the Rockies.

Signs

- Zigzag tunnels in the tops and roots of plants.
- Defoliation.

Organic Remedies

- Use row covers all season.
- Fall cultivation.
- Rotate crops.
- Encourage chickadees, bluebirds, juncos, and warblers.
- **Biological controls**: Beneficial nematodes when plants are very small.
- **Botanical controls**: Sabadilla, rotenone (5% solution).

Carrotworm

see Parsleyworm

Celery Leaftier

also known as Greenhouse Leaftier

Description

- Medium (¾"), pale green-yellow worms, with white stripe down the length of their backs, eat a host of vegetables. They web foliage together as they feed, and eventually pupate in silky cocoons inside webs.
- The adult is a small (¾"), brown nocturnal moth. It lays eggs that look like fish scales on leaf undersides. They can do great damage but don't usually appear in large numbers.
- Produces 5–6 generations per year in warm areas; 7–8 per year in greenhouses.

Plants Affected: Celery,* kale, many other plants

Where It Occurs: Widespread but worst in the Northeast and southern California. Most destructive in greenhouses.

Signs

- Holes in leaves and stalks.
- Leaves folded, and tied with webs.

Organic Remedies

- Handpick and destroy pests.
- Handpick damaged or rolled leaves where leaftier may be hiding inside.

Celeryworm

see Parsleyworm

Cherry Fruit Fly

Description

- These black fruit flies resemble small houseflies, with yellow margins on the thorax and two white crossbands on the abdomen.
- They emerge in early June and feed for 7–10 days by sucking sap. They then lay eggs through small slits in developing fruits.
- The maggots are small yellow or white worms with two dark hooks on their mouths. They feed inside fruit, then drop to the ground and pupate for 6 months 2″–3″ deep in the soil.
- Pupae overwinter in the soil.

Plants Affected: Cherry, pear, plum

Where It Occurs: Widespread, except in the Southwest and Florida.

To Monitor: Use sticky red balls or pheromone traps to monitor population levels.

* *Plant most frequently attacked by the insect.*

Signs

- Small, disfigured fruit.
- Premature fruit drop.
- Rotten flesh with maggots feeding inside.

Organic Remedies

- Fall cultivation after the first several frosts will expose pupae to predators.
- Red sticky balls. Hang these on perimeter trees at shoulder height, using about 4 balls per tree.
- Make a control trap of 1 part molasses to 9 parts water; add yeast; and pour into wide-mouthed jars. When fermentation bubbling subsides, hang the jars in trees. Use experimentally; see page 348.
- Remove and destroy badly infested fruit or feed it to animals. Drop mildly damaged fruit into water to kill maggots; these are okay for cider.
- **Biological controls:** Braconid wasps.
- **Botanical controls:** Neem.

Cherry Fruit Sawfly

Description

- Small (1/8") adult wasplike flies have two pairs of transparent wings hooked together and yellow appendages.
- Small (1/4") larvae are white with brown heads. They bore into young fruit and feed on seeds. They exit fallen fruit to pupate and overwinter in silken cocoons in the ground.

Plants Affected: Apricot, cherry, peach, plum

Where It Occurs: Pacific Coast.

Signs: Fruits shrivel and drop.

Organic Remedies

- Shallow 2" cultivation around trees to expose pupae.
- Use sticky traps to catch adults.
- Handpick eggs and larvae in spring.
- **Botanical controls:** Rotenone (1% solution); pyrethrin.

Cherry Fruitworm

Description

- The small, gray adult moth lays single eggs on cherries in May and June.
- Small (3/8") pinkish larvae hatch in 10 days and bore into fruit where they feed. They overwinter in the stubs of pruned branches or in bark crevices.
- Distinguished from the cherry fruit fly maggot by black head and caterpillar-like body.
- Produces 1 generation per year.

Plants Affected: Cherry

Where It Occurs: Prevalent in Colorado, and north and west of Colorado.

Signs

- Rotten flesh with larvae feeding inside.
- Can be very damaging.

Organic Remedies

Biological controls: Bt, sprayed in the first week of June.

Cherry Slug

see Pearslug

Chestnut Weevil

also known as Snout Weevil

Description

• The adult (5⁄16"–½") has a very long proboscis (snout), as long or longer than its body. It emerges from April to August, and in August deposits eggs into the burr through a tiny hole.
• Larvae are legless, white, plump, and curved like a crescent moon.
• Mature larvae overwinter 3"–6" below ground.
• Completes a life cycle in 1–2 years.

Plants Affected: Chestnut

Where It Occurs: Wherever Asian chestnuts are grown.

Signs: Larvae feed on kernel tissue for 3–5 weeks, sometimes hollowing out the interior, then leave the nut.

Organic Remedies

• Collect fallen nuts every day and immerse in hot water (122°F) for 45 minutes to kill all larvae. This prevents larvae from entering the soil and continuing the cycle.
• Plant away from woods or forest from which squirrels and other rodents can bring in infested nuts.
• Fall cultivation. Also, chickens can help clean up larvae.

Click Beetle

see Wireworm

Codling Moth

Description

• Large (1") larva is white tinged with pink, has a brown head, and a voracious appetite.
• Adult moths (¾") are gray-brown with dark brown markings on lacy forewings, fringed back wings, and dark brown edging on all wings. They lay flat white eggs, singly, on twigs or upper leaf surfaces.
• Eggs hatch in 6–20 days, and larvae tunnel into and out of fruit. Pupae overwinter in tough cocoons built in cracks of loose bark, fences, buildings, or garden debris.
• Produces 2 generations per year; the first generation attacks immature fruit; the second attacks mature fruit.

Plants Affected: Almond, apple,* apricot, cherry, peach, pear, walnut (English)

Where It Occurs: Widespread

To Monitor: Use pheromone traps to determine population levels. More than five moths per trap per week indicates that controls may be needed.

Signs

• Holes and tunnels in fruit, with brown fecal material at the core and a brown mound at the hole opening.
• Sometimes forms cocoons in bark crevices.

Organic Remedies

• In spring, band tree trunks with several thicknesses of 6"-wide corrugated cardboard. Exposed ridges must be at least 3⁄16"-wide and must face tree. This gives larvae a place to spin cocoons when they

* *Plant most frequently attacked by the insect.*

leave fruit. Remove and kill larvae once a week in warm weather, once every 2 weeks in cool weather. Continue through harvest. Burn cardboard.

- Use sticky bands to catch larvae.
- Make a trap for larvae of 2 parts vinegar to 1 part molasses in a wide-mouth jar — use experimentally; see page 348. Hang three to four traps per tree, 8" below the limb. Clean and replenish daily.
- In spring, scrape off all loose rough bark from trunk and limbs. Catch scrapings and destroy. Seal pruning wounds.
- Apply soap and lime spray, or fish oil spray, to entire tree before leaves appear in late winter, and, later, weekly to tree trunk and base.
- Apply horticultural oil spray before buds open. Be sure to cover all surfaces of the tree.
- Encourage woodpeckers in winter with one suet ball per tree.
- **Biological controls**: Two or three sprays of Bt, applied 3 to 4 days apart at peak egg-laying time, is very effective. Three timed releases of trichogramma wasps — the first at petal fall, the second 3–8 weeks later, and a final release 5–8 weeks later in early fall — are also helpful. Beneficial nematodes help most when sprayed on wet tree trunks and nearby soil in late winter.
- Pheromone-based mating disruption devices should be commercially available soon but may not be effective in small orchards.
- **Botanical controls**: Ryania, sabadilla, pyrethrin.
- **Allies**: Cover crops of the buckwheat, clover, and daisy families; dill; garlic; wormwood. (See chart on pages 420–437.)

Colorado Potato Beetle

Description

- Small (1/3"), yellow, hard-shelled beetles have orange heads with black dots and black stripes down their backs. They lay bright yellow eggs on leaf undersides.
- Eggs hatch in 4–9 days and become plump red larvae with black spots and black heads.
- Produces 1–3 generations per year.

Plants Affected: Eggplant,* pepper, potato,* tomato*

Where It Occurs: Widespread, but mostly a problem in the eastern United States. Rarely a problem in southern California through Texas, Louisiana, and Georgia.

Signs

- Skeletonized leaves and complete defoliation.
- Adults and larvae chew foliage.

Organic Remedies

- Use row covers in early season.
- Handpick immediately, when sighted, and crush adults and egg masses — a very effective control.
- Apply thick organic mulch to impede the movement of overwintered adults to plants. Beetles walk more than fly during the early season. Potatoes can be started in thick mulch above ground.
- Fall cultivation.
- Where different potato and tomato varieties are grown, resistant varieties will be less susceptible. If only one resistant vari-

** Plant most frequently attacked by the insect.*

ety is grown, however, even this will be consumed. All eggplants are susceptible.

• Time plantings to avoid beetles. Plant potatoes as early as possible to allow sufficient plant growth to withstand attack of overwintered adults.

• Research in Ontario shows that potatoes can serve as an effective trap crop for tomatoes, and that crop rotation of potato and tomato crops can help reduce more overwintering Colorado Potato Beetles.

• Foliar spray of fish wastes was shown to repel these bugs by University of Maine studies. They used Biostar.

• A University of Delaware study showed that ground tansy — diluted 1:100 (leaf to water weight ratio) and used as a foliar spray — dramatically reduced bug feeding.

• Garlic, onion, and pepper sprays, applied directly on beetles, are irritants — use experimentally; see pages 319, 320–321.

• Foliar sprays of hydrogen peroxide (1 tablespoon/1 gallon water) applied directly on active adults provide fair control. Do not apply in direct sunlight in the heat of the day.

• A dusting of diatomaceous earth dries out beetles. Soap and lime spray is also believed to dry out beetles, but generally soap is not effective against hard-bodied insects.

• Sprinkle fine-milled bran on leaves at the first sign of beetles — use experimentally; see page 348. Found effective by an Ohio gardener.

• In fall, let chickens into the garden to eat beetles and larvae.

• Encourage predators like songbirds, toads, and ground beetles.

• **Biological controls:** Ladybugs, lacewings, and *Edovum puttleri* eat the eggs. Beneficial nematodes. Two-spotted stink bug (*Perillus bioculatus*). New Bt strains (M-One, Trident, Foil) are very effective if applied in the larval stage. Once the larvae are too big, it doesn't help much.

• **Botanical controls:** Sabadilla, rotenone (5% solution), neem.

• **Allies:** Bean, catnip, coriander, dead nettle, eggplant, flax, horseradish, nasturtium, onion family, and tansy. Because beans are allegedly noxious to this bug, and potatoes repel the Mexican Bean Beetle, they might be companions. (See chart on pages 420–437.)

Corn Borer

see European Corn Borer

Corn Earworm

also known as Vetchworm, Cotton Bollworm, Tomato Fruitworm, Tobacco Budworm

Description

• Large (1½" to 2"), light yellow, green, red, or brown caterpillars; striped, with "spines" at bands; feed first on leaves and corn silk. They feed on kernels and exit through the husk to pupate. Feeding lasts about 1 month, when they drop to the ground and pupate 3"–5" deep in the soil.

• The adult (1½") grayish-brown moth feeds on flower nectar. It lays 500–3,000 tiny, single, ribbed, dirty white eggs on host plants.

- This pest is most numerous 2–3 weeks after a full moon. Some say corn should ideally silk during the full moon. Its numbers are reduced by cold winters and wet summers.
- Extremely destructive.
- Produces 2–3 generations per year.

Plants Affected: Bean, corn (sweet),* pea, peanut, pepper, potato, squash, tomato

Where It Occurs: Widespread, but primarily a problem in southern and central states.

Signs

- Ragged holes in tender leaves.
- Eaten tassels and damaged pods on developing fruits.
- Chewed silk and damp castings near the silk.
- Damaged kernels, often at the ear tip.

Organic Remedies

- Mineral oil, applied just inside the tip of each ear, suffocates the worms. Apply only after silk has wilted and started to brown at the tip, or pollination will be incomplete. Use ½ of a medicine dropper per small ear, ¾ of a dropper per large ear. You might add red pepper to the oil to see if it increases the effectiveness. Apply two or more applications of oil, spaced at 2-week intervals.
- Fall cultivation. In spring, cultivate the top 2 inches of soil.
- Handpick worms after silks brown.
- Time plantings to avoid worms. In northern states, early plantings that silk before mid-July often avoid attack.
- Plant resistant varieties with tight husks. Or clip husks tightly with clothespins. You can also try covering ears with pantyhose.
- The USDA recently showed petunia leaves contain a natural repellant.
- Spray light or superior horticultural oil (2–3 percent solution). Bt added to the oil spray will increase its effectiveness.
- **Biological controls:** Minute pirate bugs. Lacewings. Trichogramma wasps and Tachinid flies lay eggs in the moth eggs and prevent hatching. Inject beneficial nematodes into infested ears; they seek out and kill worms in 24 hours. Bt can be applied to borers before they move into the stalks. Then wettable Bt, applied every 10–14 days, is effective.
- **Botanical controls:** Ryania, rotenone (1% solution), pyrethrin. Use pheromone traps to identify moth flight paths before spraying. Spray moths before they lay eggs.
- **Allies:** Corn, marigold, and soybean. (See chart on pages 420–437.)

Corn Maggot

also known as Seedcorn Maggot

Description: Small (¼"), yellow-white larvae, with long heads tapering to point, tunnel into larger vegetable seeds. The adult is a small gray-brown fly that lays eggs in April through May in the soil and on seeds and seedlings. Produces 3–5 generations per year.

Plants Affected: Bean, corn (sweet),* pea

Where It Occurs: Widespread. Worst injury occurs early in cold, wet soil that is high in organic matter.

** Plant most frequently attacked by the insect.*

Signs
- Damaged seeds that fail to sprout.
- Poor, stunted plants.

Organic Remedies
- Plant seeds in shallow furrows to speed emergence.
- Delay planting until soil is warm; avoid planting early in soils high in organic matter and manure.
- If damage is heavy, replant immediately. Seeds will germinate before the next generation of adults emerges.
- **Allies**: Rye. (See chart on pages 420–437.)

Corn Rootworm (Northern and Western)

Description
- Adult yellow-green beetles are small (¼") and emerge in late July and August. They feed on corn silks and other plants. They lay eggs in the ground in late summer near corn roots.
- Eggs hatch in late spring. Small (½"), narrow, wrinkled white worms with brown heads feed only on corn and burrow through corn roots.
- Produces 1 generation per year.

Plants Affected: Corn (sweet)

Where It Occurs: The western variety is active in the upper Midwest, and east to Pennsylvania and Maryland. The northern variety is active in New York to Kansas and South Dakota.

Signs: Weak plants with damaged silks and brown tunnels in the roots.

Organic Remedies: Crop rotation; if beetles are found feeding on silks then plant sweet corn in a different location next year.

Corn Rootworm (Southern)

see Cucumber Beetles (Spotted)

Cucumber Beetles (striped/spotted)

Spotted *also known as* Southern Corn Rootworm

Description
- Small (⅓"), thin white larvae with brown heads and brown ends feed for 2–6 weeks on roots and underground stems. Heavy larvae populations can reduce plant vigor and damage melon rind surfaces next to the ground.
- Adults of both types are small (¼") with black heads and yellow or yellow-green backs.
- Striped beetles have three black stripes down the back. Spotted beetles have eleven to twelve black spots scattered across the back.
- Both types lay yellow-orange eggs in the soil near host plants. Beetles tend to congregate on one leaf or plant, so it may be possible to remove a selected plant and destroy large quantities.
- Midsummer adults feed on upper plant parts, while autumn adults feed on fruits, then weeds and trees.
- Produces 1 generation per year in cold areas; 2 generations in warm climates.

Plants Affected: Asparagus, beans (early),* corn (sweet),* cucumber,* eggplant, muskmelon,* pea, potato, pumpkin,* squash,* tomato, watermelon,* some fruit trees

** Plant most frequently attacked by the insect.*

Where It Occurs: Widespread east of the Rockies, but most serious in the South and where soils are heavy.

Signs

- Striped adults are the most destructive and eat mostly cucurbits.
- Both types eat stems and leaves of cucurbits before the first true leaves emerge. Spotted beetles also eat flowers and fruit. Adults transmit bacterial wilt of cucurbits, brown rot in stone fruit, cucumber mosaic, and wilt. Larvae feed on the root system.
- Break a cucumber plant stem, put it back together, pull it back apart and see if strings (looking like pizza cheese) form. If so, it has bacterial wilt. This test is not so easy with other cucurbits.

Organic Remedies

- Use row covers from the time of sowing or transplanting to bloom time. Lift edges during bloom time for 2 hours in early morning — just twice per week — to allow pollination; secure edges again until harvest.
- Vacuum adults with a handheld vacuum at dusk. Empty into a plastic bag immediately, or they'll crawl out.
- Select resistant varieties.
- Transplant strong seedlings.
- Fall cultivation.
- Circle plants with a 3" to 4" wide trench, 3" deep. Fill with wood ashes, moistened to prevent blowing — use experimentally; see page 348. Don't get ashes on plants.
- Handpick beetles.
- Apply thick mulch.
- Lime is thought to dry out these beetles. Apply one of the following directly on beetles: soap and lime spray, lime dust (hydrated or plasterer's lime), or a spray containing equal amounts of wood ashes and hydrated lime mixed in water. Soap alone isn't very effective against hard-bodied insects.
- Spray of hot pepper and garlic — use experimentally; see pages 319, 320–321.
- Sprinkle onion skins over plants — use experimentally; see page 348.
- Time plantings to avoid bugs.
- Plant zucchini or yellow squash to trap early infestations — use experimentally; see page 348. Studies found that beetles feed intensely on the squash (*Cucurbita maxima*) cultivar "NK530," so this may be a particularly good trap plant.
- Encourage songbirds.
- **Biological controls:** Lacewings and ladybugs eat eggs. To kill adults, apply beneficial nematodes in the seed furrows, around roots, and in mulch.
- **Botanical controls:** Sabadilla, rotenone (1% solution), pyrethrin.
- **Allies:** Broccoli, catnip, corn, goldenrod, marigold, nasturtium, onion skins, radish, rue, tansy. (See chart on pages 420–437.)

Cutworm

Description

- Large (½") soft-bodied larvae are gray or brownish with bristles. They curl into a circle when disturbed. They feed at night for several weeks and burrow into the soil during the day. The last generation overwinters as naked brown pupae in the soil.
- Adults are night-flying moths with ragged blotches like paint drips on their

wings. They lay egg masses on leaves, tree trunks, fences, and buildings. Eggs hatch in 2–10 days.

Plants Affected: Beans, brassicas, corn (sweet), cucumber, eggplant, lettuce, melon (seedlings), pea, peanut, pepper, potato, radish, tomato, all seedlings

Where It Occurs: Widespread

Signs

- Severed stems, straight across, at or below soil surface.
- Plants wilt and collapse.
- Buds, leaves, and fruit may also be eaten by the variegated cutworm.

Organic Remedies

- Stem collars.
- Plant transplants inside ½-gallon milk cartons, with bottoms cut out and rims emerging about 1" out of the soil.
- Diatomaceous earth sprinkled around the base of each plant and worked slightly into the soil, is very effective.
- Sprinkle cornmeal or bran, ½ teaspoon per plant, around stem in a circle leading away from the stem — use experimentally; see page 348. Worms eat this and die.
- Make a trap of equal parts of sawdust (pine is best) and bran, add molasses and a little water. At dusk, sprinkle several spoonfuls near plants. The sticky goo clings to their bodies and dries, making them food for prey — use experimentally; see page 348.
- Create barriers by making a 3"- to 4"-wide trench, 2"–3" deep; fill with wood ashes (moistened to prevent blowing) crushed egg shells, oak leaves, or cornmeal or bran (½ teaspoon per plant).
- Fall cultivation. Following fall cultivation, allow chickens into the garden to clean out exposed pests.
- Handpick at night with light.
- Encourage birds and toads.
- University of British Columbia student Greg Salloum claims cutworms will starve before eating plants treated with extracts of pineapple weed or sage brush.
- **Biological controls:** Apply beneficial nematodes at a rate of 50,000 per plant. Apply Bt Lacewing, braconid, and trichogramma wasps. Tachinid flies.
- **Allies:** Shepherd's purse and tansy. (See chart on pages 420–437.)

Deer

Description: Large quadruped mammal.

Plants Affected: Azalea, fruit trees, holly, juniper, rose, saplings, vegetables

Where It Occurs: Widespread

Signs: Deer have two large and two small toes on each foot, a distinct print.

Organic Remedies

- Fences: Serious deer fences need to be electrified with seven strands, at 8", 16", 24", 40", 50", and 60".
  - An alternative is to use a one-wire electric fence, but only if you bait the fence. All electrified fences work best if baited. Bait with attached pieces of aluminum foil smeared with peanut butter; this attracts deer, shocks them, and trains them to stay away.
  - An effective nonelectric fence is two 4'

high fences spaced 5' apart, with bare ground between. Deer aren't broad jumpers and realize that, once they get into the middle section, there isn't enough room to clear the second fence.

– An alternate nonelectric is a fence inclined horizontally. Research at the Institute of Ecosystem Studies, New York Botanical Garden, in Millbrook, New York, shows that deer won't jump a horizontal fence.

– An Extension Horticulturist from the University of Georgia Cooperative Extension Service said that a fence of fishing line with flagging tape at every 2- to 3-foot interval will startle and deter deer.

- Tie bags with 1 ounce of human or dog hair, dried blood meal, or fish heads, to orchard trees and along perimeters of melon and sweet potato patches — use experimentally; see page 348. This allegedly provides protection for 10 months.
- Make a spray of 2 egg yolks in 1 quart of water; spray fruit tree foliage — use experimentally; see page 348. One farmer claims trees weren't bothered the entire season.
- Lay chicken wire squares wherever you find deer droppings; deer will avoid the area — use experimentally; see page 348.
- Hanging soap in outer tree limbs is reported by some to work — use experimentally; see page 348. Some suggest Camay, some Safeguard, some say any brand will work. Be aware, however, that groundhogs love soap.
- Hinder, a deer-repellent, reputedly works when sprayed on leaves and branches. Commercial deer repellant was found unreliable in intense browsing areas in a Virginia study, however.
- Maintain dogs to keep deer away.
- Use automated blinking lights at night to keep deer away — use experimentally; see page 348.
- Old, smelly shoes are reputed to be effective deterrents when placed around the garden perimeter — use experimentally; see page 348.
- Plant a border of plants that deer won't eat, such as Fritillaria.
- For trees, translucent tubes apparently accelerate growth of the tree seedlings while protecting them from rabbits and deer, although some trees are sensitive to leaf scorch and overheating. England has produced the Netlon Shelter Guard, a polyethylene mesh bonded to a double-skin translucent plastic film laminate.
- Repel, a spray-on liquid containing Bitrex, a bitter-tasting substance, is claimed to repel domestic pets and wild mammals, including deer.

Diamondback Moth

Description: Small (⅓"), green-yellow larva with black hairs chews leaves. When disturbed, it wriggles and drops to the ground. Gray-brown adult moth (¾") has fringed back wings, with a diamond that shows when the wings are at rest. Usually a minor pest.

Plants Affected: Brassicas

Where It Occurs: Widespread

Signs: Small holes in outer leaves.

Organic Remedies
- Apply soap and lime spray directly on worms; before harvest, spray 3 days in a row to kill new worms.
- Southernwood is an herbal repellant — use experimentally; see page 348.
- **Biological controls:** Lacewings and Trichogramma wasps.
- **Botanical controls:** Sabadilla, rotenone, and pyrethrum work against larvae.
- **Allies:** Cabbage, tomato. (See chart on pages 420–437.)

Earwig

Description
- Nocturnal, slender, brown, beetle-like insect (¾") with sharp pincers at its tail. It usually crawls but can fly if it takes off from a high place. Hides under and in things during the day.
- As a beneficial, it can scavenge larvae, slow-moving bugs, and aphids.
- As a pest, it feeds on soft plant tissue such as foliage, flowers, and corn silks.
- It can be particularly damaging to seedlings.

Plants Affected: Bean (seedling), beet (seedling), cabbage (Chinese), celery,* corn (sweet),* flowers, lettuce, potato,* strawberry*

Where It Occurs: Widespread, but primarily a pest in the West, particularly the San Francisco Bay Area and northern California.

Signs: Round holes in the middle of leaves.

Organic Remedies
- Probably best left alone in the areas where they're not a serious problem.
- In areas where they're a serious problem, trap and kill. Good trap locations are moist, tight areas where they spend the day. Try rolled-up moistened newspapers, rolls of moistened cardboard, or bamboo. Collect and dispose of pests in the morning.
- Purdue University entomologists are exploring spraying away the diamondback moth with overhead irrigation as a method for reducing damage to cabbage. See if it works for you.
- If very bad, use commercial earwig bait.

Eelworm

see Nematode

European Apple Sawfly

Description
- Small wasplike adult flies are brown and yellow with two pairs of transparent wings hooked together. They emerge at blossom time.
- Larvae are white worms with seven forelegs that bore into fruit and leave chocolate-colored sawdust on the fruit surface. Worms then drop to the ground, where they pupate through the winter in the ground.
- Produces 1 generation per year.

Plants Affected: Apple, pear, plum

Where It Occurs: Present in Connecticut, Massachusetts, New Jersey, New York, Rhode Island.

** Plant most frequently attacked by the insect.*

Signs

- Premature fruit drop.
- Brown scars on fruit skin.

Organic Remedies

- Clean up fallen fruit and destroy larvae by placing them in a sealed black bag in a sunny location.
- **Botanical controls:** Ryania. Rotenone (1% solution), applied at petal fall and again 1 week later.

European Corn Borer

also known as Corn Borer

Description

- The gray-pink or flesh-colored caterpillar (1") has brown spots on each segment and a dark brown head.
- The adult, night-flying, yellowish moth (½") has dark wavy bands across its wings and lays clumps of white eggs on leaf undersides. Eggs hatch in up to 1 week.
- Larvae overwinter in corn stubble.
- Pupae are reddish brown grubs.
- Bores into ears and feeds on kernels at both tips and butt ends.
- Bores into the stem and fruit parts of pepper, potato, and green beans.
- Can attack the stems, foliage, and fruits of more than 260 different plants.
- Produces 1–3 generations per year, depending on the climate.
- Extremely destructive.

Plants Affected: Bean (green), chard, corn (sweet),* pepper, potato, tomato

Where It Occurs: Widespread, except in the Southwest and far West.

Signs

- Broken tassels and bent stalks.
- Sawdust castings outside small holes.

To Monitor

- Use blacklight traps to monitor populations before spraying moths.
- Catches of more than 5 moths per night warrant sprays applied every 4–5 days, starting at full tassel.

Organic Remedies

- Remove and destroy all plant debris.
- Fall cultivation destroys larvae. Plants must be destroyed and turned under at least 1" beneath the soil line.
- Avoid early plantings, which are more susceptible to larvae attack.
- Handpick by slitting damaged tassels and removing borer.
- Plant resistant varieties. Small-stemmed, early season types are less tolerant of borer injury.
- Cover ears with pantyhose.
- Encourage predators such as toads, downy woodpeckers, phoebes, swallows.
- **Biological controls:** Trichogramma wasps parasitize eggs. Braconid wasps. Tachinid flies. Ladybugs and lacewings eat the eggs. Release Trichogramma when adults are first caught in a monitoring trap. Bt is also effective.
- **Botanical controls:** Ryania, sabadilla, rotenone (1% solution), pyrethrin.
- Apply Bt and botanical controls at tassel emergence if moth activity is high. Repeat sprays every 4–5 days until silks turn brown.
- **Allies:** Clover, peanuts. (See chart on pages 420–437.)

** Plant most frequently attacked by the insect.*

European Red Mite

see Mite

Fall and Spring Armyworm

Description

- The spring species is a large (1½") tan, brown, or green caterpillar found early in the season in the whorl leaves of corn.
- The fall armyworms have three light yellow hairline stripes from head to tail, and on each side a dark stripe below which is a wavy yellow stripe marked with red. Heads have a prominent V or Y.
- Both species usually feed on cloudy days and at night, but the fall species also feeds during the day.
- The adult gray moth of the spring species has one yellow or white spot on each forewing. Each lays up to 2000 eggs on corn and grasses. Eggs hatch in 7 days, and the cycle continues. Further generations feed through late summer.
- Produces 3–6 generations per year.

Plants Affected: Bean, cabbage, corn (sweet)*

Where It Occurs: The spring species is widespread, but most serious in warm climates, where there is more generation turnover.

The fall species overwinters in the South and migrates to Northern states every year in June and July.

Signs: Chewed leaves, stems, and buds. The fall species also bores into ears and feeds on kernels; it can be very destructive to late plantings.

Organic Remedies

- Encourage natural predators such as birds, toads, and ground beetles. Skunks also prey on these, but you may not want to attract them.
- In case of a serious problem, dig a steep trench around the garden — use experimentally; see page 348. Armyworms will be trapped inside and can be destroyed by putting them in boiling water or in water laced with insecticidal soap (don't use kerosene — it is toxic to the soil and difficult to dispose of safely).
- Alternate rows of corn with sunflowers to discourage population movement — use experimentally; see page 348.
- **Biological controls:** Lacewings, ladybugs, and other insect predators eat eggs and young larvae. The new Hh strain of beneficial nematodes (*Neoplectana carpocapsae*) is effective against the larval stage. Ichneumon and braconid wasps. Tachinid flies. Bt is effective against the larvae.

Fall Webworm

Description

- This yellowish caterpillar (1¼") has long, light brown or whitish hairs, a dark stripe down its back, and is dotted with small black spots. It builds gray cocoons in secluded sites, fences, bark, or garden debris. White or green eggs are layered in clusters of 200–500 on leaf undersides.
- Adult white moths, with black or brown spotted wings, emerge in spring and late summer.
- Produces 2 generations per year.

* *Plant most frequently attacked by the insect.*

Plants Affected: Fruit trees, pecan

Where It Occurs: Widespread

Signs

- Large silken tents or nests on ends of branches.
- Tents include foliage, unlike tentworms.
- Very damaging.
- Trees can be stripped and die.

Organic Remedies

- Cut off and destroy nests. Burn where permitted. Otherwise destroy nests by putting them in boiling water or water laced with insecticidal soap (don't use kerosene — it is toxic to the soil and difficult to dispose of safely).
- Pick off leaves with eggs and destroy.
- **Biological controls:** Trichogramma wasps eat eggs. Bt is also effective.
- **Botanical controls:** Neem.

Filbert Mud Mite

Description

- Tiny white arachnids (0.3 mm) difficult to see without a magnifying lens.
- Adults invade new buds of current year's shoots, which become deformed as early as September.
- Adults breed in fall and winter, and in spring females leave deformed buds to invade young leaves and lay eggs.

Plants Affected: Filbert

Where It Occurs: Widespread

Signs

- Buds swell through summer and fall to 2–3 times their normal size. They will be easily recognized by December, with thousands of mites feeding per bud.
- Swollen buds open, dry, and drop prematurely in spring.
- Yields are seriously diminished.
- Tiny mites migrate to newly developing buds before old buds fall in spring.

Organic Remedies

- Plant resistant cultivars. This is the best control.
- Early-spring aphid foliar sprays may help.
- You might try pruning out infested buds in early winter.

Filbert Weevil

also known as Hazelnut Weevil

Description

- Small (¼"–⅜") adult beetle is light brownish-yellow with a long snout about half the length of its body. It emerges May to June and lays eggs in late June or early July in the green shell.
- White larvae feed first on the green shell then on the kernel. They exit through small holes and drop to the ground, where they overwinter 2"–8" below the soil surface.

Plants Affected: Filbert, western oak

Where It Occurs: Throughout United States.

Signs

- Deformed green shells.
- Hollowed out kernels and shells.

Organic Remedies: Similar to chestnut weevil and plum curculio (see pages 359 and 396, respectively).

- **Botanical controls:** Neem.

Filbertworm

Description

- Small (½"-long) adult moth varies in appearance. Its forewings are reddish-brown, with a broad coppery band down the center.
- Larvae (½") are white with yellowish heads. They feed on nut kernels. Larvae overwinter in cocoons on the ground and pupate in spring.

Plants Affected: Almond, chestnut, filbert, oak, Persian walnut

Where It Occurs: Throughout the United States.

Signs: Small holes in the chestnut.

Organic Remedies

- Early harvest of nuts and immediate destruction of infested nuts.
- See your Extension agent for further help.

Flathead Borer

also known as Flathead Appletree Borer

Description

- Adults (½") are dark bronze beetles with a metallic sheen and are usually found on the warm side of trees, where they lay yellow, wrinkled eggs in cracks in the bark of unhealthy or injured trees. May and June are the worst times.
- Eggs develop into long (1½") yellow-white, U-shaped larvae that have swollen areas just in back of their heads. They bore tunnels into the tree.
- Produces 1 generation per year.

Plants Affected: Many trees: apple, ash, beech, boxelder, dogwood, hickory, maple, oak, pecan, fruit trees

Where It Occurs: Widespread, but particularly in the South and Midwest. Oak is a prime target in the West; maple and fruit trees are prime targets in the East.

Signs

- Sunken areas in the bark, which indicate feeding tunnels, are filled with a dry powdery substance known as frass, a mixture of droppings and sawdust.
- Minor foliage damage.
- The bark turns dark and may exude sap.
- Sunny sides of the tree are attacked most.

Organic Remedies

- Shade the trunks of young trees with some kind of shield.
- Protect newly transplanted trees by wrapping trunks with burlap or cardboard from the soil level up to the lower branches. Or, cover with a thick coat of white exterior latex paint.
- Keep young trees pruned to a low profile.
- Seal wounds with tree paint.
- Encourage predators such as crows, wasps, woodpeckers, predatory beetles, and vireos.

Flea Beetle

many species; also known as Corn Flea Beetle, Potato Flea Beetle

Description

- These tiny (1/6") dark brown or black beetles jump when disturbed. They can have white or yellow markings.
- Adults lay eggs in the soil and favor the sun. The larvae feed on plant roots and can also damage tubers.
- Adults transmit Stewart's Wilt to sweet corn and Early Blight to potatoes.
- Flea beetles stop feeding and hide in wet weather.
- Usually produce 2 generations per year.

Plants Affected: Bean, brassicas, corn, eggplant, lettuce, muskmelon, pepper, potato, radish, spinach, sweet potato, tomato, watermelon

Where It Occurs: Widespread.

To Monitor: Use white sticky traps for an early warning device.

Signs

- Numerous small holes in leaves in early summer.
- Worst after mild winters and cool, wet springs.

Organic Remedies

- Use row covers. (Pantyhose can be used over small cabbages.)
- Plant as late as possible.
- Seed thickly until the danger of infestation is past.
- Attach sticky bands around the base of wintering plants.
- Use yellow sticky traps for control.
- Frequent fall and spring cultivation will expose eggs to predators.
- Sprinkle wood ashes (moistened to prevent blowing) around the plant base; or, mix equal parts of wood ashes and lime in small containers and place around the plants — use experimentally; see page 348.
- A dusting of diatomaceous earth dries up beetles.
- Vacuum with handheld vacuum. Empty immediately into plastic bag.
- Plant after a trap crop of radish or pak choi — use experimentally; see page 348.
- Sprinkle crushed elderberry or tomato leaves on vulnerable plants — use experimentally; see page 348.
- Mulch with chopped clover — use experimentally; see page 348.
- Companion plant with cover crops of clover or annual ryegrass to reduce populations — use experimentally; see page 348.
- Interplant with shading crops — use experimentally; see page 348.
- Hot pepper or garlic spray — use experimentally; see pages 319, 320–321.
- **Biological controls**: Apply beneficial nematodes in mulch or seed furrows.
- **Botanical controls**: Sabadilla, rotenone (1% solution), pyrethrin.
- **Allies**: Candytuft, catnip, mint, shepherd's purse, tansy, tomato. (See chart on pages 420–437.)

Fruit Tree Leaf Roller

see Leaf Roller

Gall Wasp

Description

- The small (⅛") adult wasp emerges in late May-early June and lays eggs inside buds.
- Eggs hatch in late July and larvae begin to grow.
- Larvae overwinter inside the bud.
- Produces 1 generation per year.

Plants Affected: Chestnut

Where It Occurs: Prevalent in the Southeast, particularly Georgia.

Signs

- Vegetative buds and shoot growth are hindered by galls.
- Buds are turned into ⅓"–½" rose-colored balls that often hang onto the branch for several years.
- Buds may have some parts of leaf or stem growth.
- Trees lose vigor and may die.

Organic Remedies: Prune and destroy infested shoots.

Garden Symphylan or Centipede

Description

- Small (½"), white worm-like creature with twelve pairs of legs thrives in damp soil, leaf mold, and manure piles. It feeds on root hairs.
- After harvest, it burrows down 12" into the soil, where it lays clusters of white spherical eggs.
- This pest is not a true centipede, which is beneficial and grows to 3" long.

Plants Affected: Asparagus, cucumber, lettuce, radish, tomato

Where It Occurs: Warm climates and greenhouses. Particularly damaging to asparagus in California.

Signs

- Stunted plants that slowly die.
- Destroyed root hairs.

Organic Remedies

- Inspect soil. Two or more of these pests per shovel means you can expect damage. Avoid planting where you find damaging populations.
- Pasteurize potting soil.
- Solarize outdoor growing beds, where possible.
- In California, flood asparagus field for 3 weeks in late December to early January, to a depth of 1–3 feet.

Garden Webworm

Description

- Tan-brown moths (1"), with gray markings, appear in spring and lay clusters of eggs on the leaves of host plants.
- Caterpillars (1") are greenish with small dark spots and hairy. They hatch, feed, and spin webs for shelter. When disturbed, they drop to the ground or hide inside silken tubular shelters on the ground. They feed for about 1 month before pupating. The last generation overwinters in the soil in the pupal phase.

Plants Affected: Bean,* beet, corn (sweet),* pea, strawberry*

** Plant most frequently attacked by the insect.*

Where It Occurs: Widespread, but only serious in parts of the South and Midwest.

Signs

- Holes in leaves and stems.
- Folded leaves held together with fine webs.
- Defoliation.
- Produces several generations per year.

Organic Remedies

- Cut off and destroy webbed branches.
- Handpick caterpillars.
- Remove infested leaves and stems.
- Crush worms inside silken tubes on the ground.
- **Botanical controls:** Dust with rotenone as a last resort, or use a rotenone/pyrethrin mix; neem.

Gopher and Groundhog

Description: Gophers eat the roots of vegetables, fruit trees, and grasses. They also dine on bulbs and tubers. Unlike groundhogs, gophers can't climb fences and, unlike the solitary mole, they invade in numbers.

Plants Affected: Flowers, fruit trees, vegetables

Where It Occurs: Widespread

Signs

- Burrows and chewed vegetation.
- Fan or crescent-shaped mounds of dirt next to holes.

Organic Remedies

- Trapping or shooting are the only sure controls. Determine all openings to the tunnel system, then fumigate.
- For gophers, install ¼" hardware cloth so that it extends 2' down into the soil and 1' above the ground.
- Gophers don't like scilla bulbs (also known as squills) — use experimentally; see page 348. These spring flowering bulbs are minimal care and are best for borders or rock gardens.
- Plant the animal's favorite foods (e.g., alfalfa and clover) at a distance — use experimentally; see page 348.
- Place empty narrow-mouth bottles in a hole; wind vibrating in the bottles repels gophers, groundhogs, and moles — use experimentally; see page 348.
- Dog manure placed in animals' holes is reported to drive them away — use experimentally; see page 348.
- Hinder is a product reputed to repel groundhogs. Spray on foliage, borders, and animal paths.
- Sprinkle human hair throughout the garden (you can get a plentiful supply from your local hairdresser) — use experimentally; see page 348. This is unsightly but found to be effective by some.

Grape Berry Moth

Description

- Adult moths emerge in late spring and lay flat round eggs on stems, flowers, and grapes.
- Larvae pupate in cocoons attached to bark, in debris, or in fallen leaves.
- Produces 2 generations per year.

Plants Affected: Grape

Where It Occurs: Prevalent in northeastern states, west to Wisconsin and Nebraska, south to Louisiana and Alabama.

Signs: Young grapes are webbed together, fail to mature, brown, and fall to the ground.

Organic Remedies

- One month before harvest (late summer), hoe around grape vines. Create a wide, flat ridge to seed with a winter cover crop.
- Turn under any cocoons on the soil surface. Water well to compact the soil and seal in the cocoons, where they'll be smothered.

Grasshopper

also known as Locust

Description: The large (1"–2") brown, gray, yellow, or green heavy-shelled adult has large hind legs, big jaws, and antennae. It lays eggs in weeds or soil. Grasshoppers can pollute well water and reservoirs. Larvae overwinter in the soil.

Plants Affected: All plants

Where It Occurs: Primarily in grassland areas, particularly in dry seasons.

Signs

- Chewed leaves and stems.
- Defoliation.

Organic Remedies

- Repeated fall cultivation.
- Use row covers.
- Encourage natural predators like birds, cats, chickens, field mice, skunks, snakes, spiders, squirrels, and toads. Also, blister beetle larvae prey on grasshopper eggs, so unless they're a pest themselves, leave these alone. If numerous, the larvae can eat up to 40–60 percent of the area's grasshopper eggs.
- Fill a jar with a mixture of molasses and water. Bury it up to its mouth in the soil; clean and refresh as needed — use experimentally; see page 348.
- Spray insecticidal soap, mixed with beneficial nematodes, directly on the grasshopper. Apply in evening hours. Soap alone is not very effective against hard-bodied insects — you can also try mixing in hot peppers.
- **Biological controls:** Preying mantis. *Nosema lucustae,* a beneficial protozoan often sold as Grasshopper Attack, controls most grasshopper species. It must be applied in early spring — before grasshoppers grow to more than ¾" — or they won't eat enough for it to be effective. It lasts several years, and its effects are greatest the summer following its application.
- **Botanical controls:** Neem.

Greenhouse Leaftier

see Celery Leaftier

Gypsy Moth

Description

- This hairy gray or brown caterpillar (1⁄16"–2") has five pairs of blue and six pairs of red spots. Larvae crawl to top of trees and dangle on silky threads to be blown to another tree. When more mature (2½") they hide during the day and feed at night.

- The adult female white moth doesn't fly, but lays tan eggs in 1" long masses on trunks, branches, and under rocks. The adult male moth is large (1½") and brown. Adults live less than 2 weeks.
- Larvae pupate and overwinter in dark cocoons or tie themselves to a branch with silken threads.
- Produces 1 generation per year.

Plants Affected: Apple, apricot, basswood, birch, linden, oak, peach, pear, willow

Where It Occurs: Prevalent primarily in the east, but moving south and west. Worst attacks follow a dry fall and warm spring.

Signs

- Rapid defoliation.
- Masses of worms feeding.
- Large holes in leaves.
- No tents.
- Defoliation in two successive years may kill deciduous trees.

Organic Remedies

- **August to April:** To destroy egg masses, paint with creosote or drop into water laced with insecticidal soap. (Don't use kerosene — it is toxic to the soil and difficult to dispose of safely.)
- **Late April to early June:** Attach a 12"-wide burlap strip to a tree, about chest height, by draping it over a string. Caterpillars will hide under the cloth during the day. Every afternoon, using gloves, sweep worms into soapy water. Apply sticky bands; remove them in mid-July.
- **Fall:** Check lawn furniture, woodpiles, walls, and outbuildings for egg masses. Also check all vehicles for egg masses to avoid transporting eggs to new areas. Destroy all egg masses.
- Use gypsy moth pheromone traps to monitor and control populations.
- **Biological controls:** Bt is effective when applied every 10–14 days starting in April and continuing through mid-June until the caterpillars are 1" long. Trichogramma and chalcid wasps provide limited control. Lacewings, tachinid flies, predaceous ground beetles, white-footed mice can all help. Also try *Glyptapanteles flavicoxis* and *Cotesia melanoscelus*. Use beneficial nematodes at the tree base to prevent migration.
- **Botanical controls:** Ryania, pyrethrin, neem.

Harlequin Bug

also known as Cabbage Bug, Calicoback, Terrapin Fire Bug

Description

- This small (¼"), flat and shield-shaped bug has a shiny black and red-orange back. It sucks leaf juices and smells bad.
- The adult female lays two neat rows of black-ringed white eggs on leaf undersides. Eggs hatch in 4–7 days.
- Nymphs suck leaf juices, causing leaf blotches.
- The adult overwinters in cabbage stalks.
- Very destructive.
- Produces several generations per year.

Plants Affected: Brassicas,* eggplant, radish

Where It Occurs: Appears in southern half of the United States, from coast to coast.

** Plant most frequently attacked by the insect.*

Signs

- Wilting plants, especially seedlings.
- Yellowish or black spots on leaves.
- White blotches.

Organic Remedies

- Handpick adults and eggs and destroy.
- Insecticidal soap. Lace the spray with isopropyl alcohol, which helps it penetrate the shells of hard-bodied insects. Soap alone is not very effective against hard-bodied insects.
- Fall cultivation.
- Mustard greens or turnips can be used as trap crops — use experimentally; see page 348.
- In spring, place old cabbage leaves in the garden to attract the bugs; destroy them when collected — use experimentally; see page 348.
- **Biological controls**: Praying mantis might help.
- **Botanical controls**: Sabadilla, rotenone (5% solution), pyrethrin.

Hickory Shuckworm

Description

- Large (⅔") caterpillars are cream-colored with brownish heads. They overwinter in shucks on the ground or in the tree.
- Small (½") dark brown-black adult moths emerge when nuts begin to develop and continue emerging throughout the summer. They lay eggs on foliage and nuts.
- Very damaging.
- Produces 1–4 generations per year.

Plants Affected: Hickory, pecan

Where It Occurs: Prevalent in eastern Canada and United States, south to Florida, west to Missouri, Oklahoma, and Texas.

Signs: Premature nut drop, poor-quality kernels, dark-stained spots on shells.

Organic Remedies

- Blacklight traps can reduce populations (one per three trees).
- Collect all prematurely dropped nuts; at harvest, collect all shucks. If legal, burn. Otherwise, destroy larvae by dropping shucks and nuts in water laced with insecticidal soap or boiling water. (Don't use kerosene — it is toxic to the soil and difficult to dispose of.)
- **Biological controls**: Trichogramma wasp — release 2–3 times per season (mid-April; 2 weeks later; and again 2 weeks following).

Imported Cabbageworm, Moth, and Butterfly

Description

- This large (1¼"), velvety smooth worm, light to bright green, has one yellow stripe down its back. It feeds on foliage and pupates by suspending itself by silken threads from plants or objects.
- The adult (2") is a day-flying, white to pale yellow butterfly, with grayish tips and 3–4 black spots on each wing. Butterflies drop hundreds of single, light yellow-green eggs on leaf undersides, which hatch in 4–8 days.
- Produces 3–6 generations per year.

Plants Affected: Brassicas,* radish

Where It Occurs: Widespread.

Signs

- Huge ragged holes in leaves with bits of green excrement.
- Tunnels inside broccoli, cabbage, and cauliflower heads.

Organic Remedies

- Use row covers all season. Also try nylon stockings over cabbage heads. Nylon stretches, allowing sun, air, and water through but keeping the butterfly out.
- Sprinkle damp leaves with rye flour; worms eat it, bloat, and die — use experimentally; see page 348.
- Handpick worms in early morning. Handpick eggs off the undersides of leaves every few days.
- Fall cultivation; repeated again in the spring.
- Any green mulch deters this moth.
- Make a viral insecticide (see directions under Organic Remedies for Cabbage Looper on page 353).
- Use butterfly nets to catch moths; each one caught means 200–300 worms destroyed.
- Encourage bluebirds, chickadees, English sparrows.
- **Biological controls:** Bt, applied every 10–14 days until heads form, is effective. Lacewings. Trichogramma and braconid wasps.
- **Botanical controls:** Sabadilla.
- **Allies:** Celery, dill, garlic, hyssop, mint, onion, rosemary, sage, tansy, thyme, tomato, and more. (See chart on pages 420–437.)

Inchworm

see Cankerworm

Japanese Beetle

Description

- This medium-large (½") beetle is shiny metallic green with copper-brown wings. The adult lays eggs in the soil and eats and flies only during the day, often up to 5 miles.
- The grub, similar to our native white grub, is gray-white with a dark brown head and two rows of spines. It's smaller, about 1", and lies curled in the soil. It overwinters deep in the soil.
- Produces 1 generation per year.

Plants Affected: Asparagus, basil, bean,* corn (sweet),* grape,* grasses, okra, onion, peach, potato, raspberry, rhubarb, rose,* tomato; most fruit trees*

Where It Occurs: Mostly in the eastern United States, but this beetle is slowly moving farther west. Eradication near Sacramento, California, was attempted in the early 1960s.

Signs

- Lacy, skeletonized leaves.
- Beetles also feed on fruit and corn silk.
- Grubs feed on grass roots.

Organic Remedies

- Handpick in early morning by shaking tree limbs or branches. Catch them on a

** Plant most frequently attacked by the insect.*

sheet spread on ground. Drop bugs into water laced with insecticidal soap. (Don't use kerosene — it is toxic to the soil and difficult to dispose of.)
• High soil pH discourages grubs.
• Make a bait of water, sugar, mashed fruit, and yeast. Place it at least 1' off the ground in sunny spots on the periphery of the garden, not in the middle. Strain out beetles every evening — use experimentally; see page 348.
• Use commercially available yellow pheromone traps. Place at a distance from the crop so that existing beetles won't be attracted to the crop and new beetles from other areas won't be drawn in.
• Encourage starlings, the only bird that eats adult beetles. Other birds eat the grubs.
• Time plantings to avoid beetle peak.
• Four-o'clocks *(Mirabilis)*, larkspur, and geraniums all poison the beetle. These are good interplantings.
• Try trap crops of African marigold, borage, evening primrose, four-o'clocks, soybeans, white roses, white and pastel zinnias — use experimentally; see page 348.
• **Biological controls:** Apply beneficial nematodes at a rate of 50,000 per square foot of lawn to prevent chewing damage. *Bacillus popilliae,* or milky spore disease, can be applied to lawns and orchard grasses. It attacks grubs and is effective for 15–20 years after just one application.
• **Botanical controls:** Rotenone (5% solution), pyrethrin, neem.
• **Allies:** Catnip, chives, garlic, rue, and tansy. (See chart on pages 420–437.)

June Beetle

also known as May Beetle

Description: Night-flying adults are large brown beetles. They emerge in May or June. They lay eggs in midsummer (see White Grub). White grubs, slightly larger than the Japanese beetle grub, feed for 2 years before pupating. Has 1 emergence per year.

Plants Affected: Corn (sweet), potato, strawberry

Where It Occurs: Causes serious damage in the midwestern and southern United States.

Signs

• Damaged leaves of berry plants (adult feeding).
• Sudden wilting, especially in May or June, and roots and underground stems are chewed or severed (grub feeding).

Organic Remedies

• Fall and spring cultivation.
• Don't plant susceptible crops on areas just converted from untilled sod, as grubs will come up through the grass.
• Allow chickens to pick over garden following fall and spring cultivations.
• Encourage birds.
• **Biological controls:** Beneficial nematodes in spring or early summer, as mulch or dressing. Milky spore disease, which takes a few years to spread enough to be effective.
• Handpick beetles and drop into water laced with insecticidal soap. (Don't use kerosene — it is toxic to the soil and diffi-

cult to dispose.)

- Shake beetles from the tree or shrub early in the morning while they're sluggish, letting them fall onto a sheet beneath; collect and destroy beetles.
- **Botanical controls:** Rotenone (5% solution).

Lace Bug

many species; also known as Eggplant Lace Bug

Description

- Nymphs and adults suck plant juices from leaves and stems. They feed in groups on leaf undersides.
- Small adults lay black eggs on leaf undersides in either fall or spring. Eggs hatch in spring. Brown-black nymphs start feeding immediately.
- Adults overwinter in garden trash.

Plants Affected: Eggplant, potato, tomato

Where It Occurs: Prevalent in the southern half of the United States, from coast to coast.

Signs

- Pale, discolored (bronzed), and curled leaves.
- Plants may die.
- Leaves may be dotted with dark, shiny droppings.

Organic Remedies

- Check for eggs every 7–10 days and destroy all egg clusters.
- Insecticidal soap spray.

Leaf Beetle

see Bean Leaf Beetle

Leaf-footed Bug

Description

- This large (¾") bug resembles the squash bug, but its hind legs are expanded and look like leaves. It is dark brown with a yellow band across its body. When handled it emits a distinctive odor.
- Adults lay barrel-shaped eggs on leaves of host plants. Nymphs look like adults.
- Adults hibernate in winter, especially in thistle, and emerge in early summer.
- Produces 1 generation per year.

Plants Affected: Almond, bean,* nuts, potato,* tomato*

Where It Occurs: Widespread, but primarily a problem in the South and west to Arizona.

Signs

- Chewed leaves in vegetables.
- In almond, there may be poorly developed misshapen nuts or premature nut drop.

Organic Remedies

- Handpick and destroy bugs.
- Handpick and destroy eggs.
- **Botanical control:** Sabadilla.

Leafhopper

many species; Beet Leafhopper *is also known as* Whitefly *in the West*

Description

- Small (¼"–⅓"), green, brown, or yellow

** Plant most frequently attacked by the insect.*

slender bugs suck juices from leaves, stems, and buds. Nymphs move sideways.

- The Beet Leafhopper (⅓") is yellow-green, jumps quickly into the air, and looks like a Whitefly.
- The Potato Leafhopper (⅓") is green with white spots.
- The Six-spotted Leafhopper (⅛") is yellowish with six black spots.
- Adults lay eggs in early spring on leaf undersides. The second generation of eggs, 2 weeks later, may be laid in plant stems. Adults overwinter in garden trash and weeds.
- Produces 1–5 generations per year.

Plants Affected: Aster, bean, beet, carrot, celery, chard, citrus, corn, eggplant, fruit trees, grape, lettuce, potato, raspberry, rhubarb, rose, spinach, squash, tomato

Where It Occurs: Widespread. For specific regions, see Beet Leafhoppers (page 351) and Potato Leafhoppers (page 397).

Signs

- White or yellow mottled, curled leaves die and drop.
- Excreted honeydew attracts ants and supports black sooty mold.
- Leafhoppers transmit Curly Top, where mature leaves roll upward, turn yellow with purple veins, and become stiff and brittle.
- Especially damaging in potatoes due to decreased yields.
- Nymphs suck plant juices.

Organic Remedies

- Use row covers in early spring.
- Avoid planting, if possible, in wide open space.
- Insecticidal soap spray. If the infestation is bad, add isopropyl alcohol to the spray.
- Sprinkle diatomaceous earth or wood ashes, moistened to prevent blowing, around the plant base. Both will dry out leafhoppers.
- Remove weeds and afflicted plants.
- Fall cultivation.
- A reflective mulch such as aluminum foil will repel leafhoppers.
- Keep beets and spinach far from tomatoes. Also avoid planting carrots, asters, and lettuce together if your garden has a problem with the six-spotted leafhopper.
- Plant resistant varieties.
- Black lights trap adults — use experimentally; see page 348.
- Boil 1 pound tobacco in 1 gallon of water; strain; and use as a spray — use experimentally; see page 348.
- Encourage songbirds.
- **Biological controls:** Lacewings eat the eggs.
- **Botanical controls:** Sabadilla, rotenone (5% solution), pyrethrin, neem.
- **Allies:** Bean, blackberry, clover, goosegrass, red sprangletop, rye, and vetch. Geraniums and petunias allegedly repel leafhoppers. (See chart on pages 420–437.)

Leaf Miner

many species; symptoms of the spinach species are given here

Description

- Larvae (⅓") are pale green or whitish.

They mine between the upper and lower edges of leaves, causing a scorched, blotched, and blistered appearance.

- Adults are tiny (1⁄6"–1⁄4") black or gray flies that lay eggs on leaf undersides. Eggs are tiny, white, and lined up in groups of four to five. If hatched, the leaf will have a grayish blister.
- Produces several generations per year.

Plants Affected: Bean, beet greens, blackberry, cabbage, chard, chestnut, lamb's quarters, lettuce, oregano, pepper, radish, spinach, Swiss chard, turnip

Where It Occurs: Widespread

Signs

- White-brown tunnels or blotches on leaves.
- Yellowed, blistered, or curled leaves.
- Stem damage below the soil surface.
- Leafminers are disease vectors for black leg and soft rot.

Organic Remedies

- Use row covers.
- Cut out the infested parts of leaves with grayish blisters.
- Remove all lamb's quarters unless used as trap crop.
- Handpick and destroy eggs.
- Plant fall crops to avoid the insect.
- Rotate crops.
- Apply superior horticultural oil.
- Encourage chickadee, purple finch, robin.
- Controls are usually not warranted on the chestnut.
- **Biological controls:** Beneficial nematodes give some control. Ladybugs and lacewings may eat leafminer eggs.
- **Botanical controls:** Neem.

Leaf Roller

also known as Fruit Tree Leaf Roller

Description

- Small (3⁄4") green caterpillars usually have dark heads. They feed for about 4 weeks, then spin webs around leaves and sometimes around the fruit. They pupate in these rolled leaves.
- The adult is a light brown moth that lays eggs with a camouflage coating in clusters of 30–100 in midsummer on twigs and bark. Eggs overwinter and hatch into caterpillars in spring.
- Produces 1 generation per year.

Plants Affected: Apple; other fruit trees

Where It Occurs: Widespread, but most damage occurs in northern United States and southern Canada.

To Monitor: Use pheromone traps to monitor moth populations and to know when to spray to prevent egg-laying.

Signs

- Rolled leaves with fine webbing that holds the leaves shut.
- Chewed fruit, leaves, and buds.

Organic Remedies

- Handpick eggs in the winter.
- Spray light or superior horticultural oil in early spring before buds appear.
- Garlic Barrier, a product containing only garlic oil and water, is supposed to repel leaf rollers.

- **Predators:** Trichogramma wasps will eat the caterpillars.
- **Biological controls:** Dipel Bt is an effective control; be sure to spray inside the rolled leaf.
- **Botanical controls:** Rotenone (5% solution).

Locust

see Grasshopper

Mealy Bug

Description

- These minuscule bugs suck plant sap.
- They lay tiny, yellow, smooth eggs where leaves join the stem.
- Produces numerous generations (each cycle takes 1 month).

Where It Occurs: Widespread, but particularly a problem in warm climates.

Plants Affected: Fruit trees, greenhouses, houseplants, rosemary, vegetables

Signs

- Cotton tufts on the leaf underside.
- Honeydew excretions attract ants and support black sooty mold.
- Dwarfed plants.
- Wilt.
- Premature fruit drop.

Organic Remedies

- Direct strong water jets at the undersides of leaves.
- Destroy mealybugs with cotton swabs dipped in alcohol.
- Spray light or superior horticultural oil before buds appear, to smother eggs.
- Insecticidal soap spray, especially in the early spring dormant stage.
- Use sticky bands to trap ants.
- **Biological controls:** Green lacewings. *Crytolaemus* ladybug (Australian and uncommon) is a predator and also works in the greenhouse. Pauridia parasites prey on bugs. In the greenhouse, PFR (*Paecilomyces fumosoroseus*) is a beneficial microorganism available in granule form that attacks whiteflies, mites, aphids, thrips, mealybugs, and certain other greenhouse pests, and is claimed to be safe for humans and beneficials.
- **Botanical controls:** Neem.

Measuring Worm

see Cankerworm

Mexican Bean Beetle

Description

- The small (¼") copper, round-backed adult beetle, with sixteen black spots in three horizontal rows on its back, looks like an orange ladybug.
- The female adult lays orange-yellow eggs in groups of 40–60 on leaves. Eggs hatch in 5–14 days.
- Most beans can tolerate 10–20 percent defoliation (more prior to bloom) without loss in yields.
- Produces 2 generations per year in cold climates, and 3–5 in warmer regions.

Plants Affected: Bean,* kale, squash

* *Plant most frequently attacked by the insect.*

Where It Occurs: Widespread, but particularly a problem in the East and Southwest.

Signs

- Lacy, skeletonized leaves. Pods and stems are eaten in bad infestations.
- Orange-yellow fuzzy larvae (⅓") are longer than the adult and attach themselves to leaves, usually on the underside, or inside a curled leaf.

Organic Remedies

- Handpick and destroy beetles, larvae, and egg clusters.
- Use row covers.
- A reflective mulch like aluminum foil will repel them.
- A spray of crushed turnips with corn oil — use experimentally; see page 348.
- Spray cedar sawdust or chips boiled in water — use experimentally; see page 348.
- **Fall:** Pull up infested plants as soon as the main harvest is over but while pests are still present. Stuff vines in a plastic bag, tie, and leave in the sun for 10–14 days to kill the bugs.
- Fall cultivation.
- Early planting.
- Encourage insect-eating birds.
- **Biological controls:** Spined soldier bug is excellent control. Predatory mites. Pediobius foveolatus wasps parasitize larvae; they are excellent controls, but expensive.
- **Botanical controls:** Rotenone (1% solution), pyrethrins, neem.
- **Allies:** Garlic, marigold, nasturtium, petunia, potatoes, rosemary, and savory. (See chart on pages 420–437.)

Millipede

Description

- This caterpillar-like worm (½"–1") has a hard-shelled body divided into multiple segments and 30–400 pairs of legs.
- It moves slowly by contracting and stretching, feeds at night on decaying vegetation, and sometimes transmits diseases.
- It lays sticky sacs of hundreds of eggs on or in the soil in summer.
- Adults live 1 to 7 years.
- Produces 1 generation per year.

Plants Affected: Lettuce, root vegetables, rose, seeds

Where It Occurs: Widespread, but most damaging in the South and West.

Signs

- Ragged holes in stems and roots, especially seedlings.
- Fungal disease may be present.

Organic Remedies

- Peat compost is more hostile to millipedes than other types such as leaf or manure compost.
- Place window-screen wire under plants.

Mite

many species

Description

- Mites are very tiny red, green, yellow, black, or brown arachnids that are difficult to see without a magnifying lens. Some are beneficial. Females lay numerous eggs on webbing under leaves. Mites overwinter in the soil.

• Mites suck chlorophyll out of plants and inject toxins. They can lower chlorophyll by as much as 35 percent. They are worst in hot, dry conditions.
• Produce up to 17 generations per year; each life cycle takes 7–14 days.

Plants Affected: Apple, apricot, asparagus, bean, blackberry, blueberry, brassicas, celery, chestnut, cucumber, eggplant, grape, herbs, muskmelon, peach, peanut, pepper, raspberry, strawberry

Where It Occurs: Widespread. Many types exist only in specific locales, but general characteristics remain the same.

To Monitor: To detect, hold a white paper underneath and tap the leaves to see if mites, the size of salt grains, are dislodged.

Signs

• Yellowed, dry leaves, with yellow or red spots or blotches and small white dots.
• Veins yellow or turn reddish-brown first.
• Fine webbing between leaves and across undersides.
• Poorly developed fruit that drops early.

Organic Remedies

• Spray forceful jets of water in early morning, 3 days in a row, or every other day (three times).
• Spray insecticidal soap at least three times, every 5–7 days.
• Mix ¼ pound of glue in a gallon of water; let stand overnight. Spray on twigs and leaves — use experimentally; see page 348. When dried it will flake off, taking trapped mites with it. Spray three times, every 7–10 days.
• Vacuum plants with handheld vacuum. Empty bag immediately into plastic bag or mites will crawl back out.
• **Fruit trees:** High nitrogen fertilizers can increase mite populations, so avoid them.
• Spray fruit trees with horticultural oil late in the dormant period, right at bud break, when mite eggs are most vulnerable.
• Cooking oil spray mix, as recommended by USDA.
• Garlic Barrier, a product containing only garlic oil and water, is supposed to repel spider mites.
• Ensure adequate water.
• Cinnamon oil sprays may be highly effective for controlling mites, aphids, and even fungi that attack greenhouse-grown plants with minimal concerns about health and safety. The Mycogen Corporation recently registered a product called Cinnamite.
• **Biological controls:** Predatory mites are good outdoors and in greenhouses. Green lacewings and ladybugs feed on the mites. In the greenhouse, PFR (*Paecilomyces fumosoroseus*) is a beneficial microorganism available in granule form that attacks whiteflies, mites, aphids, thrips, mealybugs, and certain other greenhouse pests, and is claimed to be safe for humans and beneficials.
• **Botanical controls:** Sabadilla, pyrethrins (apply twice, 3–4 days apart).
• **Allies:** Alder, bramble berries, coriander, dill, rye mulch, sorghum mulch, wheat mulch. (See chart on pages 420–437.)

Mole

Description

- Moles are solitary, unlike gophers.
- They tunnel extensively, often using each tunnel only once.
- They eat grubs and beetles, which isn't bad, but also feed on earthworms, their favorite food.
- Their tunnels can harm the root systems of young plants.

Plants Affected: Many vegetables

Where It Occurs: Appear throughout the United States but are a major problem in the West.

Signs

- Extensive tunnels.
- Main runways are usually 6"–10" below ground with frequent mounds of soil heaped above ground.

Organic Remedies

- The best control is traps. Set out traps at the first sign of tunnels. Tunneling can occur anytime in most western states. In cooler climates, tunneling occurs in spring. Determine which runs are active before setting traps, then set the traps, mark the spots, and check again in 2 days. If spots are raised, the tunnel is active. Spear or harpoon-type traps are considered the easiest to use.
- Plant 4 or 5 poisonous castor bean plants (*Euphobia lathyris*), also known as *mole plants*, nearby — use experimentally; see page 348. (*Note:* This plant can become a pest itself in areas of the West and South. It is also poisonous to humans.) Mole-Med contains castor oil as its active ingredient.
- Scatter red pepper or tobacco dust to repel moles — use experimentally; see page 348.
- Wearing gloves to prevent human scent, place Juicy Fruit gum in tunnels. Moles are alleged to love it, eat it, and die from it. Use experimentally; see page 348.
- Control grubs, food for moles.
- Plant favorite foods (e.g., alfalfa and clover) at a distance.
- A New York gardener reports that kitty litter placed in main tunnel openings and along the borders of areas is an effective way to get rid of moles. Use experimentally; see page 348.

Mouse

many species, including Voles

Description

- The house mouse is uniformly gray.
- The vole, or field mouse, is silver-bellied with gray-brown fur on top.
- The deer mouse, or white-footed mouse, resembles the vole but has white feet.
- Mice will move into mole tunnels and feed on crop roots.

Plants Affected: Apple, fruit trees, greenhouses, onion (storage), strawberry

Where It Occurs: Widespread

Signs

- Chewed tree trunks at ground level.
- Gnawed roots (by pine mice).
- In strawberry beds, you may find nests in mulch and destroyed roots.

Organic Remedies

- At planting, install hardware cloth girdles 6" in diameter and 18" high around tree trunks. Set them at least 6" into the soil, preferably in coarse gravel.
- Remove protective cover for mice by removing all vegetation within a 3' radius of trunk. In winter, pull mulch at least 6" away from tree trunks. Keep grass mowed.
- For strawberries, wait to mulch until mice have made winter homes elsewhere, when the ground has developed a frosty crust.
- Make gravel barriers around garden plots, at least 6"–8" deep and 12" or more wide. This prevents rodents from tunneling and, if kept free of weeds, from crossing the area.
- VoleBloc, a lightweight, porous, sharp-edged gravel, is claimed to protect bulbs and roots against voles when it is placed in planting holes. The same gravel is marketed as PermaTill for aerating heavy soil.
- Encourage owls and snakes.
- Don't mulch perennials until a few frosts have occurred.
- Sprinkle mint leaves in garden. Use experimentally; see page 348.
- **Allies:** Wormwood. (See chart on pages 420–437.)

Navel Orangeworm

Description: Worms are yellow or dark gray with dark heads. They pupate in cocoons within the fruit. The gray adult moth has crescent-shaped dots along its outer margins.

Plants Affected: Almond, citrus,* walnut

Where It Occurs: Prevalent in Southwestern United States and California.

Signs: Worms burrow into fruit and nuts on the tree and in storage.

Organic Remedies

- Maintain good orchard sanitation.
- Early harvest.
- Nuts can be fumigated before storage with methyl bromide.
- Remove and destroy nuts left on the tree in winter.
- **Biological controls:** *Goniozus legneri* and *Pentalitomastix plethoricus* both parasitize the pupae. Release them in early spring or after harvest.

Nematode

many species; also known as Roundworm, Eelworm

Description: Nematodes are blind and usually microscopic. Not all are harmful; some are beneficial (see page 316).

Plants Affected: Apple, bean, carrot, celery, cucumber, dill, eggplant (in South), garlic, mustard, okra, onion, parsley, pea, potato, raspberry, strawberry, sweet potato, tomato; many others

Where It Occurs: Widespread, especially in warm areas and areas with sandy or loamy soils. Not very common in clay soils. The eggs of some nematode species remain viable in soil for years.

Signs

- Malformed flowers, leaves, stems, and roots.

* *Plant most frequently attacked by the insect.*

• Stunted, yellowing leaves.
• Leaves may wilt during the day.
• Dwarfed plants, with poorly developed roots, leaves, and flowers.
• Dieback.
• Root-knot nematodes cause galls, knots, or branched root crops.
• Lesion nematodes cause root fungal infections and lesions on roots.

Organic Remedies

• Plant resistant varieties.
• Solarize soil.
• Increase soil organic matter. Heavy mulch or compost is one of the best deterrents. Compost is host to saprophytic nematodes and predacious fungi that destroy harmful nematodes. Compost also releases fatty acids toxic to nematodes. Leaf mold compost is especially effective, particularly pine needles, rye, and timothy grasses.
• Disinfect tools used in infected soil.
• Long crop rotation.
• Pent-A-Vane, a fertilizer mixture of 70 percent fish emulsion and 30 percent yucca extract, reduces nematodes.
• Kelp meal and crab shell meal (chitin, sold as a *Chitosan*) stimulate beneficial fungi that prey on nematodes. Dig these into the soil 1 month before planting to reduce populations.
• Sprinkle an emulsion of 1 part corn oil to 10 parts water — use experimentally; see page 348.
• Research in India indicates that spraying a solution containing boric acid (concentration not specified) on the leaves of container-grown okra plants that are infested with root-knot nematodes (*Meloidogyne incognita*) results in better growth than that seen in untreated controls.
• A solution of water hyacinth leaves or flowers macerated in water (1:3 on a weight basis) has been shown to kill nematodes (1989 study by botanists in India). Also, tomato and eggplant roots soaked in this solution for 80 minutes prior to planting grew three times faster than unsoaked controls.
• **Botanical controls:** A nematicide (DMDP) derived from the Costa Rican tropical tree, *Lonchocarpus costariecensis,* will probably be the first commercially available botanical effective against plant-pathogenic nematodes. British Technology Group is cooperating with a conservation organization in Costa Rica to develop methods for extracting DMDP.
• Laboratory trials have revealed root-knot nematode is highly sensitive to the essential oils derived from several plants in concentrations of 1,000 micrograms per liter of water (plus a small amount of ethanol). The following oils caused immobilization of juvenile root-knot nematodes within two days of exposure (oils marked with asterisks also reduced hatching of nematode eggs to under 5%): *Artemisia judaica,** *Carum carvi* (caraway),* *Coridothymus capitatus* (wild thyme), *Cymbopogon citratus* (lemongrass), *Foeniculum vulgare* (fennel),* *Mentha rotundifolia* (applemint),* *Mentha spicata* (spearmint),* *Micromeria fruticosa,** *Origanum syriacum* (Syrian oregano),* *O. vulgare.*
• **Allies:** Asparagus, barley, corn, garlic, hairy indigo, marigold (*Tagetes* sp.),

** Plant most frequently attacked by the insect.*

mustard (white or black). See chart on pages 420–437. (*Note:* Research shows that marigolds suppress root lesion nematodes for 3 years when planted in the entire infested area for a full season. Spot plantings are not as effective and may reduce yields of nearby crops.)

- Also, 'Commodore' radishes grown as a green manure have been shown to greatly reduce root-knot nematode populations in potato plots. The radishes can be sown in spring, before potatoes are planted, or in fall, following potato harvest, and tilled into the soil after a couple of months.
- Cover crops of barley, castor bean, corn, cotton, joint vetch, millet, rye, sesame, or wheat reduce populations. All plants must be turned under to be effective. Winter rye, when tilled under in spring, produces an organic acid toxic to nematodes. (Note: The only rye that can control nematodes is cereal rye [*Secale cereale*].)
- **Companion planting:** Louisiana Agricultural Experiment Station showed that planting cucumbers (for which no nematode-resistant cultivar has yet been developed) with nematode-resistant tomatoes such as 'Celebrity' had significantly higher yields than the same cucumber grown with nematode-susceptible tomatoes.
- White and black mustard exude an oil hostile to nematodes.
- Plant tomatoes near asparagus; asparagus roots are toxic to tomato nematodes.

Nut Curculio

Description

- The small (3⁄16") adult is black with reddish-brown blotches and a long, curved snout of curculios.
- It emerges in May and June, occasionally feeds on foliage, and deposits eggs when the burr cracks to expose the nut.
- Larvae feed on nut kernels for about 3 weeks, then emerge through small (3⁄16") holes.
- Produces 1 generation per year.

Plants Affected: Chestnut, oak

Where It Occurs: Southeast

Signs

- Premature nut drop.
- Circular cavities in nuts and shells.

Organic Remedies

- Hang white sticky traps (8"×10") in trees at chest height for both monitoring and control. Use several per tree. Remove after 3–4 weeks.
- Starting at blossom time, every day in the early morning spread a sheet or tarp under the tree and knock branches with a padded board or pole. Shake collected beetles into a bucket of water laced with insecticidal soap. (Don't use kerosene — it is toxic to the soil and difficult to dispose.)
- Remove all diseased and fallen fruit immediately. Destroy larvae by burning fruit or burying it in the middle of a hot compost pile.
- Keep trees pruned; curculios dislike direct sun.
- Encourage the chickadee, bluebird, and purple martin. Domestic fowl will also eat these insects.
- Drop infected nuts into water to kill larvae.

Onion Eelworm

see Nematode

Onion Fly Maggot

also known as Onion Maggot

Description

• Small (⅓"), legless, white worm that tapers to a point at the head. It feeds on stems and bulbs.
• The adult fly resembles a hairy housefly. It lays eggs at the base of plants, near the bulb or neck, or in the bulb.
• Produces 3 generations per year; the last generation attacks harvest and storage onions.

Plants Affected: Onion,* radish

Where It Occurs: Particularly prevalent in northern and coastal regions, where cool, wet weather abounds.

Signs

• Rotting bulbs in storage.
• Destroyed seedlings.
• Faded and wilted leaves.
• Lower stems of onion near the bulb are damaged or destroyed.
• Worse damage than that caused by the cabbage maggot.

Organic Remedies

• Avoid close spacing and planting in rows. This discourages maggot movement between plants.
• Red onions are the least vulnerable, followed by yellow, and then white varieties.
• Sprinkle diatomaceous earth around the plant base and work it slightly into the soil. If this isn't available, then sprinkle wood ashes (moistened to prevent blowing) near the base of the plants.
• Don't store damaged bulbs.
• Encourage robins and starlings.
• **Biological controls:** Beneficial nematodes.

Oriental Fruit Moth

Description

• These small larvae (½") are yellow-pink with brown heads and are very active.
• When small, they tunnel into tender shoots and later enter fruit through the stem end. They don't tunnel to the core of apples, but in peaches they feed close to the pit. On the outside, fruit may not look damaged.
• Larvae emerge from fruit and pupate in silken cocoons attached to tree trunks, weeds, or garden debris.
• The adult moth (½") is gray, and lays white eggs on leaves and twigs.
• Several generations per year.

Plants Affected: Apple,* peach,* pear, plum, quince

Where It Occurs: Prevalent east of the Mississippi and in the upper Northwest.

To Monitor: Pheromone traps are available for early detection. Captures exceeding 5 to 10 moths per trap per week warrant control action.

Signs

• Wormy fruit.
• Terminals of rapidly growing shoots wilt, turn brown in a few days, and die.

** Plant most frequently attacked by the insect.*

• Gummy exudate; holes in fruit and in fruit stems.

Organic Remedies

• If this is a consistent pest, plant early-ripening cultivars to starve the last generation.
• Prune trees annually to avoid dense growth.
• Plant early maturing varieties of peach and apricot.
• Inspect tree trunks and destroy all cocoons.
• Pheromone-based mating disruption lures are commercially available but may not be effective in small orchards.
• **Biological controls**: Import the Braconid wasp *Macrocentrus ancylivorus* and release according to instructions from supplier.
• **Botanical controls**: Ryania, rotenone, pyrethrins.
• **Allies**: Goldenrod, lamb's quarters, strawberries. (See chart on pages 420–437.)

Parsleyworm

also known as Carrotworm, Celeryworm, Black Swallowtail

Description

• This long (2"), stunning green worm has a yellow-dotted black band across each segment. It emits a sweet odor and projects two orange horns when disturbed.
• The adult is the familiar black swallowtail butterfly. Its large black forewings (3"–4" across) have two rows of parallel yellow spots. Its rear wings have a blue row of spots with one orange spot.
• Adults lay white eggs on leaves; they hatch in 10 days.
• Produces several generations per year.

Plants Affected: Carrot,* celeriac, celery, dill, parsley,* parsnip

Where It Occurs: East of the Rockies, and a similar species west of the Rockies.

Signs

• Chewed leaves, often down to bare stems.
• Damage is usually minor because of low populations.

Organic Remedies

• Handpick in early morning.
• Encourage songbirds.
• **Biological controls**: Lacewing larvae, parasitic wasps.

Peach Tree Borer (Greater)

Description

• Large (1¼"), yellow-white larvae have dark brown heads. They feed under the bark at or below the soil surface all winter. The highest they'll usually go is about 12" above the soil.
• Adult female moths (1") are blue, with a wide orange stripe around the abdomen and look like wasps; the male is gray with yellow markings and clear wings. They emerge in the North in July and August, and in the South in August and September. They lay brown-gray eggs at the trunk base in late summer to fall.
• Produce 1 generation per year.

* *Plant most frequently attacked by the insect.*

Plants Affected: Apricot, cherry, nectarine, peach,* plum

Where It Occurs: Appear where vulnerable plants are grown, particularly in eastern states and the lower half of the United States from coast to coast.

Signs

- Brown gummy sawdust (frass) on the bark, usually near the ground.
- Very damaging.
- Trees can die.

Organic Remedies

- Insert stiff wires into holes to kill larvae. Do *not* remove gummy exudate, which helps to seal wounds.
- Use sticky bands, from 2" below the soil line to 6" above. Destroy and replace the bands each week (see page 324).
- Encourage birds.
- **Tobacco dust ring:** Encircle the trunk with a piece of tin, 2" away from the bark. In mid-May fill with tobacco dust. Repeat ever year.
- Tie soft soap around the trunk from the soil line up to the crotch; soap drips repel moths and larvae — use experimentally; see page 348.
- In the near future, pheromone dispensers should be available. These confuse males and stop mating but may not be effective in small orchards.
- **Biological controls:** Spray or inject beneficial nematodes into the holes. Spray Bt every 10 days or syringe it into the holes.
- **Allies:** Garlic. (See chart on pages 420–437.)

Peach Tree Borer (Lesser)

Description:

- Larvae look the same as Greater Peach Tree Borer.
- Adult female and male moths are blue with clear wings and yellow markings.
- Life cycle is same as Greater Borer, except the Lesser Borer lays eggs higher in the limbs of the tree in rough bark or cracks.
- Complete about 1½ life cycles per year, slightly more than the Greater Borer.
- Produces 2–3 generations per year.

Plants Affected: Apricot, cherry, nectarine, peach,* plum

Where It Occurs: Appear where vulnerable plants are grown, particularly in eastern states and the lower half of the United States from coast to coast.

Signs

- Brown gummy sawdust (frass) on the bark, usually in upper limbs.
- Invade through wounds created by such things as winter injury, pruning, and cankers.

Organic Remedies

- Insert stiff wires into holes to kill larvae. Do *not* remove gummy exudate, which helps to seal wounds.
- Use sticky bands on affected limbs. Destroy and replace the bands each week.
- **Tobacco dust ring:** Encircle the trunk with a piece of tin, 2" away from the bark. In mid-May fill with tobacco dust. Repeat ever year.

* *Plant most frequently attacked by the insect.*

- Encourage birds.
- **Biological controls:** Spray or inject beneficial nematodes into the holes. Spray Bt every 10 days or syringe it into the holes.
- **Allies:** Garlic. (See chart on pages 420–437.)
- Take measures to minimize winter injury (see Sunscald on pages 345–346), pruning wounds, or other entry sites.

Peach Twig Borer

Description: Small (less than ½") red-brown larvae construct cocoons under curled edges of bark. The adult is a small (½"), steel-gray moth. Produces 3–4 generations per year.

Plants Affected: Almond, apricot, peach,* plum

Where It Occurs: Widespread, but very damaging on the West Coast.

Signs

- Red-brown masses of chewed bark in twig crotches.
- Infested fruit late in the season.

Organic Remedies

- Increase organic matter in soil.
- Take measures to minimize winter injury (see Sunscald on pages 345–346), pruning wounds, or other entry sites.
- Insert stiff wires into holes to kill larvae. Do not remove gummy exudate, which helps to seal wounds.
- Tie soft soap around the trunk from the soil line up to the crotch; soap drips repel moths and larvae — use experimentally; see page 348.
- In the near future, pheromone dispensers should be available. These confuse males and stop mating but may not be effective in small orchards.
- Encourage birds.
- Pheromone traps are available that catch males (not the same as mating-disruption lures). These can be used to monitor population levels and, in small areas, to help control populations.
- Spray dormant lime-sulfur spray (diluted 1:15) before the pink stage and after petal fall.

Pear Psylla

Description

- Nymphs are very tiny and yellow, green, or light brown. They feed on the top sides of leaves until only the veins remain.
- Adults are tiny (1⁄10"), light brown to dark orange-red, and have clear wings. Before buds open, adults lay tiny yellowish-orange eggs in cracks and crevices, in the base of terminal buds and in old leaf scars. Also check tender growing tips of the highest shoots. Nymphs hatch in 2–4 weeks. Most hatch by petal fall.
- Nymphs and adults suck plant juices, and can transmit fire blight and pear decline.
- Produces 3–5 generations per year.

Plants Affected: Pear

Where It Occurs: Eastern (east of Mississippi) and northwestern states.

Signs

- Yellow leaves.

** Plant most frequently attacked by the insect.*

• Leaf drop.
• Low vigor.
• Honeydew excretion attracts yellow jackets and ants and supports black soot mold. Blackened leaves and fruit.
• Scarred and malformed fruit.
• Very damaging.

Organic Remedies

• Spray insecticidal soap, in high pressure, as soon as females emerge and when leaf buds are just beginning to turn green. Continue through the season as needed.
• Dust tree and leaves with limestone.
• Destroy all debris, which harbors the eggs. Adults and eggs spend winter on or near tree.
• Spray light horticultural oil in fall and again in spring at the green tip stage; continue to spray every 7 days until larvae emerge.
• Plant tolerant varieties. (Bartlett and D'Anjou are the most susceptible.)
• **Botanical controls**: Rotenone (5% solution).

Pearslug

also known as Cherryslug

Description

• Larvae (½") are dark green-orange, covered with slime, tadpole-shaped, and look like small slugs with large heads.
• Larvae feed for 2–3 weeks on upper leaf surfaces. Larvae in apples feed just under fruit skin until about one-third grown, then bore through the fruit. After feeding they drop to the ground to pupate.
• The adult is a small black-and-yellow sawfly, a little larger than a housefly, with two sets of transparent wings hooked together. At bloom time, the adult emerges from a cocoon in the soil. The sawfly inserts its eggs into leaves or into the skin of fruit.
• Produces 2 generations per year.

Plants Affected: Apple, cherry, pear,* plum

Where It Occurs: Widespread

Signs

• Pink-brown patches on upper surface of leaves.
• Lacy skeletonized leaves.
• Defoliation.
• Streaks of chocolate-covered sawdust on apples.

Organic Remedies

• Shallow cultivation, no more than 2" deep, at the tree base right before full bloom. This exposes cocoons to predators.
• Dust trees with wood ashes, slightly moistened to prevent blowing, which will dry out larvae and kill them — use experimentally; see page 348. Wash trees with water after 5 days.
• Pick up fallen fruit every day.
• **Botanical controls**: A pyrethrin/rotenone/ryania blend.

Pecan Casebearer

Description

• This small (½") caterpillar is olive gray to jade green with a yellow-brown head. When buds open, caterpillars feed on buds and shoots.

* *Plant most frequently attacked by the insect.*

• Adult moths emerge when the nut forms. They lay single, light-colored eggs on the blossom end of the nut. Larvae then feed on developing nuts.
• Produces 2 generations in northern climates, 4 generations in southern climates

Plants Affected: Pecan

Where It Occurs: Present wherever pecans are grown.

Signs
• Small cocoons at the base of the bud indicate overwintering larvae.
• Tunneled shoots.
• Signs of eaten nuts in the early maturation phase.
• Trees will experience little damage, but nut yields are reduced.

Organic Remedies
• Blacklight traps (one for every three trees) will reduce populations.
• Collect all prematurely dropped nuts; at harvest time, collect all shucks. If legal, burn. Otherwise, destroy in hot water or water laced with insecticidal soap. (Don't use kerosene — it is toxic to the soil and difficult to dispose.)
• **Biological controls:** Trichogramma wasps — release 2–3 times per season (once in mid-April; again 2 weeks later; and a third time 2 weeks following).

Pecan Weevil

Description
• Similar to the chestnut weevil. The adult (¾") is light brown, and as it ages it may become dark brown. Its snout is needle-thin and can be as long as its body. Adults feed on husks and young nuts.
• Grubs are white with reddish-brown heads. They feed on kernels, emerge, and pupate in "cells" as much as 12" below the soil surface. Adults may not emerge for 2–3 years.
• Has 1 emergence per year.

Plants Affected: Hickory, pecan

Where It Occurs: Present wherever pecan and hickory are grown.

Signs: Premature nut drop and hollowed kernels.

Organic Remedies
• Use same controls as for chestnut weevil and plum curculio (see pages 359 and 396, respectively).
• Burlap bands wrapped around trees. (See Sticky Bands on page 324.)
• **Botanical controls:** Neem

Phylloxera Aphid

see Aphid

Pickleworm

Description
• This medium (¾") caterpillar is green or copperish. The first generation emerges midsummer.
• Damage is worst late in the season when the broods are largest.
• The adult nocturnal moth is yellowish. It lays eggs that develop into small caterpillars that are pale yellow with black dots. These change color as they mature.

• Produces 3–4 generations per year.

Plants Affected: Cucumber, melon, squash

Where It Occurs: Prevalent primarily in the southeast and the Gulf states, particularly Florida and Louisiana.

Signs

• Holes in buds, blossoms, and fruits of most cucurbits.
• Masses of rotting green excrement.

Organic Remedies

• Plant susceptible crops as early as possible to avoid pest.
• Destroy all plant debris after harvest.
• Cultivate the soil deeply in early fall, after harvest.
• Plant resistant varieties.
• **Botanical controls:** Apply rotenone (5% solution) at petal fall and every 10–14 days through September.

Pill Bug

also known as Roly-poly

See Sow Bug, page 404. (Unlike sow bugs, pill bugs roll up into tight balls about the size of a pea.)

Plum Curculio

Description

• This small (¼") dark brown beetle has a long down-curving snout and four humps on its back. It emerges at blossom time when temperatures climb above 70°F. When disturbed, it folds its legs and drops to the ground. It lays eggs in a crescent-shaped fruit wound.
• Grubs are gray-white with brown heads and curled bodies. They feed at the fruit center for 2 weeks and emerge only after fruit falls to ground. They pupate in the ground. Adults emerge in about 1 month.
• Produces 1–2 generations per year.

Plants Affected: Apple, apricot, blueberry, cherry, peach, pear, plum*

Where It Occurs: East of the Rockies.

Signs

• Eaten leaves and petals.
• Crescent-shaped cavities in fruits.
• Sap exudate from apple fruit dries to a white crust.
• Misshapen fruit.
• Premature fruit drop.
• Brown rot disease.

Organic Remedies

• Hang several white sticky traps (8"×10") per tree at chest height for monitoring and control. Remove after 3–4 weeks.
• Starting at blossom time, every day in the early morning spread a sheet or tarp under the tree and knock branches with a padded board or pole. Shake collected beetles into a bucket of water laced with insecticidal soap. (Don't use kerosene — its toxic to the soil and difficult to dispose.)
• Remove all diseased and fallen fruit immediately. Destroy larvae by burning fruit or burying it in the middle of a hot compost pile.
• Keep trees pruned; curculios dislike direct sun.
• Encourage the chickadee, bluebird, and purple martin. Domestic fowl will also eat these insects.

** Plant most frequently attacked by the insect.*

Plume Moth

Description: Larvae bore into the fruit, blemish the scales, and tunnel into the heart. Nocturnal adult brown moths (1") have plumed wings and fly near the plant. They lay eggs on leaf undersides.

Plants Affected: Artichoke

Where It Occurs: Lives on the Pacific and Texas Coasts.

Signs

- Irregular holes in stems, foliage, and bud scales.
- Small worms are on bud scales and new foliage.
- Damage is year-round but worst in the spring.

Organic Remedies

- Pick and destroy all wormy buds.
- Remove and destroy all plant debris in the fall.
- Remove all nearby thistles.
- **Biological controls**: Bt is effective.

Potato Bug

see Colorado Potato Beetle

Potato Flea Beetle

see Flea Beetle

Potato Leafhopper

also known as Bean Jassid. *See also* Leafhopper.

Description

- This leafhopper causes "hopperburn:" a triangular brown spot appears at leaf tips, leaf tips curl, yellow, and become brittle.
- Reduced yields in potatoes.
- Produces 2 generations in mid-Atlantic states.

Plants Affected: Apple,* bean (South), peanut, potato (East and South)*

Where It Occurs: Lives primarily in eastern and southern United States. It migrates south for winter, and returns north in spring to feed first on apples, then on potatoes.

Organic Remedies

- See Leafhopper on page 381.
- Potato leafhoppers are said to be trapped by black fluorescent lamps — use experimentally; see page 348.

Potato Tuberworm

also known as Tuber Moth

Description

- The gray-brown adult moth is small with narrow wings. It lays eggs on the leaf undersides or in tubers.
- Larvae (¾") are pink-white with brown heads. They tunnel into stems and leaves.
- They pupate on the ground in cocoons covered with soil, and can emerge in storage to pupate.
- Produces 5–6 generations per year.

Plants Affected: Eggplant, potato,* tomato

Where It Occurs: Present in the South from coast to coast, and northward to

* *Plant most frequently attacked by the insect.*

Washington, Colorado, Virginia, and Maryland. Worst in hot, dry years.

Signs

- Wilted shoots and stems.
- Dieback.

Organic Remedies

- Plant as early as possible.
- Keep soil well cultivated and deeply tilled.
- Cut and destroy infested vines before harvesting.
- Destroy all infested potatoes.
- Screen storage areas and keep storage area cool and dark. (Darkness discourages moth activity.)

Rabbit

Description: A small white or gray mammal with long ears and short tail that travels primarily by hopping.

Where It Occurs: Widespread

Plants Affected: Bean, carrot, lettuce, pea, strawberry, tulip shoots, bark of fruit trees

Signs: Rabbits eat vegetables, herbs, flowers, and chew on young fruit trees.

Organic Remedies

- Sprinkle any of these around plants: Blood meal, moistened wood ashes, ground hot peppers, chili or garlic powder, crushed mint leaves, talcum powder (use experimentally; see page 348). Replenish frequently, especially after rain. You can also sprinkle black pepper right on the plants, which gives rabbits sneezing fits and keeps them away.
- Cover seedlings with plastic milk jugs that have the bottoms cut out. Anchor them well in the soil, and keep the cap off for ventilation. This can also be used as a season extender in the spring.
- Place old smelly leather shoes on the garden periphery — use experimentally; see page 348.
- Place fake snakes near garden — use experimentally; see page 348.
- Wrap base of fruit trees with hardware cloth.
- Set rabbit traps.
- Encourage owls and sparrow hawks with nest boxes.
- Hinder is reputed to be an effective repellant. Spray on foliage, borders, and animal paths. When painted on bark it prevents tree girdling by rabbits.
- Fine-woven fences are effective deterrents.
- Garlic, marigold, and onion are said to deter rabbits — use experimentally; see page 348.

Raspberry Caneborer

Description

- Medium-sized (½") adult is a long-horned black-and-yellow beetle. It lays one egg between a double row of punctures on the stem near the cane tip.
- A small worm hatches and burrows 1"–2" deep near the base of the cane to hibernate.
- This is a major cane pest.
- Produces 1 generation per year.

** Plant most frequently attacked by the insect.*

Plants Affected: Blackberry, raspberry

Where It Occurs: Present in Kansas and eastward.

Signs
- Sudden wilting of tips.
- Two rows of punctures about 1" apart.

Organic Remedies
- Cut off cane tips 6" below the puncture marks, and burn or destroy.
- **Botanical controls:** Rotenone (1% solution). Apply two treatments 7 days apart, timed to coincide with adult emergence during June. Consult your county Extension agent for the local adult activity period.

Raspberry Root Borer

also known as Raspberry Crown Borer

Description
- Larvae are small, white, and hibernate at the soil level near canes. They tunnel into cane crowns and bases.
- The adult moth has clear wings and a black body with four yellow bands.
- Produces 1 generation.

Plants Affected: Blackberry, raspberry

Where It Occurs: Prevalent in eastern United States.

Signs: Wilting and dying canes in early summer, usually when berries are ripening.

Organic Remedies: Cut out infected canes below the soil line.

Raspberry Sawfly

see Blackberry Sawfly

Rednecked Cane Borer

see Caneborer

Root Fly and Root Maggot

see Corn Maggot *and* Cabbage Maggot

Rose Chafer

also known as Rose Bug

Description
- This slender, adult beetle (⅓"–½") is tan, with a reddish-brown head and long spiny, slightly hairy, legs.
- It emerges in late May to early June and feeds for 3–4 weeks, attacking flowers first, then fruit blossoms and newly set fruit.
- Eggs laid in the soil hatch in 1–2 weeks.
- Larvae (¾") feed on grass roots, then pupate 10"–16" deep in the soil.
- Produces 1 generation annually.

Plants Affected: Grape,* peony, rose,* other fruits and flowers

Where It Occurs: Prevalent east of the Rockies; primarily a pest in sandy soils and north of New York City.

Signs
- Chewed foliage.
- Destroyed grass roots.

Organic Remedies
- Handpick adults.
- Cheesecloth fences, stretched higher than the plants, will deter the beetle as it doesn't fly over barriers.

** Plant most frequently attacked by the insect.*

• Do not allow chickens to clean garden because these beetles are poisonous to chickens.
• Ohio State University research shows that commercially available white traps are most effective with this pest (and no chemical attractants are necessary).

Roundworm

see Nematode

Sap Beetle

many species; also known as Corn Sap Beetle, Strawberry Sap Beetle

Description

• This small (3⁄16"), black, oblong beetle invades ears through the silk channel or via holes in the husk caused by other pests. This is often associated with corn borer or corn earworm damage. A smaller, brown, oval beetle attacks small fruit.
• White, maggot-like larvae eat inside kernels and infested fruit. They scatter when exposed to light.

Plants Affected: Corn, raspberry, strawberry

Where It Occurs: Widespread

Signs

• Brown hollowed-out kernels at the corn ear tip. Sometimes individual damaged kernels are scattered throughout.
• Round cavities eaten straight into ripe strawberries.
• Feeding injury between the raspberry and stem. Injury predisposes small fruit to secondary rot organisms.

Organic Remedies

• Clean and complete harvesting of small fruit and removal of damaged, diseased, and overripe berries helps to reduce populations.
• Deep plowing in fall or early spring reduces overwintering populations of the corn sap beetle.
• Destroy alternate food sources, such as old vegetable crops beyond harvest.
• Mineral oil, applied just inside the tip of each ear, suffocates the worms. Apply only after silk has wilted and started to brown at the tip, or pollination will be incomplete. Use half of a medicine dropper per small ear, and three-fourths of a dropper per large ear. You might add red pepper to the oil to see if it increases the effectiveness. Apply two or more applications of oil, spaced at 2-week intervals.
• Fall cultivation. In the spring, cultivate the top 2 inches of soil.
• Plant resistant varieties with tight husks. Or clip husks tightly with clothespins. You can also try covering ears with pantyhose.
• **Biological controls:** Minute pirate bugs. Lacewings. Trichogramma wasps and Tachinid flies lay eggs in the moth eggs and prevent hatching. Inject beneficial nematodes into infested ears; they seek out and kill worms in 24 hours. Bt can be applied to borers before they move into the stalks. Then wettable Bt, applied every 10–14 days, is effective.
• **Botanical controls:** Ryania, rotenone (1% solution), pyrethrin. Use pheromone traps to identify moth flight paths before spraying. Spray moths before they lay eggs.

- **Allies:** Corn, marigold, and soybean. (See chart on pages 420–437.)

Scale

many species

Description

- Scales are extremely small and suck plant nutrients from bark, leaves, and fruit.
- You may see the insect's "armor," which is either a part of its body or, in some species, is made up of old skeletons and a waxy coating that shields them from attacks. This armor may look like flaky, crusty parts of the bark.
- The immature insect (1⁄16") crawls to a feeding spot, where it stays for the remainder of its life.
- Ants carry scales from plant to plant.
- Produces 1–3 or more generations per year.

Plants Affected: Fruit trees, nut trees, shrubs

Where It Occurs: Different species occur throughout United States.

Signs

- Small spots of reddened tissue on leaves or branches.
- Fine dusty ash.
- Hard bumps on fruit.
- Dead twigs or branches.
- Limbs lose vigor, leaves yellow, and the plant dies, usually from the top down. Honeydew excretions support black fungus and attract ants.

Organic Remedies

- Spray insecticidal soap.
- Apply horticultural oil spray late in spring, right at bud break. One spray should be sufficient.
- Mix ¼ pound of glue in a gallon of water; let stand overnight. Spray on twigs and leaves. (Use experimentally; see page 348.) When dried it will flake off, taking trapped mites with it. Spray three times, every 7–10 days.
- Scrape scale off plants or touch them with an alcohol-soaked cotton swab. Repeat every 3 or 4 days until scale falls off.
- **Biological controls:** Use *Comperiella bifasciata* for yellow scale; the chalcid wasp *A. luteolus* for soft scale in California; *A. melinus* for red scale; the *Metaphycus helvolus* wasp for soft scale or black scale; the *Vedalia* ladybug for cottony cushion scale; and the *Chilococorus nigritis* ladybug for all scales.

Seedcorn Maggot

see Corn Maggot

Skipjack

see Wireworm

Slug (many species) and Snail

Description

- Slugs are large (½"–10" long), slimy wormlike creatures that resemble snails without shells. They are mollusks but have no outer protective shells. Their eyes are at the tips of two tentacles. They come in all colors— brown, gray, purple, black, white, and yellow — and can be spotted. They feed mainly from 2 hours after sundown to

2 hours before sunrise. Females lay clusters of twenty-five oval white eggs in damp soil, which hatch in about 1 month.

• Snails, also in the mollusk family, have hard shells and scrape small holes in foliage as they feed and lay masses of eggs. Eggs are large (⅛") and look like clusters of white pearls. Snails can go dormant in periods of drought or low food supply.

Plants Affected: Artichoke, asparagus, basil, bean, brassicas, celeriac, celery, chard, cucumber, eggplant, greens, lettuce, onion, pea, pepper (seedling), sage, squash, strawberry; most fruit trees

Where It Occurs: Widespread, but they rank as one of the top pests (if not the top one) in the western United States. They thrive in temperatures below 75°F.

Signs

• Large, ragged holes in leaves, fruits, and stems, starting at plant bottom.

• Trails of slime on leaves and soil.

Organic Remedies

• Reduce habitat by removing garden debris, bricks, boards, garden clippings, and weeds. Use mulches that are slug irritants, such as shredded bark, crushed rock, or cinders.

• Make the habitat less conducive to this pest by watering in the morning instead of at night. Swiss researchers found lettuce leaf consumption by slugs with morning watering was less than one-fifth of the consumption with evening watering.

• Cultivation or spading in spring or times of drought will help destroy dormant slugs and eggs.

• A spray of two parts vinegar and one part water can be effective against snails and slugs, but may result in phytotoxicity; a ratio of one-to-one was found to be an effective control with less plant damage.

• Place stale beer in shallow pans, the lip of which must be at ground level — use experimentally; see page 348. Replace every day and after rain, and dispose of dead slugs. Because slugs are attracted to the yeast, an even more effective solution is made by dissolving 1 teaspoon of dry yeast in about ¼ cup of water. Where slugs are too numerous for this solution (e.g., in the West), try the following other traps.

• Use the weed Plantain as a trap crop, but first be sure it will not become a pest in your area. Use experimentally; see page 348.

• Spray with wormwood tea. Use experimentally; see page 348.

• Researchers in India suggest that both garlic powder and raw garlic contain the toxic substance allicin that can kill snails.

• **Board trap:** Set a wide board about 1" off the ground in an infested area. This provides daytime shelter for slugs and easy collection for you.

• **Gutter trap:** Set aluminum gutter around the garden beds and coat them with Ivory soap. The slugs will get trapped in this. Empty and kill them regularly.

• **Seedlings:** Keep seedlings covered until they're 6" high, particularly pea seedlings. Try plastic jugs with the bottoms cut out, anchored firmly in the soil, with the caps off for ventilation. Make sure no slugs are inside.

- **Asparagus:** Plant crowns in wire baskets anchored several inches down in the ground.
- **Barriers for raised beds:**
  - Strips of aluminum screening about 3" high, pushed about 1" into the soil. Bend top of the screen outward, away from bed, and remove two strands from the edge so it's rough.
  - A 2" strip of copper flashing tacked around the outside of beds, 1" from top of bed. This carries a minor electrical charge that repels slugs. Copper bands around tree trunks works similarly. Snail Barr is available commercially.
  - Crushed eggshells around plants.
  - Strips of hardware cloth tacked onto top edges of bed, extending 2" above edge. Make sure there are sharp points along the top edge.
- Sprinkle diatomaceous earth around the plant base and work slightly into the soil. If not available, sprinkle moistened wood ashes around the plant base. Avoid getting ashes on plants.
- An Oregon master gardener devised a moat that reportedly protects her young seedling brassicas. Seedling pots are placed on plywood supported by bricks in a shallow plastic tub filled with soapy water. Use experimentally; see page 348. She reports no more losses.
- Two products (Concern Slug Stop and Slug Out) contain the active ingredient of biodegradable soap derived from coconuts. Slug Stop is formulated as a paste, while Slug Out is sold as flakes; both are intended to create barriers.
- Don't apply mulch until soil has warmed to 75°F, which is warmer than slugs like.
- Destroy all eggs. Find them under rocks, pots, and boards.
- Encourage predators such as birds, ducks, lightning bug larvae, ground beetles, turtles, salamanders, grass, and garter snakes. Ducks have a voracious appetite for these pests.
- Chemists at Texas Southern University report that uscharin, a glycoside compound found in the latex-like sap of *Calotropis procera* (Mudar, native to the Middle East and grown as green houseplant in temperate areas), is extremely toxic to the white garden snail *(Theba pisana)*. Home experimenters can lay parts of this plant around mollusk-susceptible plants.
- Pestoban, a new herbal concoction from India, consists of three medicinal herbs dissolved in a non-ionic emulsifier. Zoologists at the University of Gorakhpur found that extremely low doses of Pestoban were lethal.
- An Oregon recipe for slug bait combines 3 cups of water, 1 tablespoon of granulated yeast, and 2 tablespoons of sugar. Slugs are attracted to the yeast. Put in a pan with the edge at least ½" above the soil surface, to keep beneficial ground beetles out. Use experimentally; see page 348.
- Flowers said to be "distasteful to slugs" are Achillea, Ageratum, Alyssum, Arabis, Armeria, Aster, Astilbe, Calendula, Campanula, Cosmos, Dianthus, Dicentra, Eschscholzia, Galium, Hemerocallis, Iberis, Kniphofia, Lobelia, Mentha,

Nasturtium, Paeonia, Penstemon, Phlox, Portulaca, Potentilla, Ranunculus, Rudbeckia, Saxifraga, Sedum, Thymus, Verbena, Vinca, Viola, Zinnia. Use experimentally; see page 348.
- **Biological controls:** Predatory snail, *Ruminia decollata*, will feed on brown garden snails. They can't be shipped to most places, like northern California or the Northwest, because they kill native snails.
- **Allies:** Fennel, garlic, and rosemary. (See chart on pages 420–437.)

Southern Corn Rootworm

see Cucumber Beetles

Sow Bug

also known as Dooryard Sow Bug

Description

- These bugs are crustaceans. They are small (¼"–½") oval, hump-backed bugs with fourteen legs. Sow bugs scurry when disturbed and generally hide under leaves and debris. The Pill Bug will roll itself into a ball for protection.
- Produces 1 generation per year.

Plants Affected: Seedlings

Where It Occurs: Widespread. Common in gardens but not usually a severe problem.

Signs: Chewed seedlings, stems, and roots.

Organic Remedies

- Apply wood ashes or an oak-leaf mulch.
- Water plants with a very weak lime solution. Mix 2 pounds per 5 gallons, and let sit for 24 hours before using.
- Bait with one-half potato placed cut-side down on the soil surface. In the morning collect and kill bugs. Use experimentally; see page 348.
- Botanical controls: If the problem is severe, rotenone (1% solution).

Spider Mite

see Mite

Spinach Flea Beetle

see Flea Beetle

Spittlebug

many species; also known as Froghopper

Description

- Adults (⅓") are dull brown, gray, or black, sometimes with yellow markings. They hop and look like short, fat leafhoppers.
- They lay eggs in plant stems, and in grasses between the stem and leaf sheath. The foam covers the eggs, which overwinter.
- In spring when nymphs hatch, they produce even more froth for protection.
- Adults and nymphs suck plant juices from leaves and stems.
- Produces 1 or more generations.

Plants Affected: Corn; many others

Where It Occurs: Widespread, but worst in the Northeast, Oregon, and high humidity regions.

Signs

- Foamy masses ("frog spit"), usually at

stem joints.
• Faded, wilted, curled, or discolored leaves.
• Stunted, weakened, and sometimes distorted plants.

Organic Remedies
• Many species are not a serious problem, so you may want to leave them alone.
• If seriously damaging, cut out plant parts with "spittle" or simply remove the foamy mass. Destroy egg masses or nymphs contained inside.

Spotted Asparagus Beetle

Description
• This slender reddish brown-orange beetle has six black spots on each side of its back.
• It lays single greenish eggs on leaves, which develop into orange larvae in 1–2 weeks.
• Larvae bore into the berry and eat seeds and pulp, then pupate in the soil.
• Produces 1 generation per year.

Plants Affected: Asparagus,* aster, cucurbits,* zinnia

Where It Occurs: East of the Mississippi.

Signs
• Defoliation and misshapen fruit.
• It usually appears in July.

Organic Remedies
• Handpick in the morning when the beetle can't fly, due to cool temperatures.
• See Organic Remedies for Asparagus Beetle, page 351.

Squash Bug

Description
• These medium (⅝"), brown-black bugs feed by sucking plant sap and injecting toxins. When crushed at any age they emit a foul odor. They like moist, protected areas and hide in deep, loose mulch like hay or straw, as well as in debris or under boards.
• They lay clusters of yellow, red, or brown eggs on the undersides of leaves along the central vein. Eggs hatch in 7–14 days.
• Young bugs have red heads and bright green bodies that turn to gray as they mature.
• Bugs overwinter under vines, in boards and buildings, and under dead leaves.
• Produces 1 generation per year.

Plants Affected: Cucumber, muskmelon, pumpkin,* squash,* watermelon

Where It Occurs: Widespread

Signs
• Rapidly wilting leaves dry up and then turn black.
• No fruit development.

Organic Remedies
• Use row covers. When female blossoms (fruit blooms) open, lift edges of covers for 2 hours in the early morning, twice a week, until blossoms drop. This permits pollination.
• Handpick bugs.
• Sprinkle a barrier of wood ashes that have been moistened to prevent blowing around the plant base. (Use experimen-

** Plant most frequently attacked by the insect.*

tally; see page 348.) Don't get ashes on the plant. Renew periodically.

- In fall, leave a few immature squash on the ground to attract the remaining bugs. Destroy squash covered with bugs.
- Trellised plants are less susceptible to this bug.
- Pull vulnerable plants as soon as they finish bearing, place in a large plastic bag, tie securely, and place in direct sun for 1 to 2 weeks. This destroys all eggs.
- Spray insecticidal soap, laced with isopropyl alcohol to help it penetrate the shell. Soap, however, is not very effective against hard-bodied insects.
- Time plantings to avoid bugs.
- Rotate crops.
- Use heavy mulch materials. Avoid aluminum and white or black plastic mulch, which increase squash bug populations, according to Oklahoma State University studies.
- Place boards in garden where they will hide and be easy to catch. Use experimentally; see page 348.
- Plant resistant varieties.
- Encourage birds.
- **Biological controls:** Praying mantises eat eggs and nymphs. Tachinid flies are natural predators.
- **Botanical controls**: Sabadilla, rotenone (1% solution).
- **Allies:** Borage, catnip, marigold, mint, nasturtium, radish, and tansy. (See chart on pages 420–437.)

Squash Vine Borer

Description

- This long (1") dirty-white worm, with brown head and legs, bores into stems, where it feeds for 4–6 weeks. It overwinters 1"–2" below the soil surface.
- The adult wasplike moth (1½") has clear copper-green forewings, transparent rear wings, and rings on its abdomen colored red, copper, and black. It lays rows or clusters of individual tiny, longish, brown or red eggs near the plant base on the main stem.
- Produces 1 generation per year in the North, and 2 generations per year in the southern and Gulf states.

Plants Affected: Cucumber, gourd, muskmelon, pumpkin, squash

Where It Occurs: East of the Rockies.

Signs: Sudden wilting of plant parts. Moist yellow sawdustlike material (frass) outside small holes near the plant base.

Organic Remedies

- Use row covers. When female blossoms (fruit blooms) open, lift cover edges for just 2 hours in early morning, twice a week, until blossoms drop. This permits pollination.
- Reflective mulch such as aluminum foil will repel them. Use experimentally; see page 348.
- Stem collars can prevent egg-laying.
- If not grown on a trellis, pinch off the young plant's growing tip to cause multistemming. For trailing squash stems, bury every fifth leaf node to encourage rooting.

When one part becomes infested, cut it off and remove. Leave other sections to grow. Mound soil over vines up to blossoms.

- Slit stem vertically to remove and destroy borers. Mound soil around the slit stem to encourage new rooting. Remove and destroy damaged stems.
- Handpick the eggs.
- Sprinkle wood ashes, crushed black pepper, or real camphor around plants to deter borer. Use experimentally; see page 348.
- Time planting to avoid borer. In the North, plant a second crop in midsummer to avoid borer feeding.
- Plant resistant varieties.
- Remove plants as soon as they finish bearing, place in a plastic bag, tie securely, and place in direct sun for 1–2 weeks to destroy borers and eggs.
- **Biological controls:** Inject beneficial nematodes into infected vines at 4" intervals over bottom 3' of vine, using 5000 nematodes per injection; use them in mulch around vines as well. Bt is another effective control; inject it into the vine after the first blossoms and again 10 days following. Clean syringe between injections. Trichogramma wasps attack borer eggs. Lacewings are also predators.
- **Allies:** Borage, nasturtium, radish (plant radishes around the plant base). See chart on pages 420–437.

Squirrel

Description: One of the most physically variable mammals. Fur on head may range from red to black; long furry tail; tufts of hair on ears. All except those with black fur have white underbellies.

Plants Affected: Fruit, nuts, seeds

Where It Occurs: Widespread

Signs: The presence of squirrels is obvious: you'll see the squirrel digging in the garden or scurrying in and around the bottom of your fruit and nut trees. They steal seeds, fruit, and nuts.

Organic Remedies

- Before planting corn try mixing 1 teaspoon of pepper with 1 pound of corn seed. Use experimentally; see page 348.
- Gather fruits and nuts every day. You may need to harvest fruit slightly underripe, for they can clean out every single ripe fruit in one night.

Stink Bug

many species

Description

- The medium-sized (⅓"–¾") adult is an ugly gray, brown, green, or black. Its back is shaped like a shield.
- It emits a very unpleasant odor when touched or frightened.
- Adults lay eggs on plants in mid-spring, and hibernate in debris.
- Adults and nymphs suck sap.
- Some stink bugs prey on the Colorado potato beetle.
- Produces 1–4 generations per year.

Plants Affected: Bean,* cabbage, cucumber,* mustard, okra, pepper,* snapdragon, tomato*

** Plant most frequently attacked by the insect.*

Where It Occurs: Widespread; many different species.

Signs

- Tiny holes in leaves and stems, particularly in new growth.
- Holes are surrounded by milky spots.
- Stunted, distorted, and weak plants.

Organic Remedies

- Main control is to keep the garden well weeded.
- Insecticidal soap sprays must be laced with isopropyl alcohol to help penetrate the bug's outer shell; soap alone is not very effective against hard-bodied insects.
- Hand pick and destroy.
- **Botanical controls:** Sabadilla, rotenone (1% solution), pyrethrins.

Strawberry Clipper

see Strawberry Weevil

Strawberry Crown Borer

Description

- Small yellow grubs bore into strawberry roots and crowns, and turn pink the longer they feed.
- The small (1/5") adult brown, snout-nosed beetle has reddish patches on its wings. It feeds on stems and leaves.
- Adults overwinter just below soil surface, and lay eggs in spring in shallow holes in the crown at the base of leaf stalks.
- Produces 1 generation per year.

Plants Affected: Strawberry

Where It Occurs: East of the Rockies, particularly in Kentucky, Tennessee, and Arkansas.

Signs

- Stunted or weakened plants.
- Chewed leaves and stems.

Organic Remedies

- If a patch is infested with crown borer, plant new strawberries at least 300 yards away from the area. Beetles can't fly and won't migrate.
- Pull and destroy plants that show damage. Replacement plants can be established immediately.
- Make sure plants are healthy and fed with compost; healthy plants can usually tolerate these beetles.
- **Allies:** Borage. (See chart on pages 420–437.)

Strawberry Leaf Roller

Description

- Yellow-green-brown larvae (½") feed inside rolled leaves.
- Small (½") adult moths, gray to reddish-brown, emerge in May to lay eggs on leaf undersides.
- Produces 2 or more generations per year.

Plants Affected: Strawberry

Where It Occurs: Northern United States, Louisiana and Arkansas.

Signs

- Rolled-up leaves.
- Skeletonized leaves that turn brown.
- Withered and deformed fruit.

Organic Remedies

- For minor infestations, remove and burn leaves.
- For larger infestations, mow or cut plants off 1" above the crowns. Burn or destroy plants.
- Handpick eggs in the winter.
- Spray light or superior horticultural oil in early spring before buds appear.
- Garlic Barrier, a product containing only garlic oil and water, is supposed to repel leaf rollers.
- **Predators:** Trichogramma wasps will eat the caterpillars.
- **Biological controls:** Dipel Bt is an effective control; be sure to spray inside the rolled leaf.
- **Botanical control:** Rotenone (5% solution).

Strawberry Root Weevil

many species

Description

- Beetles are small to large (1/10"–1"), shiny brown, gray to black. Females lay eggs on the soil surface in spring.
- Small, white to pinkish, thick-bodied, legless curved grubs feed on roots and then hibernate in soil.
- Produces 1 generation annually.

Plants Affected: Strawberry

Where It Occurs: Prevalent in northern United States.

Signs

- Adults eat notches into leaves.
- Larvae eat roots and crown, which weakens, stunts, and eventually kills the plants.

Organic Remedies

- Remove and destroy plants showing damage.
- Planting annual crops can, in some areas, help avoid weevil damage.
- **Botanical controls:** Generally ineffective because of the difficulty in timing sprays to coincide with adult emergence.

Strawberry Weevil or Clipper

Description

- These small (1/10") reddish-brown beetles, with black snouts, feed at night and hide in daylight. Females lay eggs in buds, then sever them.
- Small white grubs feed inside severed buds and emerge after fruit is picked in July.
- Produces 1 generation per year.

Plants Affected: Brambles, strawberry

Where It Occurs: Northern United States.

Signs: Holes in blossom buds and severed stems.

Organic Remedies

- Remove and destroy all stems hanging by threads. These carry the eggs.
- Remove mulch, maintain open-canopy beds, and renovate immediately after harvest to discourage new adults.
- Plant late-bearing strawberries.
- **Biological controls:** Beneficial nematodes are effective, especially the new Hh strain.

• **Botanical controls:** Sabadilla, rotenone (5% solution), and pyrethrins (apply 2 treatments 7–10 days apart, starting at early bud development); neem.
• **Allies:** Borage. (See chart on pages 420–437.)

Striped Blister Beetle

Description

• Slender adult beetles (½") are black with a yellow stripe. They swarm in huge numbers and can destroy everything in sight. They lay eggs in the soil that hatch midsummer. Other blister beetles may be less damaging.
• Larvae are heavy-jawed, burrow in the soil, and eat grasshopper eggs. They become hard-shelled as pseudopupa and remain dormant for under 1 year up to 2 years.
• Produces 1 generation per year.

Plants Affected: All vegetables

Where It Occurs: East of the Rockies.

Signs

• Eaten blossoms, chewed foliage, eaten fruit.
• Human skin that contacts a crushed beetle will blister.

Organic Remedies

• Handpick with gloves to protect skin.
• Use row covers or netting.
• **Botanical controls:** Rotenone dust.

Striped Flea Beetle

see Flea Beetle

Tarnished Plant Bug

Description

• Small (¼") highly mobile adults suck juice from young shoots and buds. They inject a poisonous substance into plant tissue and can spread fireblight.
• Generally brownish and oval, this insect can have mottled yellow, brown, and black triangles on each side of its back. It lays light yellow, long, curved eggs inside stems, tips, and leaves.
• Adults hibernate through winter under stones, tree bark, garden trash, or in clover and alfalfa.
• Nymphs are green-yellow with black dots on the abdomen and thorax.
• This bug attacks more plants than any other insect; 328 hosts have been recorded.
• Several generations per year.

Plants Affected: Bean, celery, most vegetables; peach, pear, raspberry, strawberry

Where It Occurs: Widespread

Signs

• Deformed, dwarfed flowers, beans, strawberry, and peaches.
• Wilted and discolored celery stems.
• Deformed roots.
• Black terminal shoots.
• Pitting and black spots on buds, tips, and fruit.
• Fireblight.

Organic Remedies

• Remove all sites of hibernation.
• Use row covers.
• Use white sticky traps.

- Spray insecticidal soap weekly in the early morning, though this is not very effective against hard-bodied insects.
- **Biological controls:** Try beneficial nematodes in fall to kill overwintering forms.
- **Botanical controls:** Sabadilla, rotenone (5% solution), pyrethrins (in 3 applications 2–3 days apart). Spray in the early morning when bugs are sluggish.

Tent Caterpillar

also known as Eastern Tent Caterpillar, Apple Tree Caterpillar

Description

- Large (2") caterpillars are black, hairy, with white and blue markings and a white stripe down the back. They feed in daylight on leaves outside the tent.
- Adults (1¼") are red-brown moths.
- Females lay egg masses in a band around twigs, and cover them with a foamy substance that dries to a dark shiny brown, hard finish. Eggs hatch the following spring.
- Produces 1 generation per year.
- Worst infestations come in 10-year cycles.

Plants Affected: Apple,* cherry, peach, pear

Where It Occurs: Prevalent east of the Rockies; a similar species exists in California.

Signs

- Woven tent-like nests, full of caterpillars, in tree forks.
- Defoliation.

Organic Remedies

- Destroy nests by hand. Wearing gloves, in early morning pull down nests and kill caterpillars by crushing or dropping into a bucket of water laced with insecticidal soap. (Don't use kerosene; it is toxic to the soil and difficult to dispose.)
- Use sticky burlap bands. Remove pests daily.
- In winter, cut off twigs with egg masses and burn. Check fences and buildings for eggs as well.
- Remove nearby wild cherry trees.
- Attract Baltimore orioles, bluebirds, digger wasps, and chickadees.
- **Biological controls:** Bt, sprayed every 10–14 days.
- **Botanical controls:** Pyrethrins; neem.
- **Allies:** Dill. (See chart on pages 420–437.)

Thrips

many species

Description

- Tiny (1⁄25") straw-colored or black slender insects with two pairs of slender wings edged with hairs.
- Nymphs and adults both suck plant juices and scrape and sting the plant. They transmit spotted wilt to tomatoes.
- Adults insert eggs inside leaves, stems, and fruit. Eggs hatch in 1 week.
- Produces 5–8 generations per year.

Plants Affected: Bean, corn, onion, peanut, pear, squash, tomato

** Plant most frequently attacked by the insect.*

Where It Occurs: Widespread

Signs

- Damaged blossoms, especially those colored white and yellow.
- Buds turn brown.
- Whitened, scarred, desiccated leaves and fruit.
- Pale, silvery leaves eventually die.
- Dark fecal pellets.

Organic Remedies

- Immediately remove infested buds and flowers.
- A reflective mulch like aluminum foil is reported to repel them. Use experimentally; see page 348.
- Control weeds.
- Spray insecticidal soap.
- Spray a hard jet of water in early morning, 3 days in a row.
- Dust with diatomaceous earth.
- Garlic or onion sprays — use experimentally; see page 348. Garlic Barrier, a product containing only garlic oil and water, is supposed to repel thrips.
- Spray light horticultural oil twice, 3–4 days apart, in morning.
- Ensure sufficient water supply.
- Rotate crops.
- **Biological controls:** Predatory mites (*Amblyseiulus mackenseii* and *Euseius tularensis*). Green lacewing larvae, ladybugs, and predatory thrips prey on thrips. Beneficial nematodes may work in soil control in the greenhouse. Also in the greenhouse, PFR (*Paecilomyces fumosoroseus*) is a beneficial microorganism available in granule form that attacks whiteflies, mites, aphids, thrips, mealybugs, and certain other greenhouse pests, and is claimed to be safe for humans and beneficials.
- **Botanical controls:** Rotenone (5% solution), pyrethrins, neem. A mix might work best. Sulfur and tobacco dusts also work.
- **Allies:** Carrots, corn. (See chart on pages 420–437.)

Tobacco Hornworm

also known as Southern Hornworm

Description: Virtually the same as the Tomato Hornworm (see below), except this worm has a red horn.

Plants Affected: Eggplant, pepper, tomato

Where It Occurs: Worst in the Gulf states and on ornamentals in California.

Signs: See Tomato Hornworm.

Organic Remedies: See Tomato Hornworm.

Tomato Fruitworm

see Corn Earworm

Tomato Hornworm

Description

- This very large (3"–4") green worm has white bars down both sides and a black or green horn at its tail end.
- The large (4"–5") adult moth emerges in May and June and is sometimes called the Hawk or Hummingbird Moth. It has long narrow gray wings, yellow spots on its abdomen, flies at dusk, and feeds like a hummingbird. It lays single, green-yellow

eggs on leaf undersides. Pupae (2") emerge and overwinter 3"–4" under the soil in a hard-shelled cocoon.
- Produces 1 generation per year in the North, and 2 per year in the South.

Plants Affected: Dill, eggplant,* pepper,* potato,* tomato*

Where It Occurs: Widespread

Signs
- Holes in leaves and fruit.
- Dark droppings on leaves.
- Defoliation.

Organic Remedies
- Handpick larvae and eggs. Look for green droppings under the plant.
- Do not pick worms with cocoons on their backs. Eggs of the Braconid wasp, a predator, are in the cocoons. If you see these, make the NPV control spray described under the Organic Remedies for Corn Earworm (see page 362). Also, do not pick eggs with dark streaks, which means they're parasitized by the trichogramma wasp.
- Apply hot pepper or soap and lime sprays directly on worms. Use experimentally; see page 348.
- Encourage birds.
- Blacklight traps and bug zappers are effective against the adult, as well. But these kill beneficial insects. Use experimentally; see page 348.
- Fall cultivation.
- **Biological controls:** Lacewings, braconid and trichogramma wasps, and ladybugs attack the eggs. Release at first sign of adults laying eggs. Bt (berliner/kurstake strain), sprayed every 10–14 days, is very effective.
- **Botanical controls:** Rotenone (1% solution), pyrethrins, neem.
- **Allies:** Borage and dill (both used as trap crops), and opal basil and marigold. (See chart on pages 420–437.)

Vine Borer

see Squash Vine Borer

Vole (Field Mouse)

see Mouse

Walnut Caterpillar

Description
- This black caterpillar (2"), with long white hairs, lifts its head and tail when disturbed. At night they gather at branch bases. Pupae overwinter 1"–3" below the soil surface.
- Adult moths emerge in spring. They are dark tan with four brown transverse lines on their forewings. They lay clusters of eggs on leaf undersides.
- Produces 1 generation per year in northern climates, 2 generations in southern climates.

Plants Affected: Hickory, pecan, walnut.

Where It Occurs: Prevalent in eastern United States south to Florida, and west to Texas and Wisconsin.

Signs
- Defoliation, usually in large branches first.

* *Plant most frequently attacked by the insect.*

• Black walnuts are prime targets.

Organic Remedies

• In late evening, brush congregating caterpillars into water laced with insecticidal soap. (Don't use kerosene — it is toxic to the soil and difficult to dispose.)
• Spray with dormant horticultural oil. Make sure all parts of the tree are covered.

Walnut Husk Fly and Maggot

Description

• Small larvae feed on the outer husk, then drop to the ground. Pupae hibernate under the trees in the ground in hard brown cases.
• Adult flies, the size of houseflies, emerge in midsummer. They are brown with yellow stripes across their backs and have transparent wings. Females lay eggs inside husks.
• Produces 1 generation per year.

Plants Affected: Peach (in West), walnut

Where It Occurs: Various species occur throughout the United States.

Signs

• Dark liquid stain over the walnut shell, and sometimes the kernels. This is a byproduct of larvae feeding.
• Kernels may have an off taste.

Organic Remedies

• Destroy worms in infested nuts by dropping husks into a pail of water. Remove dead maggots when removing the husk.
• Fall cultivation.
• Plant late-maturing cultivars in eastern United States.
• For large orchard sprays, consult your Extension agent or Richard Jaynes, *Nut Tree Culture in North America* (Hamden, Conn.: Northern Nut Growers, 1979).

Webworm (Garden)

see Garden Webworm

Weevil

many species

Description

• Family of hard-shelled, snout-nosed, tear-shaped small beetles. Usually brown or black, they feed at night and hide during the day.
• Small, white larvae feed inside fruit, stems, or roots. Adults usually lay eggs on the plant, sometimes inside.
• Bean and pea weevil larvae feed in young seed and emerge when beans are in storage. They can do extensive damage in storage.
• Usually produces 1 generation per year.

Plants Affected: Apple, bean, blueberry, brassicas, carrot, celeriac, cherry, pea, peach, pear, pepper, plum, raspberry, strawberry, sweet potato

Where It Occurs: Widespread

Signs: Zigzag paths in roots, fruit, stems.

Organic Remedies

• Heat beans and peas before storing. Beans at 135°F for 3–4 hours, peas at 120°–130° for 5–6 hours. Store in a cool, dry place.

- Clean cultivation is essential.
- Deep cultivation exposes larvae.
- **Pea weevil:** Plant crops early.
- **Sweet potato weevil:** Rotate crops, and use certified disease-free slips.
- **Brassicas:** Rotate crops.
- Try dusting with lime when the leaves are wet or dew-covered.
- Encourage songbirds.
- See Organic Remedies under Carrot Weevil, Plum Curculio, and Strawberry Clipper and Strawberry Root Weevil for additional controls.
- **Biological controls:** Beneficial nematodes are helpful when applied in early spring near planting time.
- **Botanical controls:** Sabadilla, rotenone (5% solution), pyrethrins, neem.
- **Allies:** Radish, summer savory, and tansy. (See chart on pages 420–437.)

Whitefly

Description

- Tiny (1⁄16") insects with white wings suck plant juices from the undersides of leaves, stems, and buds. They lay groups of yellow, conical eggs on leaf undersides. Nymphs hatch in 4–12 days and are legless white crawlers.
- Produces several generations per year.

Plants Affected: Greenhouses, most fruits, most vegetables, rosemary

Where It Occurs: Widespread

Signs

- Leaves yellow and die.
- Black fungus.
- Honeydew excretions coat leaves and support black fungus.

Organic Remedies

- Spray insecticidal soap, every 2–3 days for 2 weeks.
- Use sticky yellow traps. In greenhouses, place them at plant canopy height and shake plants.
- Mix 1 cup alcohol, ½ teaspoon Volck oil or insecticidal soap, and 1 quart water. Spray twice, at 1-week intervals, to the point of runoff. This suffocates whiteflies but doesn't harm plants.
- Use forceful water jet sprays, in early morning at least 3 days in a row.
- Hot pepper and garlic sprays — use experimentally; see pages 319, 320–321. Garlic Barrier, a product containing only garlic oil and water, is supposed to repel whiteflies.
- Cooking oil spray mix, as recommended by USDA.
- Check phosphorus and magnesium levels; whitefly may be a sign of a deficiency. Magnesium may be applied by mixing ½ cup Epsom salts in 1 gallon of water and thoroughly soaking the soil with the solution.
- Increase air circulation.
- Marigold root secretion is alleged to be absorbed by nearby vegetables and repel whiteflies. Use experimentally; see page 348.
- **Biological controls:** Ladybugs and green lacewings. Trichogramma and chalcid parasitic wasps. Whitefly parasites, *Encarsia formosa,* can be used in greenhouses. In the greenhouse, PFR *(Paecilomyces fumosoroseus)*

is a beneficial microorganism available in granule form that attacks whiteflies, mites, aphids, thrips, mealybugs, and certain other greenhouse pests and is claimed to be safe for humans and beneficials.

- **Botanical controls:** Ryania, neem, pyrethrins.
- **Allies:** Mint, nasturtium, thyme, and wormwood. (See chart on pages 420–437.)

White-fringed Beetle

Description

- Larvae (½") are yellow-white, curved, legless, and emerge in May, with the greatest numbers in June and July. They feed for 2–5 months and may travel ¼ to ¾ mile. They overwinter in the top 9" of the soil and pupate in spring.
- Adult beetles (½") are brownish-gray with broad, short snouts, have short pale hairs all over, and nonfunctional wings banded with white. They feed in large numbers.
- Produces 1–4 generations per year.

Plants Affected: Virtually all vegetation

Where It Occurs: Appear in the Southeast (Alabama, Arkansas, Florida, Georgia, Kentucky, Louisiana, Missouri, North Carolina, South Carolina, Tennessee, Virginia), but are moving north and have been seen in New Jersey.

Signs

- Severed roots.
- Chewed lower stems, root tissue, tubers.
- Plants yellow, wilt, and die.
- Extremely damaging.

Organic Remedies

- Large-scale government quarantine and eradication measures have eliminated this in some areas, but home gardeners have limited methods of dealing with the beetle. If you experience this pest, notify your Extension agent. The Cooperative Extension Service should be alerted to the movement of this pest and may be able to help you with control measures.
- Deep spading in spring can help destroy overwintering grubs.
- Dig very steep-sided ditches, 1' deep, to trap crawling beetles. Capture and destroy.

White Grub

larvae of June and Japanese Beetles

Description

- Medium to large (¾" to 1½") plump, white, curved worms have brown heads and several legs near the head. They feed on plant roots. The adult is usually either the May or June beetle.
- It takes them 10 months to several years to complete one life cycle.
- Has 1 emergence per year.

Plants Affected: Apple (young), blackberry, corn,* grain roots,* lawns,* onion, potato,* strawberry*

Where It Occurs: Widespread

Signs

- Sudden wilting, especially in early summer.
- See Signs under June Beetle, page 379.

Organic Remedies

- Fall and spring cultivation.

** Plant most frequently attacked by the insect.*

• Don't plant susceptible crops on areas just converted from untilled sod, as grubs will come up through the grass.
• Allow chickens to pick over garden following fall and spring cultivations.
• Encourage birds.
• **Biological controls:** Beneficial nematodes in spring or early summer, as mulch or dressing. Milky spore disease, which takes a few years to spread enough to be effective.
• **Botanical controls:** Neem.

Wireworm

many species

Description

• Large (1½"), slender, fairly hard-shelled worms, with three pairs of legs near the head; feed underground; do not curl when disturbed. They core into roots, bulbs, and germinating seeds. Corn, grasses, and potatoes may be badly damaged.
• The adult beetle is also known as the Click beetle or Skipjack. It flips into the air with a clicking sound when placed on its back. It can't fly well or long. The egg-adult cycle takes 3 years — 2 of which are in the larval feeding phase.
• Overlapping generations are present at all times.

Plants Affected: Bean,* cabbage, carrot,* celeriac, corn,* lettuce,* melon, onion,* pea, potato,* strawberry,* sweet potato, turnip

Where It Occurs: Widespread, but particularly a problem in poorly drained soil or recently sod soil.

Signs

• Plants wilt and die.
• Damaged roots.
• Thin and patchy crops.

Organic Remedies

• Frequent, at least once weekly, fall and spring cultivation to expose worms to predators.
• Don't grow a garden over grass sod. Plow or till soil once every week for 4–6 weeks in fall before beginning garden.
• Alfalfa is said to repel wireworms; white mustard and buckwheat are alleged to repel wireworms; clover and timothy (and other grass hays) are said by some to repel worms — others say they attract them. Use experimentally; see page 348. You might try it as a trap crop, away from your garden.
• **Potato trap:** Cut potato in half, spear with a stick, and bury 1" to 4" in soil, with stick as a handle above ground. Use experimentally; see page 348. Set 3' to 10' apart. Pull potatoes in 2 to 5 days. Destroy potato and all worms. Some gardeners reported capturing 15–20 worms in one potato.
• Put milkweed juice on the soil around affected plants; this supposedly repels worms. Use experimentally; see page 348.
• **Biological controls:** Apply beneficial nematodes 2 months before planting.
• **Allies:** Alfalfa, clover. (See preceding tip on alfalfa and clover; also see chart on pages 420–437.)

Woolley Aphid

see Aphid

** Plant most frequently attacked by the insect.*

Chapter Six

allies & companions

What they do, & how & where they do it

Allies are said to actively repel insects or to enhance the growth or flavor of the plant they help, the target plant. Companions, by contrast, are said to share space and growing habits well but do not necessarily play an active role in each other's pest protection or growth. Allies can be and are considered companions, but companions are not necessarily allies.

There is considerable controversy concerning allies and companions. Efforts to test claims scientifically with proper controls have been few, and results are often difficult to interpret. For example, in field trials a particular species of marigold was shown to repel nematodes but only in mass plantings. In small quantities planted next to the target crop, they decreased yields. As such, in the chart that follows every effort has been made to include information from as many scientific trials as possible, among them those where a putative ally was shown to carry with it some negative effects. Though an ally may effectively deter one insect pest, it may simultaneously attract other insect pests.

Plant Ally	Plant Enhanced	Pest Controlled
Alder	Fruit trees	Red Spider Mite
Alfalfa	Barley	Aphid
	Corn	Wireworm
Anise	Most vegetables	Aphid
	General	
Asparagus	Some vegetables	Nematodes: Asparagus roots contain a toxin
Barley (as cover crop)	Many vegetables	
	Soybean	Soybean pests
Basil	Asparagus	Asparagus beetle
	Most vegetables	Flies
	Tomato	Fruit flies
	Tomato	
	Tomato	Tomato Hornworm (with Opal Basil)
Beans		
French beans	Brussels sprouts	Aphid
All beans	Corn	Leaf Beetle, Leafhopper, Fall Armyworm
	Cucumber	
	Eggplant	Colorado Potato Beetle

The chart lists a plant ally; the plant(s) enhanced; the pest(s) it controls; the method of control, if known (e.g., visual masking); the test site where the data on its effects were gathered, if known; and its other benefits (e.g., repels insects, aids growth). Potential drawbacks are italicized and enclosed in parentheses.

Some sources claim to have evidence supporting the efficacy of a certain ally but fail to mention the nature of the evidence. *Evidence* constitutes a one-time occurrence for some, whereas others maintain a more rigorous, scientific standard, requiring multiple occurrences. When there is no clear scientific evidence for the effectiveness of an ally or when the source of the claim is unidentified, the method of control is designated *anecdotal.* Such claims may derive from folklore, word-of-mouth, or personal observation and may or may not be valid.

Approach companion planting with skepticism, curiosity, and a healthy experimental attitude. Conduct your own trials. The success of allies and companions depends greatly on your microclimate, soil conditions, and cropping history, to name just a few variables.

Method of Control (Test Site)*	Other Benefits
Attracts beneficial predators (United Kingdom)	
Attracts beneficial predators (Czechoslovakia)	
Anecdotal	
Anecdotal	
Chemical repellant contained in plant (essential oil mixed with sassafras *[Sassafras albidum]*)	General insecticide
Anecdotal	
Anecdotal	Reduces nematode populations
Attracts beneficial predators (Virginia)	
Anecdotal	Improves growth
Anecdotal	Improves growth and flavor
Chemical repellent contained in plant (essential oil)	
Anecdotal	Improves growth and flavor
Anecdotal	
Physical interference so pest can't reach target (England)	
Attracts beneficial predators; physical interference so pest can't reach target (Tropics)	
Anecdotal	Adds nutrients
Anecdotal	

Plant Ally	Plant Enhanced	Pest Controlled
All beans *(cont'd)*	Potato	Colorado Potato Beetle, Leafhopper
Green beans	Potato	
Snap beans		
Bee Balm *(Monarda didyma)*	Tomato	
Beet	Onion	
Blackberry	Grape	Leafhopper (Pierce's disease)
Black Salsify (Oyster plant)	Various vegetables	Carrot Rust Fly
Borage (as trap crop)	Squash	Squash Vine Borer
	Strawberry	Strawberry Crown Borer
	Tomato	Tomato Hornworm
Bramble Berries (e.g., Raspberry)	Fruit trees	Red Spider Mite
Brassicas	Peas	Root Rot (Rhizoctonia)
Broccoli	Beet (sugar)	Green Peach Aphid
	Cucumber	Striped Cucumber Beetle
Buckwheat (as cover crop)	Fruit trees	Codling Moth
Cabbage	Celery	
	Tomato	Flea Beetle
	Tomato	Diamondback Moth
Candytuft	Brassicas	Flea Beetle
Caraway	Fruit trees	
	Gardens	
	Onion	Thrip
	Peas	
	Pepper	
Catnip	General	
	Beans	Flea Beetle (plant in borders)
	Broccoli	Cabbageworm, Flea Beetle
	Cucumber	Cucumber Beetle
	Pepper	Green Peach Aphid

Method of Control (Test Site)*	Other Benefits
Anecdotal	*(May reduce potato yield)*
Anecdotal	Repels insects
Anecdotal	
Anecdotal	Improves growth and flavor
Provides alternate host to plant (temperate climate)	Improves growth
Anecdotal	
Anecdotal	
Anecdotal	Improves growth and flavor
Anecdotal	Improves growth and flavor
	Improves growth and flavor
Attracts beneficial predators (England)	
Anecdotal	
Attracts beneficial parasitic wasps (Washington)	
Physical interference so pest can't reach target (Michigan)	
Attracts beneficial parasites	
Anecdotal	Repels insects
Masking by chemical repellent confuses pest (temperate climate)	
Masking by chemical repellent confuses pest (Tropics)	
Masking by chemical repellent confuses pest (New York)	
Anecdotal	Attracts beneficial insects
Anecdotal	Loosens soils
Visual masking (Africa)	Improves growth
Anecdotal	Improves growth
Anecdotal	Improves growth
Chemical repellent contained in plant (nepetalactone in essential oil of catnip found comparable to DEET)	General insecticide
Anecdotal	*(May also increase whiteflies on snap beans)*
Anecdotal	*(Some evidence suggests that catnip increases Cabbageworms and decreases cabbage yields)*
Anecdotal	
Anecdotal	*(Catnip may compete with pepper)*

Plant Ally	Plant Enhanced	Pest Controlled
Catnip *(cont'd)*	Potato	Colorado Potato Beetle
	Squash	Squash Bug
	All vegetables	Aphid, Flea Beetle, Japanese Beetle
Celery	Beans	
	Brassicas	Cabbageworm
Chamomile	Beans	
	Brassicas	
Chervil	Radish	
Chive	Carrot	
	Celery	Aphid
	Lettuce	Aphid
	Peas	Aphid
	All vegetables	Japanese Beetle
Clover		
Red	Barley	Aphid
Clover, White	Brussels sprouts	Aphid, Cabbage Butterfly, Root Fly
Unspecified	Brussels sprouts	Aphid
Red and white	Cabbage and cauliflower	Aphid, Imported Cabbage Butterfly
Unspecified	Cabbage	Cabbage Root Fly
Unspecified	Corn	Corn Borer
Unspecified	Fruit trees	Aphid, Codling Moth
Unspecified	Fruit trees	Reduces Leafhopper population
New Zealand white	Oats	Fruit Fly
Dutch white	Turnip	Cabbage Root Maggot
Unspecified	Many vegetables	Repels Wireworm
Coriander	All fruit trees	
	Eggplant	Colorado Potato Beetle
	Potato	Colorado Potato Beetle
	Tomato	Colorado Potato Beetle
	Many vegetables	Aphid
	Many vegetables	Spider Mite
	Various vegetables	Carrot Rust Fly

Method of Control (Test Site)*	Other Benefits
Anecdotal	
	(Squash may not grow as large)
Anecdotal	
Anecdotal	Improves growth
Anecdotal	
Anecdotal	Improves growth
Anecdotal	Improves growth and flavor
Anecdotal	Improves growth and flavor
Anecdotal	Improves growth and flavor
Anecdotal	
Anecdotal	
Anecdotal	
Anecdotal	
Attracts beneficial predators (Czechoslovakia)	
Visual masking (England)	
Physical interference so pest can't reach target (England)	
Physical interference so pest can't reach target; attracts beneficial predators	
Attracts beneficial predators (Ireland)	
Physical interference so pest can't reach target (England)	
Attracts beneficial parasites	
Anecdotal	*(Can also attract Leafhoppers, which will then damage susceptible crops.)*
Physical interference so pest can't reach target (England)	
Masking by chemical repellent confuses pest (Pennsylvania)	
Anecdotal	
Anecdotal	Attracts beneficial insects
Anecdotal	
Anecdotal	
Anecdotal	
Chemical repellent contained in plant (essential oil)	
Anecdotal	
Anecdotal	

Plant Ally	Plant Enhanced	Pest Controlled
Corn	Beans	
	Cucumber	Striped Cucumber Beetle
	Cucurbits	
	Peanut	Corn Borer
	Pumpkin	
	Soybean	Corn Earworm
	Squash	Cucumber Beetle
	Squash	Western Flower Thrips
	Many vegetables	Nematode (plant corn as cover crop)
Corn spurry (*Spergula arvensis*)	Cauliflower	Aphid, Flea Beetle, Cabbage Looper
Cover grass	Brussels sprouts	Aphid
Cucumber	Radish	
Dead Nettle (*Lamium*)	Potato	Colorado Potato Beetle
Dill	Brassicas	Cabbage Looper, Imported Cabbageworm
	Cabbage	Spider Mite, Caterpillars
	Fruit trees	Codling Moth, Tent Caterpillar
	Tomato	Tomato Hornworm (use dill as trap crop)
	General	
Eggplant	Potato	Colorado Potato Beetle (use eggplant as trap crop)
Fennel	Most vegetables	Aphid
	General	
Flax	Carrot, Onion	
	Onion	Colorado Potato Beetle
	Onion	
Garlic	Beet	
	Brassicas	Cabbage Looper, Maggot, Worm
	Celery	Aphid
	Fruit trees	Codling Moth
	Lettuce	Aphid
	Peach tree	Peach Borer
	Raspberry	

Method of Control (Test Site)*	Other Benefits
Anecdotal	Improves growth
Physical interference so pest can't reach target (Michigan)	
Anecdotal	Improves growth
Visual masking (temperate climate)	
Anecdotal	Improves growth
Attracts beneficial parasitic wasps (Georgia)	
Physical interference so pest can't reach target (Tropics)	
Attracts beneficial predators (California)	
Anecdotal	
Attracts beneficial predators (California)	
Physical interference so pest can't reach target (England)	
Anecdotal	Repels insects
Anecdotal	
Anecdotal	Improves growth and flavor
Anecdotal	Improves growth and vigor
Anecdotal	
Anecdotal	
Anecdotal	
Chemical repellent contained in plant (essential oil)	General insecticide
Anecdotal	
Anecdotal	
Chemical repellent contained in plant (essential oil)	General insecticide
Anecdotal	Improves growth and flavor
Anecdotal	
Anecdotal	Improves growth and flavor
Anecdotal	Improves growth and flavor
Anecdotal	Improves growth and flavor
Anecdotal	
Anecdotal	
Anecdotal	
Anecdotal	
Anecdotal	Improves growth and health

Plant Ally	Plant Enhanced	Pest Controlled
Garlic *(cont'd)*	Rose	
	Many vegetables	Japanese Beetle, Mexican Bean Beetle, Nematodes, Slug, and Snail
	General	Aphids, flies, mosquitoes
Goldenrod	Peach tree	Oriental Fruit Moth
	Various vegetables	Cucumber Beetles
Goosegrass *(Eleusine indica)*	Beans	Leafhopper
Hairy Indigo	Various vegetables	Nematodes
Horseradish	Potato	Potato Bug (planted in patch corner)
Hyssop	Cabbage	Cabbage Looper, Moth, Worm
	Grape	
Johnson Grass *(Sorghum halepense)*	Grape	Pacific Mite
	Grape	Williamette Mite
Lamb's Quarters	Collards	Green Peach Aphid
	Cauliflower	Imported Cabbage Butterfly
	Peach tree	Oriental Fruit Moth
Lettuce	Carrot	Carrot Rust Fly
	Radish	
Marigold (*Tagetes* sp.)[c]	Asparagus	Asparagus Beetle
	Beans	Mexican Bean Beetle
	Eggplant	Nematode
	Lima Bean	Mexican Bean Beetle
	Lima Bean	Nematode
	Rose	Aphid
	Squash	Beetles, Nematode
	Tomato	Aphid, Tomato Hornworm
	Many vegetables	Cabbage Maggot
Marjoram	Vegetables	
Mint	Brassicas	Cabbage Looper, Moth, and Worm
	Broccoli	Ants
	Peas	

Method of Control (Test Site)*	Other Benefits
Anecdotal	Improves growth and health
Anecdotal	
Chemical repellent contained in plant (essential oil)	
(temperate climate)[a]	
Anecdotal	
Masking by chemical repellent confuses pest (Tropics)	
Anecdotal	
Anecdotal	
Anecdotal	
Anecdotal	Increases yields
Attracts beneficial predators (temperate climate)	*(While Johnson Grass may help with Pacific Mites, one Arkansas grower noted that it killed his vines, so you may want to make sure the grass is not allowed close to the vines.)*
Attracts beneficial predators (California)	
Attracts beneficial predators (Ohio)	
Attracts beneficial predators (California)	
(temperate climate)[b]	
Anecdotal	
Anecdotal	Improves growth
Anecdotal	
Anecdotal	
Roots excrete toxic substances (Connecticut)	
Anecdotal	
Roots excrete toxic substances (Connecticut)	
Anecdotal	
Anecdotal	
Anecdotal	
Anecdotal	
Anecdotal	Improves flavor
Anecdotal	Improves growth and flavor
Anecdotal	
Anecdotal	Improves growth and flavor

Plant Ally	Plant Enhanced	Pest Controlled
Mint *(cont'd)*	Squash	Squash Bug
	Tomato	
	Many vegetables	Whitefly
Mirabilis (Four O'clocks)		Japanese Beetles
Mustard (white or black)	Many vegetables	Nematodes
Nasturtium	Asparagus	Carrot Rust Fly
	Beans	Mexican Bean Beetle
	Brassicas	Aphid, Beetles, Cabbage Looper and Worm
	Celery	Aphid
	Cucumber	Aphid, Cucumber Beetle
	Fruit trees	
	Pepper	Green Peach Aphid
	Potato	Colorado Potato Beetle
	Radish	
	Squash	Beetles, Squash Bug
	Many vegetables	Whitefly
Onion family	Beet	Insects
	Brassicas	Cabbage Looper, Maggot Worm
	Carrot	Carrot Rust Fly
	Potato	Colorado Potato Beetle
	Swiss Chard	
	Many vegetables	Aphid
Oregano	Beans	
	Cucumber	
	Squash	
Parsley	Asparagus	Asparagus Beetle
	Tomato	
Peas	Carrot, Corn	
	Turnip	
Pennyroyal	Various vegetables	Carrot Rust Fly
	General	Fleas, mosquitoes, gnats, and ants
***Phacelia* sp.** (herbs)	Apple	Aphid, San Jose Scale

Method of Control (Test Site)*	Other Benefits
Anecdotal	
Anecdotal	Improves growth and flavor
Anecdotal	
Acts as trap crop	
Anecdotal	
Anecdotal	
Anecdotal	
Anecdotal	
Anecdotal	
Anecdotal	
Anecdotal	Provides general protection (under tree)
Anecdotal	
Anecdotal	
Anecdotal	Provides general protection
Anecdotal	
Anecdotal	
Anecdotal	
Anecdotal	
Masking by chemical repellent confuses pest (UK, Africa)	
Anecdotal	
Anecdotal	Improves growth
Anecdotal	
Anecdotal	Improves flavor and growth
Anecdotal	Deters pests
Anecdotal	General pest protection
Anecdotal	Helps growth
Anecdotal	Improves growth
Anecdotal	Improves growth and flavor by adding nutrients to the soil
Anecdotal	Improves growth
Attracts beneficial parasitic wasps (Soviet Union)	
Chemical repellent contained in plant (essential oil)	
Attracts beneficial predators (Ohio)	

Plant Ally	Plant Enhanced	Pest Controlled
Pigweed (*Amaranthus* sp.)	Collards	Green Peach Aphid (*A. retroflexus*)
	Corn	Fall Armyworm (*A. hybridus*)
	Corn, Onion, and Potato	
Potato	Beans	Mexican Bean Beetle
	Corn	
	Eggplant	Useful as trap plant
Radish	Brassicas	Cabbage Maggot
	Cucumber	Striped Cucumber Beetle
	Lettuce	
	Squash	Squash Bug, Vine Borer (use radish as trap crop)
	Sweet Potato	Sweet Potato Weevil
Ragweed	Collards	Flea Beetle
Giant	Corn	Corn Borer
Normal	Peach tree	Oriental Fruit Moth
	Peach tree	Oriental Fruit Moth
Red Sprangletop (*Leptochioa filliformis*)	Beans	Leafhopper
Rosemary	Beans	Mexican Bean Beetle
	Brassicas	Cabbage Moth
	Carrot	Carrot Rust Fly
	Many vegetables	Slug, Snail
Rue	Cucumber	Cucumber Beetles
	Raspberry	Japanese Beetle
	Rose	Japanese Beetle
	Many vegetables	Flea Beetle
	General	Fleas
Rye (as cover crop)	Fruit trees	Aphid
	Fruit trees	Leafhopper
	Many vegetables	Nematode (turn cover crop under)
	Soybean	Seedcorn Maggot
Rye (mulch)	Fruit trees	European Red Mite

Method of Control (Test Site)*	Other Benefits
Attracts beneficial predators (Ohio)	
Attracts beneficial predators; attracts beneficial parasitic wasps (Florida)	
Anecdotal	Brings nutrients to soil surface where available to plants
Anecdotal	
Anecdotal	Repels insects
Anecdotal	
Anecdotal	
Anecdotal	
Anecdotal	Improves growth
Anecdotal	
Anecdotal	
Masking by chemical repellent confuses pest (New York)	
Provides alternate host plant (Canada)	
Provides alternate host for parasitic wasps (Virginia)	
(temperate climate)[d]	
Masking by chemical repellent confuses pest (Tropics)	
Anecdotal	
Anecdotal	Repels insects
Anecdotal	Repels insects
Anecdotal	
Anecdotal	
Anecdotal	
Anecdotal	
Anecdotal	
Chemical repellent contained in plant (essential oil)	
Attracts beneficial predators	
Anecdotal	
Roots excrete toxic substances	
Physical interference so pest can't reach target (Ohio)	
Attracts beneficial predators (Michigan)	

Plant Ally	Plant Enhanced	Pest Controlled
Sage	Brassicas	Cabbage Looper, Maggot, Moth, and Worm
	Cabbage	White Cabbage Butterflies
	Carrot	Carrot Rust Fly
	Marjoram	
	Strawberry	
	Tomato	
Savory (Summer)	Beans	Mexican Bean Beetle
	Onion	
	Sweet Potato	Sweet Potato Weevil
Shepherd's Purse	Brassicas	Flea Beetle
	Corn	Black Cutworm
Smartweed	Peach tree	Oriental Fruit Moth
Sorghum (cover crop mulch)	Cow pea	Leaf Beetle
	Fruit trees	European Red Mite
Southern Wood	Cabbage	Cabbage Moth
Soybeans	Corn	Corn Earworm
	Corn	Cinch Bug
Strawberry	Peach tree	Oriental Fruit Moth
	Spinach	
Sudan Grass	Grape	Williamette Mite
Sweet Potato	Corn	Leaf Beetle
Tansy	Brassicas	Cabbageworm, Cutworm
	Cucumber	Ants, Cucumber Beetles, Squash Bug
	Fruit Trees	Ants, Aphid, Japanese Beetle
	Fruit Trees (especially peach)	Borers
	Potato	Colorado Potato Beetle
	Raspberry	Ants, Japanese Beetle
	Squash	Squash Bug
		Sweet Potato Weevil
	Sweet Potato	Flea Beetle, Japanese Beetle
	All vegetables	

Method of Control (Test Site)*	Other Benefits
Anecdotal	
Chemical repellant contained in plant (essential oil)	
Anecdotal	Improves growth
Anecdotal	Improves growth
Anecdotal	Improves growth
Anecdotal	Improves growth
Anecdotal	Improves growth and flavor
Anecdotal	Improves growth and flavor
Anecdotal	
Masking by chemical repellent confuses pest (New York)	
Attracts beneficial parasitic wasps (Illinois)	
(temperate climate)[e]	
Masking by chemical repellent (temperate climate)	
Attracts beneficial predators (Michigan)	
Anecdotal	
Attracts beneficial predators (Florida)	
Anecdotal	
Attracts beneficial predators (temperate climate)	
Anecdotal	Improves growth
Attracts beneficial predators (California)	
Attracts beneficial parasitic wasps (Tropics)	
	(Some evidence suggests that tansy increases Cabbageworms)
Anecdotal	
Anecdotal	
Chemical repellent contained in plant (when planted under tree)	
Chemical repellent contained in plant (essential oil)	
Anecdotal	
	(May make squash plants smaller)
Anecdotal	
Anecdotal	

Plant Ally	Plant Enhanced	Pest Controlled
Thyme	General	Mosquitoes
	Brassicas	Cabbage Looper and Worm, Insects
	Strawberry	Worms
	Many vegetables	Whitefly
Tomato	Asparagus	Asparagus Beetle
	Brassicas	Imported Cabbage Butterfly
	Cabbage	Diamondback Moth
	Collards	Flea Beetle
Turnip	Peas	
Vetch	Fruit trees	Aphid
Weedy Ground Cover	Apple	Tent Caterpillars
	Apple	Codling Moth
	Brussels sprouts	Cabbage Butterfly, Cabbageworm
	Collards	Flea Beetle
	Collards	Cabbage Aphid
	Mung bean	Beanfly
	Walnut	Walnut Aphid
Wheat (as cover crop)	Soybean	Soybean pests
	Many vegetables	Nematode
Wheat (mulch)	Fruit trees	European Spider Mite
Wildflowers	Fruit trees	
Wormseed mustard	Brassicas	Flea Beetle
Wormwood (pulverized)	Brassicas	Cabbage Maggot
	Fruit trees	Codling Moth
	Many vegetables	Mice, Whitefly
	Carrot	Carrot Rust Fly

*Note: All references to specific mechanisms of action, when associated with a specific testing site, are from Robert Kourik, *Designing and Maintaining Your Edible Landscape Naturally* (Santa Rosa, Calif.: Metamorphic, 1986). Consult his book for further information on the original research articles. Other references to mechanisms of action are drawn from a variety of sources.

[a] Clausen, C. P. *Ann Entomol Soc Am* 29 (1936): 201–223. Quoted in Kourik, *Designing and Maintaining Your Edible Landscape.*

[b] Ibid.

[c] Research at the Connecticut Agricultural Experiment Station has shown that small French (*Tagetes patula* L.) marigolds suppress meadow, or root lesion, nematodes for up to 3 years and one or more other nematodes for 1 or more years. Marigolds are effective when rotated, or grown in the entire infested area for a full season. Interplanting is not as

Method of Control (Test Site)*	Other Benefits
Chemical repellent contained in plant (citronella in lemon thyme)	
	(May lower cabbage yields)
Anecdotal	
	(In beans, may cause higher Whitefly population)
Anecdotal	
Anecdotal	
(Tropics)[f]	
Masking by chemical repellent (New York)	
Anecdotal	Improves growth
Anecdotal	
Attracts beneficial parasitic wasps (Canada)	
Attracts beneficial parasitic wasps (Canada)	
Attracts beneficial predators (England)	
Visual masking (New York)	
Attracts beneficial parasitic wasps (California)	
Physical interference so pest can't reach target (temperate climate)	
Provides alternate host for parasitic wasps (California)	
Attracts beneficial predators (Virginia)	
Anecdotal	
Attracts beneficial predators (Michigan)	
Anecdotal	Attracts beneficial insects
Masking by chemical repellent (New York)	
Anecdotal	
Anecdotal	
Anecdotal	
Anecdotal	

effective, and can reduce crop yields, but some beneficial nematicide effects may be seen the following year. To reduce competition, interplant marigolds 2 weeks or more after other plants. Two theories exist on how marigolds work: (1) they produce a chemical from their roots that kills nematodes; (2) they do not serve as a host to nematodes and, in the absence of a host, the nematode population dies.

[d] Clausen, C. P. *Ann Entomol Soc Am* 29 (1936): 201–223. Quoted in Kourik, *Designing and Maintaining Your Edible Landscape.*

[e] Ibid.

[f] Raros, R. S. "Prospects and Problems of Integrated Pest Control in Multiple Cropping." *IRRI Saturday Seminar Proc* (Los Banos, Philippines, 1973), 1–20. Quoted in Kourik, *Designing and Maintaining Your Edible Landscape.*

afterword

How can you or I, as small backyard gardeners, know whether our gardens are truly organic? As of this writing, the latest definition for the term *organic* in the United States was announced in December 2000 and was fully implemented in October 2002 by the National Organic Program. The National Standards are extensive and must be reviewed in detail if you want to become a certified organic grower. Only those growers who meet the National Standards for production and handling can sell products under the label "organic." Those who are not certified organic growers but who garden organically can also benefit by learning about what the National Standards do or do not allow.

The National Organic Program continues to build on the historical "Statement of Principles of Organic Agriculture," drafted by the Organic Farmers Association Council (OFAC) in February 1990. This statement is perhaps one of the best descriptions of the goals of organic growing, and it bears repeating here:

Organic farming practices are based on a common set of principles that aim to encourage stewardship of the earth. Organic procedures work in harmony with natural ecosystems to develop stability through diversity, complexity, and the recycling of energy and nutrients. They:

- Seek to provide food of the highest quality, using practices and materials that protect the environment and promote human health.
- Use renewable resources and recycled materials to the greatest extent possible, within agricultural systems that are regionally organized.
- Maintain diversity within the farming system and in its surroundings, including the protection of plant and wildlife habitat.
- Replenish and maintain long-term soil fertility by providing optimal conditions for soil biological activity.
- Provide livestock and poultry with conditions that meet both health and behavioral requirements.
- Seek an adequate return from their labor, while providing a safe working environment and maintaining concern for the long-range social and ecological impact of their work.

The Challenge and Promise of Organic Growing

The new definition of *organic*, as outlined by the National Standards, is not without controversy. Even as the standards are being enacted, organic growers debate aspects of the definition. Yet, the fact that a federal definition has been promulgated by the U.S. Department of Agriculture (USDA), under the auspices of the National Organic Program, suggests that organic agriculture is coming of age and finding a place in mainstream culture.

For more than 20 years, the organic standards have been honed, modified, and updated to keep pace with developments in

Overview of National Standards

The National Standards set specific guidelines and criteria for:

- Producing and handling livestock
- Producing and handling processed products
- Labeling products "organic"
- Composting. The composting process must begin with an initial carbon:nitrogen ratio between 25:1 and 40:1. Temperatures in in-vessel or aerated piles must stay between 131°F and 170°F for 3 days. Temperatures in windrows must stay between 131°F and 170°F for 15 days, during which time the materials must be turned a minimum of five times.
- Producing and handling organic crops, including using organically grown seeds, seedlings, or planting stock, with some exceptions; implementing a crop rotation system for several functions; using management practices to prevent crop pests, weeds and diseases; removing any plastic or synthetic mulches at the end of the growing or harvest season; keeping detailed records, which are specified in the standards; *not* using lumber treated with arsenate or other prohibited materials at an organic production site; and *not* using genetic engineering, ionizing radiation, or sewage sludge for fertilization.

The National List of Allowed and Prohibited Substances for organic production and handling is provided in appendix A. Additional information about the National Standards can be found at www.ams.usda.gov/nop, the Web site of the National Organic Program. Consult the Web site periodically for updates.

agriculture and science. In 1985, the Organic Foods Production Association of North America (OFPANA), now the Organic Trade Association, took the lead in drafting national guidelines. The guidelines were based on suggestions and reviews by state organic growers organizations and became the basis for the first federal standards outlined in the 1990 Organic Foods Production Act. These standards continue to be reviewed and updated by the National Organic Standards Board (NOSB).

In October 1993, a federal mandate required that anyone selling more than $5000 worth of organic produce per year secure annual certification, through an approved program, to use the label "organic." In response to this mandate, organic certification programs sprang up throughout the United States. Thus began a trend toward greater regulatory control over the term *organic,* which in the eyes of some, but not all, growers gave the term more credibility in mainstream commercial culture.

In the latest revision of the National Standard, however, significant controversy erupted over trends that some believed threatened the essence of what is meant by the term *organic*. Among them, should human waste in the form of sludge, or biosolids, be spread on organic fields? Should genetically modified seed be grown in organic soils, and should the resulting produce be called "organic"? Should antibiotics be allowed under certain conditions, or never? Could irradiation be used for storage of foods labeled "organic"? More than 300,000 public comments about the proposed standards were officially received by the federal government.

In all likelihood, the debate will continue to intensify as advances in biotechnology enable us to do things unimagined by previous generations. To date, organic farms have tended to be smaller family farms — with land numbering in the tens and hundreds rather than thousands of acres — whose stewards subscribe to the tenets of sustainable agriculture and often shun advances offered by biotechnology. Ironically, however, some advances in biotechnology may make it possible for smaller and more vulnerable farms to resist the pressures of industrialized agriculture. New developments suggest that tobacco may be an ideal host for bioengineered medicinal substances, for example. Might this assist southern rural communities and their threatened small family farms dependent on growing tobacco to continue their ways of life and livelihood? To bring this into the realm of organic, what would happen if some of these tobacco farmers wanted to grow certified organic bioengineered medicinal crops? Would this be possible? Will or must organic growers forever eschew biotechnological changes and opportunities that are not universally accepted as advances?

The implications of these and many other questions threaten to shake the foundations of our agricultural communities for decades. A search for answers will require much soul-searching and discussion among those involved in farmland sustainability. Because we cannot possibly anticipate the myriad questions that will need to be asked by the end of this century, our best hope for addressing the difficult questions relating to farmland sustainability today is to bring together the best minds from all fields to work toward solutions.

By law, the federal standard of approved and prohibited substances for organic production must be reviewed and updated at least every 5 years. This mandated review reflects an understanding that the definition of *organic* will necessarily evolve as field and laboratory research provide new information about the effects certain substances have on soil and food crops. The National List of approved substances, therefore, should not be considered the final word on what is organic, nor should the "approved methods" be considered the final word on what we should and should not do if we want to grow organic. We must remain open to new information and new ways of doing things, while recalling that new ways may not always be better — perhaps the most difficult lesson of the "green revolution" of the twentieth century.

According to U.S. Secretary of Agriculture Dan Glickam in 2000, "The need for these standards rose out of the exponential growth of organic agriculture. It is a sector that is here to stay — growing from $78 million in 1980 to about $6 billion today, with continuing growth of 20 percent per year."

Federal statistics from December 2000 suggest that the number of organic farms is increasing by about 12 percent per year, with an estimated 12,200 organic farms nationwide, most of them small-scale producers. Additionally, a USDA study indicates that certified organic cropland more than doubled from 1992 to 1997 and that in the livestock sectors, eggs and dairy increased at an even faster pace. Growth in organic production has also produced what can be called, for the first time, a significant organic industry. Large corporations such as General Mills, Archer Daniels Midland, Con Agra, and Dole are actively buying, packaging, and marketing all fashions of things organic.

Still, the seemingly explosive growth of things organic during the past decade must be put into context. In raw numbers, approximately 2 percent of farming in the United States is organic. Also, organic produce has a long way to go before it can be considered ordinary everyday fare, affordable to people at all economic levels. Most organic products still retail anywhere from 1.5 to more than 3 times the price of the their "conventional" counterparts; it could be argued that this difference in price is due, at least in part, to the artificial deflation of conventional food prices from subsidies and other federal agriculture assistance programs that benefit mostly large-scale industrial producers.

But do higher prices at the grocery store mean that organic produce is available only to those who can afford it? Hardly. People at all socioeconomic levels garden; in fact, gardening has become a favorite pastime in the United States. Some people garden because they must grow food for their table; others garden for enjoyment. The good news is that there is nothing preventing home gardeners from making their gardens organic.

Bibliography

General Gardening & Vegetable Growing

Appelhof, Mary. *Worms Eat My Garbage: How to Set Up and Maintain a Worm Composting System.* Kalamazoo, Mich.: Flower, 1982. A small, easy-to-read paperback. Appelhof tells you all you need to know about how to grow earthworms without much effort. Not as complete as Minnich's book, but most people don't need the detail that Minnich offers.

Arms, Karen. *Environmental Gardening.* Savannah, Ga.: Halfmoon, 1992. A fun and informative resource for all gardeners. A biologist by training, Arms covers general principles and numerous specifics, such as water gardening, attracting wildlife, saving water and energy, annuals, shrubs, trees, lawns, and edible plants. She accomplishes her goal of providing a guide to which gardening practices are environmentally sound and which are dangerous or unethical.

Backyard Composting. Ojai, Calif.: Harmonious, 1992. A great little gem of a handbook that makes composting super simple and easy. This is all you need to get your backyard compost pile started. It also provides information for those trying to be more environmentally sensitive.

Ball, Jeff. *Jeff Ball's 60-Minute Garden: One Hour a Week Is All It Takes to Garden Successfully.* Emmaus, Penn.: Rodale, 1985. One of my favorite books, not only because of its goal of low-time gardening, but also because it is so well written and fun to read. Practical, informative, and easy-to-follow. Includes diagrams of useful garden tools as well as shopping lists and instructions. Diagrams and construction plans are provided for a boxed bed with PVC foundations, tunnels, trellis and orchard fence, compost bin and sifter, seedling box, garden sink, and birdhouse.

———. *The Self-Sufficient Suburban Gardener: A Step-by-Step Planning and Management Guide to Backyard Food Production.* Emmaus, Penn.: Rodale, 1983. An excellent book for novice gardeners: enjoyable, practical, and informative. Includes many useful charts, including how much to plant, seed-starting tips, planting guide, interplanting guide, companion planting, succession planting, growing guide, and food storage options.

Bartholomew, Mel. *Square Foot Gardening.* Emmaus, Penn.: Rodale, 1981. Another fun book, good for both novice and experienced gardeners, based on the PBS television series. The square foot method is an interesting, practical method of intensive gardening, in some ways more formulaic (and therefore easier) than French intensive. The back of the book contains cultural notes on vegetables and square foot spacing rules for each vegetable.

Barton, Barbara. *Gardening by Mail: A Source Book,* 5th ed. Boston: Mariner Books, 1997. A unique and important resource for gardeners that provides the most complete listing anywhere of plant and seed sources, garden suppliers, professional societies and associations, magazines, newsletters, horticultural libraries, and books.

Best Ways to Improve Your Soil. Emmaus, Penn.: Rodale, 1987. A small, inexpensive, useful booklet packed with information. Useful charts include soil types, cover crop planting guide, composting materials, and nutrient profiles of common organic amendments.

Bubel, Nancy. *The New Seed-Starters Handbook.* Emmaus, Penn.: Rodale, 1988. For those who want an in-depth discussion of how to get plants off to a good start, Bubel covers everything from germinating to transplanting. She also explains how to save your own seeds. The last third of the book includes cultural briefs on how to start from seed vegetables, fruits, herbs, flowers, wildflowers, trees, and shrubs. Useful charts on soil deficiency: symptoms and treatment, and more.

Carr, Anna. *Good Neighbors: Companion Planting for Gardeners.* Emmaus, Penn.: Rodale, 1985. A great resource for learning more about companion planting and how to separate fact from lore. I especially like the appendix that suggests possible companion planting experiments for backyard gardens and introduces way to assess the results. Useful charts include read your weeds, recipes for botanical pesticides and repellants, legumes for companion planting, weeds to watch out for, and more.

Coleman, Eliot. *Four-Season Harvest: Organic Vegetables from Your Home Garden All Year Long.* Chelsea, Vt.: Chelsea Green, 1999. *See annotation below.*

———. *The New Organic Grower: A Master's Manual of Tools and Techniques for the Home and Market Gardener.* Chelsea, Vt.: Chelsea Green, 1989. I love this book, not just because one reviewer called it the perfect companion to *Gardening at a Glance* (the self-published precursor to this book), but because it really *is* the perfect companion. While *The Gardener's A–Z Guide to Growing Organic Food* is strong on information for individual crops and pest control, Coleman's book is strong on understanding, approach, and techniques. Coleman is a joy to read and learn from. The sections on green manure rotations and crop rotations are especially useful. There are many useful charts, but I especially like the rotation charts for green manure and vegetable crops.

Creasy, Rosalind. *The Complete Book of Edible Landscaping: Home Landscaping with Food-Fearing Plants and Resource-Saving Techniques.* San Francisco: Sierra Club, 1982. An excellent book on how to make your garden an attractive landscape and your landscape into an attractive garden. Useful to all gardeners. Major principles of edible landscaping are covered in this timely book. Creasy also discusses gardening techniques, culture, hygiene, diseases, and insects. The last half of the book is a useful "encyclopedia" of fruit, vegetable, herb, and nut culture; includes varieties and sources.

———. *Cooking from the Garden.* San Francisco: Sierra Club, 1988. A gem filled with valuable growing information and visual beauty. Creasy creates and explores specialty gardens of different countries, traditions, colors, and flavors. She shares a wealth of information on varieties known for such things as color, flavor, and heirloom history. Mouth-watering recipes accompany every section, including such wonders as violet vichyssoise and lavender ice cream. The book concludes with an encyclopedia on how to grow each type of vegetable.

Cutler, Karan Davis. *Burpee-The Complete Vegetable & Herb Gardener: A Guide to Growing Your Garden Organically.* New York: Macmillan, 1997.

Encyclopedia of Organic Gardening. Emmaus, Penn.: Rodale, 1978. A good general reference book that covers to a greater or lesser extent every plant one might be curious about. It also discusses general topics such as fertilizer, fruit cultivation, landscaping, and much more. Useful charts include planting dates (based on average last frost date); shrubs: recommended shrubs for the home grounds; straw: mineral value of straws; trace elements: chart on signs of deficiency and accumulator plants; wild plants; and edibles, among others.

Faeth, Paul, Robert Repetto, Kim Kroll, Qi Dai, and Glenn Helmers. *Paying the Farm Bill: U.S. Agricultural Policy and the Transition to Sustainable Agriculture.* Washington: World Resources Institute, March 1991. An important study that offers perhaps the first comprehensive comparison of conventional farming with sustainable agriculture techniques. If all environmental costs are factored in, WRI demonstrates that sustainable farming techniques not only save the environment but also save money. A Pennsylvania farm using conventional crop rotation of corn and soybeans resulted in a loss of $61 per acre over 10 years, while rotation of several crops of corn, soybean, wheat, and clover *saved* $325 per acre. To request this publication, write to 10 G Street, NE (Suite 800), Washington, DC 20002.

Fukuoka, Masanobu. *The One-Straw Revolution: An Introduction to Natural Farming.* Emmaus, Penn.: Rodale, 1978. Now considered a classic on the subject, this book discusses the importance and methods of no-till cultivation. Some of the methods may not be generally applicable in the United States, however, because they're designed for a mild Japanese climate.

Hamilton, Geoff. *The Organic Garden Book: The Complete Guide to Growing Flowers, Fruit and Vegetables Naturally.* New York: Crown, 1987. Excellent color photographs and diagrams make this book special. The photographs cover everything from different soil types to how to harvest different vegetables. Includes photos of the different methods of training fruit trees and provides good cultural notes. Concludes with a good list of gardening activities that is broken down by season and type of garden (ornamental, fruit, vegetable, greenhouse).

Hill, Lewis. *Secrets of Plant Propagation: Starting Your Own Flowers, Vegetables, Fruits, Berries, Shrubs, Trees, and Houseplants.* North Adams, Mass.: Storey, 1985. A complete guide to starting new plants.

Hirshberg, Gary and Tracy Calvan, eds. *Gardening for All Seasons: A Complete Guide to Producing Food at Home 12 Months of the Year.* Andover, Mass.: Brick House Publishing, 1983. Useful for its charts, tables, and cultural notes on vegetables. Notable charts include varieties of vegetables for greenhouse growing; pH preference of vegetables, fruits, flowers and grains; disease-resistant plant varieties; insect-resistant plant varieties; insect-repellant plant varieties; companion planting guide; planting; nitrogen, phosphorus, and potassium components of organic materials; plants that attract birds.

Hunt, Marjorie B., and Brenda Bortz. *High-Yield Gardening: How to Get More from Your Garden Space and More from Your Gardening Season.* Emmaus, Penn.: Rodale, 1986. Chock-full of useful suggestions for increasing garden yield, this book is a good general reference for beginning and experienced gardeners. Useful charts include cultural notes presented alphabetically by plant; soil type; cover crop planting guide; composting materials; nutrient profiles of common organic materials; survey of raised beds; high-yield low-space versions of popular vegetables; high-yielding varieties for vertical growing; garden time-savers; critical times for watering; succession planting;

traditional companions; space-efficient root patterns; and more.

Jeavons, John. *How to Grow More Vegetables than You Ever Thought Possible.* Berkeley, Calif.: Ten Speed Press, 1991. An excellent discussion on the importance and methods of French intensive raised-bed gardening. Jeavons is well known for advocating high-yield, intensive techniques. Here's a good, detailed description on how to double-dig. Useful charts include 4-year garden plan; companion plants; growing data on vegetables, fruits, grains; and fertilizers and their components.

Kourik, Robert. *Designing and Maintaining Your Edible Landscape Naturally.* Santa Rosa, Calif.: Metamorphic Press, 1986. For the novice and experienced gardener, this is essential reading on edible landscaping, the gardening wave of the future. Unusual, informative, and fun, Kourik's book discusses everything from aesthetics to specific gardening techniques — how to plan gardens, the pros and cons of tillage, how to prune, and much more. Useful charts include companion planting research summaries; intercropping for pest reduction; green manure plants; dynamic accumulators; seven-step rotation for fertility; multi-purpose edibles; soil indicators (plants); ripening dates for fruit and nut varieties; fruit tree rootstocks; disease-resistant trees; and more.

———. *Drip Irrigation for Every Landscape and All Climates.* Santa Rosa, Calif.: Metamorphic Press, 1992. A step-by-step guide to easy and simple drip irrigation. This book demystifies drip irrigation and includes illustrations, charts, and sources.

Lappé, Frances Moore, and Anna Lappé. *Hope's Edge: The Next Diet for a Small Planet.* New York: Tarcher/Putnam, 2002. A journey to different parts of the world to meet interesting people who are developing new ways of growing food; the book also addresses hunger and social change. Lappé demonstrates that the links between what we grow and eat, our health, and our economy extend to the very root and soul of our societies. Demonstrates the rich possibilities for solutions all around us and the power of individuals to initiate change. A must-read for those interested in sustainability, it will challenge and inspire gardeners everywhere.

Maynard, Donald N., and George J. Hochmuth, eds. *Knott's Handbook for Vegetable Growers,* 4th ed. New York: Wiley, 1997. A professional's resource book and authority on all aspects of vegetable growing. Just when you thought you had a handle on the issues involved in vegetable growing, this book can both humble and stimulate. Covers everything from hydroponic solutions to irrigation rates. A small ring binder, it's easy to carry around and fun to browse through for both trivia and essentials. Charts that might be appropriate for the nonprofessional include the following: composition of fresh raw vegetables; diagnosis and correction of transplant disorders; soil temperature conditions for germination; days required for seedling emergence; composition of organic material; key to nutrient-deficiency symptoms; practical soil moisture interpretation; disease control for vegetables; insect control for vegetables.

Minnich, Jerry. *The Earthworm Book.* Emmaus, Penn.: Rodale, 1977. The best book I've found on the subject. Covers everything you might want or need to know about earthworm cultivation. Also explains the important benefits of earthworms in your garden and how to grow them easily on a small scale.

National Gardening Association. *Gardening: The Complete Guide to Growing America's Favorite Fruits and Vegetables.* Reading, Mass.: Addison-Wesley, 1986. For novice and advanced gardeners, this book discusses how to plan and prepare the garden site. It also provides cultural information for individual vegetables and fruits.

Organic Gardening's Soil First Aid Manual. Emmaus, Penn.: Rodale, 1982. A useful collection of short articles on ways to improve your soil. Interesting reading and good background material.

Poincelot, Raymond P. *No-Dig, No-Weed Gardening.* Emmaus, Penn.: Rodale, 1986. An excellent discussion of the benefits of not tilling the soil and how to maintain such a garden. Useful cultural notes on vegetables and flowers.

Raymond, Dick. *Down-to-Earth Gardening Know-How for the '90s: Vegetables and Herbs.* North Adams, Mass.: Storey, 1991. A book full of useful charts and graphs. Helpful charts on green manures; plant diseases; herbs; garden planning, including amounts to plant.

———. *Joy of Gardening.* North Adams, Mass.: Storey, 1982. A good book on garden preparation, maintenance, and culture of vegetables and fruits. Good photographs. Good discussion of cover crops. Useful charts include vegetable planting guide (especially useful for planning amounts to plant), detailed first frost date in the fall map of the United States, mulching guide comparing advantages and disadvantages of different mulch types.

Reilly, Ann. *Park's Success With Seeds.* Greenwood, S.C.: Park Seed Co., 1978. A guide to how to start most species from seed.

Rogers, Marc. *Saving Seeds: The Gardener's Guide to Growing and Storing Vegetable and Flower Seeds.* North Adams, Mass.: Storey, 1990. This helpful guidebook to saving seeds presents general principles and specifics for each vegetable.

Sunset Books and Sunset magazine, eds. *Sunset Western Garden Book.* Menlo Park, Calif.: Lane Publishing, 1988. Often considered the Western gardener's bible, this book covers all growing areas west of the Rockies. Among other goodies, it includes a plant selection guide for different growing conditions and a huge encyclopedia of more than 6,000 plants. A great book.

Thomson, Bob. *The New Victory Garden.* Little, Brown, 1987. Based on the popular PBS television series, this book emphasizes how to achieve maximum yield per unit of effort. Perhaps most unique is its monthly guide on what to do for each vegetable. It also includes a short but useful section on fruit cultivation, as well as interesting chapters on cider-making and bird feeders and sample pages of a gardening journal.

Tilgner, Linda. *Tips for the Lazy Gardener.* North Adams, Mass.: Storey, 1998. Explains that it is possible to produce better vegetables with less work and more pleasure. The secret to successful lazy gardening is effective planning: layout of the garden, choice of plants, schedule of seasonal jobs, and maintenance.

Van Patten, George F. *Organic Garden Vegetables.* Portland, Oreg.: Van Patten Publishing, 1991. A well-researched and inexpensive guide on how to grow more than 50 vegetables, from soil preparation to harvesting. Designed for easy, quick reference.

Whealy, Kent, ed. *The Garden Seed Inventory.* Decorah, Iowa: Seed Saver Publications, 1992. As director of the Seed Savers Exchange, Whealy compiles a complete listing of all nonhybrid varieties (more than 5,000) offered by more than 200 seed companies. Each entry describes the variety, provides synonyms, provides a range of maturity dates, and lists all known sources. Known by some as the "seed savers bible."

Yeomans, Kathleen, RN. *The Able Gardener.* North Adams, Mass.: Storey, 1992. Tips, techniques, and inspiration to make gardening easier and more enjoyable, no matter your ability. Included are practical techniques like raised beds and automatic watering systems, and imaginative suggestions like fragrance and indoor gardens.

See also Smith, Wolfe under Greenhouses on page 451.

Fruits and Nuts

Baumgardt, John Philip. *How to Prune Almost Everything.* New York: Quill, 1982.

Best Methods for Growing Fruits and Berries. Emmaus, Penn.: Rodale, 1981. This small booklet provides good discussions of individual fruit cultures.

Bilderback, Diane E., and Dorothy Hinshaw Patent. *Backyard Fruits & Berries: How to Grow Them Better than Ever.* Emmaus, Penn.: Rodale, 1984. A good, very thorough book on fruit culture. Fun and informative.

Hill, Lewis. *Fruits and Berries for the Home Garden.* North Adams, Mass.: Storey, 1992. A good book on all aspects of fruit culture. Both entertaining and informative, Lewis goes the extra mile to explain the whys and hows behind so many orchard practices.

———. *Pruning Made Easy: A Gardener's Visual Guide to When and How to Prune Everything from Flowers to Trees.* North Adams, Mass.: Storey, 1997. A great guide on how to prune everything from evergreens and ornamentals to fruit and nut trees. Useful illustrations of before and after proper pruning.

James Jr., Theodore. *How To Grow Fruit, Berries, and Nuts in the Midwest and East.* Tucson, Ariz.: Fisher, 1983. A very good, thin paperback on fruit and nut culture, with excellent pictures on planting, pruning, and grafting. Other versions are available for different regions.

Jaynes, Richard A., ed. *Nut Tree Culture in North America.* Hamden, Conn.: Northern Nut Growers Association, Inc., 1979. A standard reference on nut culture, this book is essential for anyone interested in growing nut trees. Jaynes has the experts cover all aspects of growing nut trees in detail.

Otto, Stella. *The Backyard Orchardist: A Complete Guide to Growing Fruit Trees in the Home Garden.* Maple City, Mich.: OttoGraphics, 1993. A super resource and excellent guide for the fruit hobbyist. While not strictly organic in approach, Otto provides the information you need to take an organic approach if you so choose. She covers everything from site preparation to pest control, harvest, and storage. Useful charts include best fruit choices for U.S. regions; common tree fruit insect pests in various U.S. regions; periods of active insect pressure; a question-and-answer chart for troubleshooting seasonal problems; and a monthly almanac of things to do and watch for in the orchard.

———. *The Backyard Berry Book: A Hands-on Guide to Growing Berries, Brambles, & Vine Fruit in the Home Garden.* Maple City, Mich.: OttoGraphics, 1995. See annotation above.

Page, Stephen, and Joe Smillie. *The Orchard Almanac: A Spraysaver Guide.* Rockport, Maine: Spraysaver, 1986. An excellent small handbook on fruit culture, with a special emphasis on how to control insects and diseases organically. Effective presentation of what to do on a monthly basis. Useful charts on rootstocks, fertilizers, and sprays.

Southwick, Lawrence. *Planting Your Dwarf Fruit Orchard.* North Adams, Mass.: Storey, 1979. A good introduction to planting and pruning fruit trees; also provides a nice glossary of pruning terms.

A University of California Organic Apple Production Manual. Oakland: University of California, 2000. Publication No. 3403. This 88-page handbook was prepared at the Center for Agroecology & Sustainable Food Systems, University of California, Santa Cruz.

See also Creasy, *Encyclopedia of Organic Gardening,* Hamilton, Kourik (charts on rootstocks, diseases, resistant fruits, ripening dates, and excellent section on pruning), and Thomson under General Gardening, beginning on page 443.

Herbs

Garland, Sarah. *The Herb Garden.* New York: Penguin, 1985. Comprehensive coverage of how to grow fragrant herbs for culinary and other purposes.

Hartung, Tammi. *Growing 101 Herbs That Heal: Gardening Techniques, Recipes, and Remedies.* North Adams, Mass.: Storey, 2000. Tammi and her husband grow medicinal plants on an organically certified farm in southern Colorado. This book explains how to turn gardens into organic medicinal herb businesses.

Hutson, Lucinda. *The Herb Garden Cookbook.* Austin: Texas Monthly Press, 1987. A good book on how to grow and how to cook with herbs, including many recipes. Useful information for Southwestern gardeners.

Jacobs, Betty E. M. *Growing and Using Herbs Successfully.* North Adams, Mass.: Storey, 1981. Complete information about herbs.

Kowalchik, Claire, and William H. Hylton, eds. *Rodale's Illustrated Encyclopedia of Herbs.* Emmaus, Penn.: Rodale, 1987. Hands down the best and most complete book I've seen on herbs, covering everything from herb culture to how to use herbs for healing. Includes useful charts on companion planting, dangerous herbs, herbs for dyeing, herb pests, herb diseases, and more.

Magic and Medicine of Plants. Pleasantville, N.Y.: Reader's Digest, 1986. A wonderful resource for learning about the medicinal purposes of almost 300 plants, including herbs. With instructive illustrations and excellent color photographs, this book features chapters on the history of plants in magic and medicine, an exceptionally clear chapter on the anatomy of plants, and plant entries that differentiate between folk medicine and medicinal purposes that have been scientifically proved. Herb gardeners will also enjoy the chapters that offer designs for herb gardens and recipes for culinary and medicinal purposes.

Shaudys, Phyllis V. *The Pleasure of Herbs: A Month-by-Month Guide to Growing, Using and Enjoying Herbs.* North Adams, Mass.: Storey, 1986. Each month presents a new set of fun projects, from seed-starting and herbal gifts to creating culinary herbal delights. Cultural information, gardening techniques, and multiple ideas on how to store and use herbs are covered. Also included is a brief encyclopedia of herbs.

See also Bubel, Creasy and *Encyclopedia of Organic Gardening* under General Gardening, beginning on page 443.

Insects and Diseases

Ball, Jeff. *Rodale's Garden Problem Solver: Vegetables, Fruits, Herbs.* Emmaus, Penn.: Rodale, 1988. A good book specifically on organic disease and insect control for major plants. Easy to use; you can problem-solve by plants, or by the specific insect or disease.

Carr, Anna, ed. *A Gardener's Guide to Common Insect Pests*. Emmaus, Penn.: Rodale, 1989. An excellent booklet, with great color photographs of insects and short descriptions of their habits and natural controls.

Cravens, Richard H., and the Editors of Time-Life Books. *The Time-Life Encyclopedia of Gardening: Pests and Diseases*. Alexandria, Va.: Time-Life, 1977. A good background book on pests, identification, habits, and regions affected. Nice illustrations.

MacNab, A. A., A. F. Sherf, and J. K. Springer. *Identifying Diseases of Vegetables*. University Park, Penn.: Pennsylvania State University, 1983. More than 200 top-notch photographs of the most common diseases in vegetables. Excellent for field identification of vegetable crop diseases.

Mother Earth News. Healthy Garden Handbook. New York: Simon & Schuster, 1989. A good guide on organic control of insects and diseases. Discusses general methods of maintaining a healthy garden and specific pest remedies. It offers excellent color photographs of plant allies, diseases, and all developmental phases of pests. A garden remedy section helps you troubleshoot common problems for each vegetable. Includes a useful chart on peak emergence times of insects.

Olkowski, William, Sheila Daar, and Helga Olkowski. *Commonsense Pest Control: Least Toxic Solutions for Your Home, Garden, Pets and Community*. Newtown, Conn.: Taunton, 1991. The definitive word on integrated pest management, this hefty reference makes scientific advancements accessible to the nonprofessional. The authors (one horticulturist and two entomologists) have been leaders in reporting on advancements through the nonprofit Bio-Integral Resource Center (BIRC) in Berkeley, California. Discover ways to control garden, household, and community pests, and learn just about anything you ever wanted to know about pest control, from pesticide toxicity levels and effects on different organs, to different garden mulches for weed control. There are too many useful charts to list here, but organic gardeners will be particularly interested in the chart of plants that attract beneficial insects.

Become a member of BIRC and receive technical advice about specific pest problems, troubleshooting, access to workshops and trainings, and subscriptions to the quarterly *Common Sense Pest Control*, as well as the monthly *IPM Practitioner*.

Shepherd's Purse Organic Pest Control Handbook. Summertown, Tenn.: Pest Publications, 1987. A small useful booklet with good color drawings of insects and notes on their biological, cultural, and acute control. Suppliers of beneficial insects are listed.

Sherf, Arden F., and Alan A. MacNab. *Vegetable Diseases and Their Control*, 2nd ed. New York: Wiley, 1986. A professional textbook on vegetable diseases. Includes a detailed description of disease symptoms, cause, disease cycle, and control. Controls are not limited to organic methods.

Smith, Miranda, and Ana Carr. *Rodale's Garden Insect, Disease & Weed Identification Guide*. Emmaus, Penn.: Rodale, 1988. An excellent field guide to insects and diseases, with ninety-seven photos to assist with identification.

Steiner, M.Y., and D. P. Elliot. *Biological Pest Management for Interior Plantscapes*, 2nd ed. Vegreville, AB: Alberta Environmental Centre, 1987. An excellent booklet on major greenhouse insects and their cultural, biological, and chemical control. Describes for each insect the damage caused, occurrence, appearance, and life history. Useful charts include minor pest problems and their control; summary of primary predators and parasites of major plant pests; reported toxicity of common greenhouse pesticides to various biological control agents; suppliers of biological control agents.

Yepsen, Roger, Jr., ed. *The Encyclopedia of Natural Insect & Disease Control.* Emmaus, Penn.: Rodale, 1984. An excellent reference covering major insects and diseases. Excellent color photos. Page numbers for the charts are given because of the potential difficulty of locating them. Useful charts on insect- and disease-resistant vegetable varieties; insect emergence times, divided into sixteen zones.

See also Encyclopedia of Organic Gardening and Hirshberg, Kourik, Lorenz, and Raymond under General Gardening, beginning on page 443.

Harvest and Storage

Bubel, Mike and Nancy. *Root Cellaring: Natural Cold Storage of Fruits & Vegetables.* North Adams, Mass.: Storey, 1991. The only reference on root cellars I'm aware of that has been thoroughly researched by investigating what has and hasn't worked over the years. Many other books are pure theory; this one isn't. Many useful diagrams.

Organic Gardening Harvest Book. Emmaus, Penn.: Rodale, 1975. A small, inexpensive booklet with good information on different methods of harvesting vegetables and fruits. Useful charts include guide to drying vegetables, and selection and preparation of vegetables for freezing.

Stoner, Carol Hupping, ed. *Stocking Up: How to Preserve the Foods You Grow, Naturally.* Emmaus, Penn.: Rodale, 1988. An excellent guide to freezing, canning, drying, and other methods of saving the harvest. Useful charts include timetable for processing fruits, tomatoes, and pickled vegetables in boiling-water bath.

Greenhouses

Ball, Vic, ed. *Ball Red Book: Greenhouse Growing,* 14th ed. Reston, Va.: Reston, 1985. Greenhouse structures, tools, methods, insect control, mechanization, and computerization are discussed. Half the book is devoted to greenhouse culture of flowers, shrubs, and a few vegetables. Essential for all commercial greenhouse growers.

Smith, Miranda. *Greenhouse Gardening.* Emmaus, Penn.: Rodale, 1985. Covers everything from greenhouse design, soils, fertilizers, and insects and diseases to vegetable varieties.

Wolfe, Delores. *Growing Food in Solar Greenhouses: A Month by Month Guide to Raising Vegetables, Fruits and Herbs Under Glass.* Garden City, N.J.: Doubleday, 1981. A monthly schedule covers such things as microclimates, raising animals in the greenhouse, container growing, propagation, diseases, and insects. Fun and informative.

See also Hamilton and Hirshberg under General Gardening, beginning on page 443, and Steiner and Elliot under Insects and Diseases, beginning on page 449.

appendixes

A. The National Standards

The National Organic Standards adopted in 2000 are extensive and need to be reviewed in detail if you are considering becoming a certified organic grower. They are available on-line at <www.ams.usda.gov/nop>, the home page of the National Organic Program.

A word of caution to those who are seeking a list of specific allowed substances. Unlike the organic standards set in Europe, which detail each specific substance allowed in organic growing, in the United States the federal National List primarily identifies specific substances that are disallowed. Some general categories of allowed substances and allowed practices are provided, but if the National List is silent about a particular substance then it is legally allowed. Neem and other botanical remedies, for example, are not found on the National List as substances that are disallowed and therefore are legally allowed in organic growing.

Confusing and highly controversial, the vague language of the National List could permit substances not yet known or tested, and therefore not specifically disallowed, to be used in organic growing. If you are unsure about a specific substance and you cannot find it on the National List, consult the Glossary of Organic Remedies on pages 315–325 of this book to see if it qualifies as an organic remedy. If it is not found there, confer with your local Extension agent or organic certification program about its status.

B. Suggested Reading

Books

1998 Directory of Least-Toxic Pest Control Products. Berkeley, Calif: The Bio-Integral Resource Center, 1998. Lists more than 2,000 pest control items in four sections: insects and mites, plant disease, vertebrates, and weeds. Supplier addresses and cross-references are available.

Allen, Oliver E. *Gardening with the New Small Plants: A Complete Guide to Growing Dwarf and Miniature Shrubs, Flowers, Trees and Vegetables.* Boston: Houghton Mifflin, 1978. An excellent introduction to and discussion of small plants, whether in or out of rock gardens. The author suggests specific species and varieties of shrubs, flowers, trees, and vegetables, as well as sources of availability.

Bowe, Patrick. *The Complete Kitchen Garden: The Art of Designing and Planting an Edible Garden.* New York: Macmillan, 1996.

Campbell, Stu. *Let It Rot!: The Gardener's Guide to Composting,* 3rd ed. North Adams, Mass.: Storey, 1998. Explains the technical aspects of composting in simple, easy-to-understand terms, provides detailed information on selecting the right materials, and covers the mechanics clearly and comprehensively.

———. *Mulch It!* North Adams, Mass.: Storey, 2001. Explains the labor-saving way to maintain a healthy garden. Nothing beats mulch for controlling weeds, retaining moisture, and fertilizing and insulating soil, with minimum effort. Explains how to make your own mulch, when to mulch, and which ones to use in various types of gardens.

Disease-Resistant Vegetables for the Home Garden. Madison: University of Wisconsin Extension Cooperative Extension Publication A3110, 1999. Provides lists of cultivars with inbred resistance to specific diseases, and cultivars that are especially tolerant of low or high temperatures.

Editors of Garden Way Publishing. *The Big Book of Gardening Skills.* North Adams, Mass.: Storey, 1993. Comprehensive, illustrated guide to growing flowers, fruits, herbs, and vegetables, on a small or large scale. Numerous charts on planting, garden design, organic pest and disease control, succession planting, and more.

———. *Herbs, Fruits, and Vegetables.* North Adams, Mass.: Storey, 1990. Offers information about propagation, planting, maintenance, cultivation harvesting, and potential pests and predators.

Edwards, Linda. *Organic Tree Fruit Management.* Keremeos, BC: Certified Organic Associations of British Columbia, 1998.

Foster, Catherine Osgood. *Building Healthy Gardens: A Safe and Natural Approach.* North Adams, Mass.: Storey, 1989. Techniques and insights for reaping abundant harvests without chemical fertilizers, pesticides, or herbicides.

Franklin, Stuart. *Building a Healthy Lawn: A Safe and Natural Approach.* North Adams, Mass.: Storey, 1988. Written by a professional landscaper, this book offers clear directions for growing and maintaining a healthy lawn without the use of heavy chemicals. Includes chapters on mowing, watering, fertilizing, weed control, diseases, and insects, and a useful month-by-month guide to lawn care.

Gilkeson, Linda, Pam Pierce, and Miranda Smith. *Rodale's Pest & Disease Problem Solver: A Chemical-Free Guide to Keeping Your Garden Healthy.* Emmaus, Penn.: Rodale, 1996.

Hart, Rhonda Massingham. *Bugs, Slugs, and Other Thugs*. North Adams, Mass.: Storey, 1991. Explains how to stop pests without risk to the environment. Chapters are organized by predator, and each is illustrated with descriptions of habitat, life cycle, habit, favorite garden targets, and damage caused.

———. *Trellising: How to Grow Climbing Vegetables, Fruits, Flowers, Vines and Trees*. North Adams, Mass.: Storey, 1992. Explains the benefits of using trellises and ways to improve and increase yields, cut time output, and use less space. Also shows how to design and construct different types of garden supports and how to plant and grow more than two dozen varieties of vegetables, fruits, and flowering vines on trellises.

Jacobs, Betty E. M. *Growing and Using Herbs Successfully*. North Adams, Mass.: Storey, 1981.

Larksom, Joy. *The Salad Garden*. New York: Viking, 1984. A fun book for those interested in salad greens and cooking.

Lee, Andy. *Chicken Tractor: The Gardener's Guide to Happy Hens and Healthy Soil*. Shelburne, Vt.: Good Earth, 1994. This permaculture book explains in detail how chickens and bottomless cages can improve soil fertility.

Leighton, Phebe, and Calvin Simonds. *The New American Landscape Gardener: A Guide to Beautiful Backyards and Sensational Surroundings*. Emmaus, Penn.: Rodale, 1987. Numerous charts of both edible and nonedible plants for different landscapes makes this a helpful aide. General design principles and design flaws are discussed in-depth. Extremely readable. Useful charts include plants for meadows, plants for a sunspot, plants for a rock garden, plants for a winter landscape, plants for a water garden, plants for a wildlife garden, and plants for a woodswalk.

Medic, Kris. *The New American Backyard: Easy Organic Techniques and Solutions for a Landscape You'll Love*. Emmaus, Penn.: Rodale, 2001. Demonstrates how organic techniques can save homeowners money and time.

McClure, Susan. *The Harvest Gardener*. North Adams, Mass.: Storey, 1992. A compendium of tips and advice about choosing cultivars, scheduling plantings, organizing garden space, coping with the vagaries of weather and pests, harvesting and storing the crop. Also includes an encyclopedia of culture, harvest, and storage of fruits, herbs, and vegetables.

Natural Enemies of Vegetable and Insect Pests. Ithaca, N.Y.: Cornell University Resource Center, 1994. This 64-page manual, published by Cornell University Cooperative Extension, describes more than 90 beneficial species of insect predators, parasitoids, and diseases, providing details on appearance and lifestyle, pests attacked, and vegetable crops for which these natural enemies are important.

O'Keefe, John M. *Water-Conserving Gardens and Landscapes*. North Adams, Mass.: Storey, 1992. Explains how to meet the challenges of dwindling and overused water supplies, drought conditions, high water costs, and pollution by groundwater. Includes information on drip irrigation systems and alternative gardening techniques, such as raised beds, containers, and hardscapes.

Phillips, Michael. *The Apple Grower: A Guide for the Organic Orchardist*. Chelsea, Vt.: Chelsea Green, 1998.

Phillips, Roger, and Nicky Foy. *The Random House Book of Herbs*. New York: Random House, 1990. One of the best books I've seen for identifying and learning about the historical and modern uses for more than 400 herbs. Extensive and thorough, with more than 400 excellent color photographs.

Pleasant, Barbara. *The Gardener's Bug Book: Earth-Safe Insect Control.* North Adams, Mass.: Storey, 1994. An easy-to-use guide to identifying both the beneficial and harmful insects in your garden. Includes instructions for homemade pest control remedies that are safe for you and your garden.

———. *Warm-Climate Gardening.* North Adams, Mass.: Storey, 1993. Offers advice about how to recognize and exploit the cool seasons within a warm-climate gardening year. Also includes information on drought-resistant plants, summer hardiness, and scheduling maintenance chores when it's too hot to garden.

Poisson, Leandre, and Gretchen Vogel. *Solar Gardening: Growing Vegetables Year Round the American Intensive Way.* White River Junction, Vt.: Chelsea Green, 1994. A practical text filled with concrete particulars, not abstract generalities. Includes photos, plans of existing gardens, tools, and apparatus.

Prakash, Anand, and Jagadiswari Rao. *Botanical Pesticides in Agriculture.* Boca Raton, Fla.: Lewis (CRC Press), 1997.

Raymond, Dick. *Down-to-Earth Natural Lawn Care.* North Adams, Mass.: Storey, 1993. A guide to installing and maintaining a healthy lawn using natural methods. Useful charts include types of grasses and lawn maintenance schedules for all areas of the country.

Restuccio, Jeffrey P. *Fitness the Dynamic Gardening Way: A Health and Wellness Lifestyle.* Cordova, Tenn.: Balance of Nature Publishing, 1992. A marvelous one-of-a-kind book that shows gardeners how normal gardening tasks can be part of a fitness program. Helpful for professional gardeners because it shows the correct postures gardeners should use to avoid injuries. For fitness enthusiasts, it provides a new way to get a workout while doing something creative like growing a garden. The fitness information is detailed and replete with charts and diagrams.

Riotte, Louise. *Carrots Love Tomatoes: Secrets of Companion Planting for Successful Gardening.* North Adams, Mass.: Storey, 1998. Vegetables and fruits have natural preferences; the author explains how to plan a garden to take advantage of these productive relationships.

———. *Roses Love Garlic: Companion Planting and Other Secrets of Flowers.* North Adams, Mass.: Storey, 1998. The author explores companion planting with flowers and shows how to combine flower and vegetable gardens for striking display of color, form, and productivity.

———. *Successful Small Food Gardens: Vegetables, Herbs, Flowers, Fruits, Nuts, Berries.* North Adams, Mass.: Storey, 1993. A revised and updated edition of the author's classic intensive-gardening book, this text addresses the needs of the small-space garden. Includes information on companion plants, succession planting, raised beds, container growing, watering and drainage techniques, increasing soil quality, edible flowers, herbs, and shrubs.

Schultz, Warren, ed. *Natural Insect Control: The Ecological Gardener's Guide to Foiling Pests.* Brooklyn: Brooklyn Botanic Garden, 1995. A volume from the Brooklyn Botanic Garden's 21st Century Gardening Series. Includes full-color illustrations of pests, beneficial insects, pest life cycles, symptoms, management techniques, naturally derived pesticides, and cultural and physical control methods.

Shaudys, Phyllis V. *Herbal Treasures: Inspiring Month by Month Projects for Gardening, Cooking, and Crafts.* North Adams, Mass.: Storey, 1990. A month-by-month collection of herb crafts, recipes, and gardening ideas. Also included are projects with herbs, reference materials, and suppliers.

Smyser, Carol A. *Nature's Design: A Practical Guide to Natural Landscaping.* Emmaus, Penn.: Rodale, 1982. Explains how to make a contour map, how to assess the impact of water on your property, how to analyze your soil, climate conditions, and natural plant and wildlife habitats. An interesting and thought-provoking book, bursting with useful diagrams and pictures. A large section is devoted to landscape construction, including how to find native plants, and propagation techniques. Useful charts include how to evaluate your landscape in terms of energy efficiency and environmental impact; plant selection charts for ten ecoregions, listing the appropriate trees, shrubs, and forbs with their habitat, growth characteristics and any notable qualities; functional uses of plants; aesthetic uses of plants; landscape pest primer; plant associations that attract wildlife, divided into regions; build a better birdhouse, includes all pertinent information that might be needed for twelve different birds; a gardener's avian friends, including pests they eat and plants for food and shelter, broken down by region.

Stickland, Sue. *Heirloom Vegetables: A Home Gardener's Guide to Finding and Growing Vegetables from the Past.* New York: Fireside, 1998.

Troetschler, Ruth, and Alison Woodworth, Sonja Wilcomer, Janet Hoffman, and Mary Allen. *Rebugging Your Home and Garden.* Los Altos, Calif.: PTF Press, 1996. A "must have" for the home gardener who wants information on low-toxic pest control. Incredibly user friendly, with innovative decision flowcharts, this book includes information on indoor and outdoor pests and diseases, basic IPM techniques, human diseases transmitted by arthropods, and much more.

Wirth, Thomas. *Victory Garden Landscape Guide.* Boston: Little, Brown, 1984. A fun and useful month-by-month guide to landscaping. Wirth offers interesting and useful ideas and charts on everything concerning landscaping, from terraces and patios to fruit trees. He divides each month into four categories: landscaping opportunities, plants for a purpose, materials and construction, and plants by design.

Periodicals

American Vegetable Grower
800-572-7740
www.americanvegetablegrower.com
A monthly magazine covering new trends in commercial growing. One section covers greenhouse production.

Common Sense Pest Control Quarterly and IPM Practitioner (monthly)
510-524-2567
www.birc.org
Two journals offering the least-toxic solutions to pest problems of the home and garden. Available to members of the Bio-Integral Resource Center.

HortIdeas
http://users.mikrotec.com/~gwill/hi-main.htm
gwill@mis.net
A great monthly newsletter that abstracts the best and latest gardening research from both popular and technical journals. An excellent way to keep up to date on advances in the gardening, horticultural, agricultural, and related fields. Available via mail or e-mail; back issues available on CD.

Mother Earth News
800-234-3368
www.motherearthnews.com
A bimonthly publication about country living skills, including a regular section on organic gardening.

The New Farm
www.newfarm.org
The Web site and electronic magazine of The Rodale Institute, an organization that researches the connection between soil health and human health in order to educate and equip farmers to transition toward organic and regenerative agriculture.

Organic Gardening
www.organicgardening.com
ogdcustserv@rodale.com
A monthly magazine that covers all aspects of organic gardening, as well as the results of gardening trials and experiments conducted by the Rodale Research Center.

Sunset
800-777-0117
www.sunset.com
A monthly magazine that features a regular section on gardening and landscaping in the western United States. Highly recommended for western gardeners.

Taunton's Fine Gardening
800-477-8727
www.tauton.com/fg
A collector's magazine to be savored, like all other Taunton Press publications. Articles on edibles (vegetables, fruits, and herbs), landscaping, ornamentals, insects, diseases, and gardening methods. Well-researched information; superb photography.

C. Seed Companies and Nurseries

Abundant Life Seeds
Saginaw, Oregon
541-767-9606
www.abundantlifeseeds.com
Good selection of vegetables, small grains, herbs, and flowers. Organic, untreated, open-pollinated seeds.

Adams County Nursery, Inc.
Aspers, Pennsylvania
717-677-8105
www.acnursery.com
A large selection of fruit varieties for commercial and home growers, on a number of different rootstocks.

American Forestry Technology, Inc.
West Point, Indiana
765-572-1212
The only source of the patented "Purdue Number 1" seedling. Exclusive producers of the genetically superior black walnut trees developed by Dr. Walter Beineke, a leading authority on black walnut genetics.

Applesource
Chapin, Illinois
800-588-3854
www.applesource.com
A large selection of apple varieties for commercial and home growers.

Baker Creek Heirloom Seeds
Mansfield, Missouri
417-924-8917
www.rareseeds.com
A large selection of only nonhybrid seeds, a number of which are certified organic.

Bay Laurel Nursery
Atascadero, California
805-466-3406
www.baylaurelnursery.com
Over 500 bareroot varieties of edible and ornamental shrubs, trees, and vines.

Bountiful Gardens
Willits, California
707-459-6410
www.bountifulgardens.org
A nonprofit organization that sponsors educational programs and offers a good selection of untreated, open-pollinated seeds for vegetables, herbs, flowers, green manures, and grains.

W. Atlee Burpee Seed Co.
Warminster, Pennsylvania
800-333-5808
www.burpee.com
Conventional and some organic vegetables, fruits, herbs, and flowers.

Companion Plants
Athens, Ohio
740-592-4643
www.companionplants.com
Huge selection of herbs for culinary, medicinal, and other purposes.

The Cook's Garden
Warminster, Pennsylvania
800-457-9703
www.cooksgarden.com
Certified organic, untreated and open-pollinated seed. Large selection of vegetable seeds for outdoor, coldframe and greenhouse culture, especially lettuce.

Cumberland Valley Nurseries, Inc.
McMinnville, Tennessee
800-492-0022
Fruit trees, with an unusually extensive collection of peach trees.

Fall Creek Farm & Nursery, Inc.
Lowell, Oregon
541-937-2973
www.fallcreeknursery.com
A large selection of blueberries.

Fedco Seeds
Waterville, Maine
207-873-7333
www.fedcoseeds.com
Large selection of vegetable seeds and seedlings suitable for the Northeast. Also have large selection of trees. Operated as a cooperative — no retail store.

Fox Hill Farm
Parma, Michigan
517-531-3179
Good selection of herbs.

Grimo Nut Nursery
Niagara-on-the-Lake, Ontario
905-934-6887
www.grimonut.com
One of the best nurseries for high quality nut seedlings.

Harris Seeds
Rochester, New York
800-514-4441
www.harrisseeds.com
Good vegetable and flower selection.

Heirloom Seeds
West Elizabeth, Pennsylvania
412-384-0852
www.heirloomseeds.com
Offers more than 700 varieties of heirloom vegetable, flower, and herb seeds; also organic fertilizers, gardening books, and environmentally friendly pest control products.

Henry Field's Seed & Nursery Co.
Aurora, Indiana
513-354-1495
www.henryfields.com
Seeds, perennials, live vegetable plants, and trees — in business since 1892.

Henry Leuthardt Nurseries, Inc.
East Moriches, New York
631-878-1387
www.henryleuthardtnurseries.com
Offers a wide variety of espalier fruit trees. Carries grapes and other fruits, too.

High Altitude Gardens
(A Division of Seeds Trust)
Hailey, Idaho
928-649-3315
www.seedtrust.com
A bio-regional seed company dedicated to finding and saving seeds for short, cold seasons for vegetables, wildflowers, grasses and herbs. Many certified organic and organically grown, and many open-pollinated.

Horticultural Enterprises
Box 810082
Dallas, TX 75381-0082
Sells traditional Mexican vegetables and herbs including peppers, jicama, tomatoes, and tomatillo.

Indiana Berry and Plant Co.
Huntingburg, Indiana
800-295-2226
www.inberry.com
Sells huge selection of strawberries and offers a good comparison of traits. Also sells other fruits.

Irish Eyes, Garden City Seeds
Thorp, Washington
509-964-7000
www.irish-eyes.com
Large selection of potatoes, garlic, and northern vegetable varieties; open-pollinated and certified seed.

Johnny's Selected Seeds
Winslow, Maine
877-564-6697
www.johnnyseeds.com
Good selection of vegetables — especially carrots and greenhouse varieties. Most are untreated seeds for organic growing.

J. W. Jung Seed Co.
Randolph, Wisconsin
800-247-5864
www.jungseed.com
Selection of vegetables, flowers, herbs, berries, and fruit trees.

Kitchen Garden Seeds (John Scheepers)
Bantam, Connecticut
860-567-6086
www.kitchengardenseeds.com
Large selection of unusual vegetable and herb seeds.

Miller Nurseries, Inc.
Canandaigua, New York
800-836-9630
www.millernurseries.com
Large fruit and nut tree and berry selection.

Native Seeds/SEARCH
Tucson, Arizona
520-622-5561
www.nativeseeds.org
A nonprofit seed conservation organization working to preserve the traditional crops and their wild relatives of the American Southwest and northwestern Mexico. Grains, vegetables, herbs, fruit, cotton, tobacco, books, food, baskets, and dye for wool.

Nichols Garden Nursery
Albany, Oregon
800-422-3985
www.nicholsgardennursery.com
Good selection of vegetables, herbs, and flowers, often rare and unusual.

Nolin River Nut Tree Nursery
Upton, Kentucky
270-369-8551
www.nolinnursery.com
One of the few nurseries devoted exclusively to nut trees. Offers specific varieties of a broad range of nuts.

Nourse Farms, Inc.
South Deerfield, Massachusetts
413-665-2658
www.noursefarms.com
Good selections of strawberries, raspberries, and asparagus. Most everything propagated by tissue culture.

One Green World
Molalla, Oregon
877-353-4028
www.onegreenworld.com
Unusual fruits, ornamentals, fruit trees, berries, orchard & garden supplies, books, more.

Park Seed Co., Inc.
Greenwood, South Carolina
800-213-0076
www.parkseed.com
Vegetables, especially hybrids, flowers, shrubs, and ornamentals.

Pinetree Garden Seeds
New Gloucester, Maine
207-926-3400
www.superseeds.com
Extensive selection of vegetable seeds. Also have large book section in catalog.

Plants of the Southwest
Santa Fe, New Mexico
800-788-7333
www.plantsofthesouthwest.com
Good selection of plants native to the Southwest, many suitable for Western mountains and high plains.

Raintree Nursery
Morton, Washington
360-496-6400
www.raintreenursery.com
Good selection of fruit trees and berries for the Northwest. An informative catalog.

Redwood City Seed
Redwood City, California
650-325-7333
www.ecoseeds.com
Large selection of unusual heirloom varieties, especially unusual hot peppers, heirloom corn, and seeds from the Orient, Mexico, and Europe.

Richters Herb Catalogue
Goodwood, Ontario
905-640-6677
www.richters.com
Cadillac catalog of herbs for all purposes.

Sandy Mush Herb Nursery
Leicester, North Carolina
828-683-2014
www.sandymushherbs.com
Large selection of culinary, ornamental, medicinal, and other purpose herbs.

Seeds of Change Research Farm and Garden
Santa Fe, New Mexico
888-762-7333
www.seedsofchange.com
100% certified organic seeds. A huge selection of heirloom and lost varieties. This organization's mission is to collect and safeguard the genetic diversity and legacy of the earth.

Shepherd's Garden Seeds
Litchfield, Connecticut
800-503-9624
www.shepherdseeds.com
A large selection of vegetable, herbs, flowers, and some fruits; many open-pollinated heirlooms.

Sierra Gold Nurseries
Yuba City, California
800-243-4653
www.sierragoldtrees.com
A good selection of almond and fruit trees.

South Carolina Foundation Seed Association
Clemson, South Carolina
864-656-2520
www.clemson.edu/seed/
Sweet potato varieties.

Southern Exposure Seed Exchange
Mineral, Virginia
540-894-9480
www.southernexposure.com
Certified organic seed. Open-pollinated and heirloom vegetable varieties. Special varieties for solar greenhouses. Untreated seeds. Very informative catalog.

Stark Bro's Nurseries and Orchards Co.
Louisiana, Missouri
800-325-4180
www.starkbros.com
Fruit trees, berries, and nuts. Many common and some unusual varieties. Some varieties listed in this book are offered in their catalog for professional growers.

Stokes Seeds, Inc.
Buffalo, New York
800-396-9238
www.stokeseeds.com
Good selection of vegetables and flowers. Good greenhouse varieties. Some open-pollinated and untreated seeds.

Territorial Seed Company
Cottage Grove, Oregon
800-626-0866
www.territorialseed.com
Nice selection of vegetable seeds for the Northwest. Many open-pollinated varieties. Some certified organic seed, some organically grown seed.

Thomas Jefferson Center for Historic Plants
Charlottesville, Virginia
800-243-1743
www.monticello.org/shop
A small selection of historical vegetables grown by Thomas Jefferson, and a large selection of historical ornamentals.

Thompson & Morgan
Jackson, New Jersey
800-274-7333
www.thompson-morgan.com
Cadillac catalog of flowers and also some vegetables. Many unusual and interesting varieties; none are genetically altered.

Tomato Growers Supply Co.
Fort Myers, Florida
888-478-7333
www.tomatogrowers.com
Huge selection of tomatoes, most hybrids.

Trees of Antiquity
Formerly Sonoma Antique Apple Nursery
Paso Robles, California
805-467-9909
www.treesofantiquity.com
Organically grown antique apple and other fruit trees. Espaliered trees.

Van Well Nursery
Wenatchee, Washington
800-572-1553
www.vanwell.net
Retail and wholesale selection of conventional fruit trees and berries.

Vermont Bean Seed Company
Randolph, Wisconsin
800-349-1071
www.vermontbean.com
A large selection of vegetable, herb, and fruit seeds, including a great selection of bean seeds.

West Coast Seeds
Delta, British Columbia
604-952-8820
www.westcoastseeds.com
Organic and non-GMO vegetable, herb, and flower seeds.

William Dam Seeds, Ltd.
Dundas, Ontario
905-628-6641
www.damseeds.com
Untreated seeds. Good selection of vegetables, especially cold-tolerant varieties, and some cover crops and herbs.

D. Equipment and Pest Control Suppliers

Ag Biochem, Inc.
Orinda, California
925-254-0787
www.crowngall.com
Beneficial insects and other biocontrol agents.

Alsto's
Galesburg, Illinois
800-447-0048
www.alsto.com
Garden tools, supplies, and accessories.

A. M. Leonard, Inc.
Piqua, Ohio
800-543-8955
www.amleo.com
Extensive offering of garden tools, light and heavy.

ARBICO Organics
Tucson, Arizona
800-827-2847
www.arbico.com
Beneficial insects and organisms, traps, soil test kits, compost, fertilizers, more.

Biofac Crop Care
Mathis, Texas
800-233-4914
www.biofac.com
Beneficial insects.

Charley's Greenhouse & Garden
Mount Vernon, Washington
800-322-4707
www.charleysgreenhouse.com
A large selection of supplies for greenhouse growing as well as supplies for outdoor gardening.

Clyde Robin Seed Co.
Castro Valley, California
800-647-6475
www.clyderobin.com
Pest control supplies, beneficial insects, and seed mixes to attract beneficials.

Diggers
Soquel, California
831-462-6095
Wire gopher baskets: 20-gauge galvanized root guard (¾-inch mesh in 1- to 5-foot widths), and 15-gallon containers.

Gage Industries
Lake Oswego, Oregon
800-443-4243
www.gageindustries.com
Contact for your nearest distributor of DuraPots, which are made from recycled plastic.

Gardener's Supply
Burlington, Vermont
888-833-1412
www.gardeners.com
Tools, seed starting kits, row covers, organic fertilizers, greenhouses, drip irrigation, composting supplies, sprayers, bird houses, botanical and mineral products, soaps.

Gardens Alive!
Lawrenceburg, Indiana
513-354-1483
www.gardensalive.com
Great selection of organic pest and disease controls, beneficial insects for IPM, botanical and mineral products, traps, composting and irrigation supplies, pet products, more. Very informative catalog.

Harmony Farm Supply & Nursery
Sebastopol, California
707-823-9125
www.harmonyfarm.com
Excellent selection of pest controls, fertilizers, irrigation supplies, tools, lab services, horticultural supplies, books, and posters.

IFM (Integrated Fertility Management)
Wenatchee, Washington
509-662-3179
www.agecology.com
Promotes ecologically sound orchard, farm, and garden practices. Garden equipment, nutrient analysis, soil amendments, foliar sprays, green manures, pest controls, botanical and biological controls, beneficial predators, and parasites.

Natural Insect Control
Stevensville, Ontario
905-382-2904
www.naturalinsectcontrol.ca
Environmentally safe pest controls and beneficial insects.

Nitron Industries
Fayetteville, Arkansas
800-835-0123
www.nitron.com
Unusual agricultural enzymes to condition, rebuild, loosen, and detoxify soil. Irrigation equipment, organic fertilizers and amendments, more.

Peaceful Valley Farm & Garden Supply
Grass Valley, California
888-784-1722
www.groworganic.com
Tools, composting supplies, row covers, many cover crops and mixes, sprayers, soil amendments, greenhouse covering, seed thresher, and also bulbs.

Planet Natural
Bozeman, Montana
800-289-6656
www.planetnatural.com
Environmentally safe controls for grasshoppers, including Semaspore, which contains *Nosema locustae* to stop them from feeding.

Plow & Hearth
Madison, Virginia
800-494-7544
www.plowandhearth.com
Fine selection of garden tools, supplies, and accessories.

Smith & Hawken
Pueblo, Colorado
800-940-1170
www.smithandhawken.com
A cadillac selection of garden tools. Also offers irrigation supplies, composting equipment, clothing, books, and furniture.

Worm's Way
Bloomington, Indiana
800-274-9676
www.wormsway.com
A large selection of supplies for propagation, greenhouse growing, lighting, composting, and natural fertilizers.

E. Other Useful Resources

State and Local Associations

State and local organic growers associations are excellent resources. Joining your local association offers an ideal opportunity to learn more about trends in organic growing and to network and make new friends.

I urge you to contact your state and local organic growers associations for a list of locally recommended insect and disease controls. Issues of disease and insect control vary widely from region to region: pests in the Northeast may never be seen on the West Coast, in the Southwest, or in Florida, and vice versa. State and local associations will have the most accurate information for your location.

Also, be aware that many state associations have compiled lists of substances that are permitted, regulated, or prohibited in organic growing within the state. Some states may even have standards that are stricter than the federal guidelines stipulated in the National Standards (see appendix A).

Last, state and local associations may be able to direct you to cooperatives that may be able to assist with producing and marketing issues as they arise.

On the Web

If you have access to the Interent, search the World Wide Web for myriad helpful links and resources. Try searching under "organic growers associations" by state, "organic producers associations" by state, "certified growers and producers" by state, "organic farmers," "organic growers," and "sustainable agriculture."
A short list of useful Web sites includes:

California Certified Organic Farmers
www.ccof.org

Canadian Organic Growers
www.cog.ca

National Sustainable Agriculture Information Service
www.attra.ncat.org

National Gardening Association
www.garden.org

Northeast Organic Farming Association
www.nofa.org

Organic Crop Improvement Association
www.ocia.org

Organic Farming Research Foundation
www.ofrf.org

Organic Trade Association
www.ota.com

The Organic Trade Association is an excellent resource for information at the national level. Many of these Web sites offer links to other helpful resources.

F. USDA Hardiness Zone Map

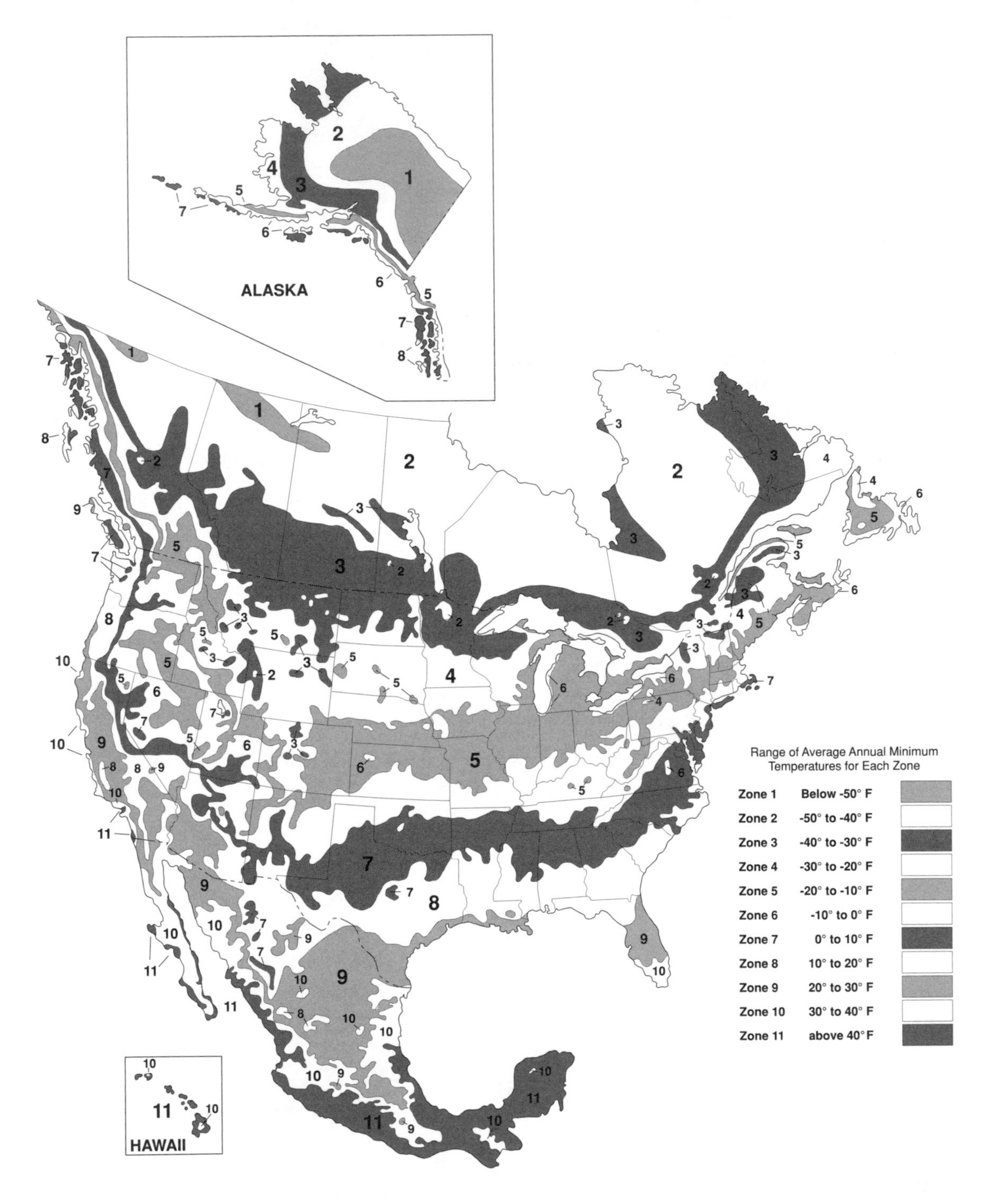

index

Note: Numbers in **boldface** indicate plant entries and charts.

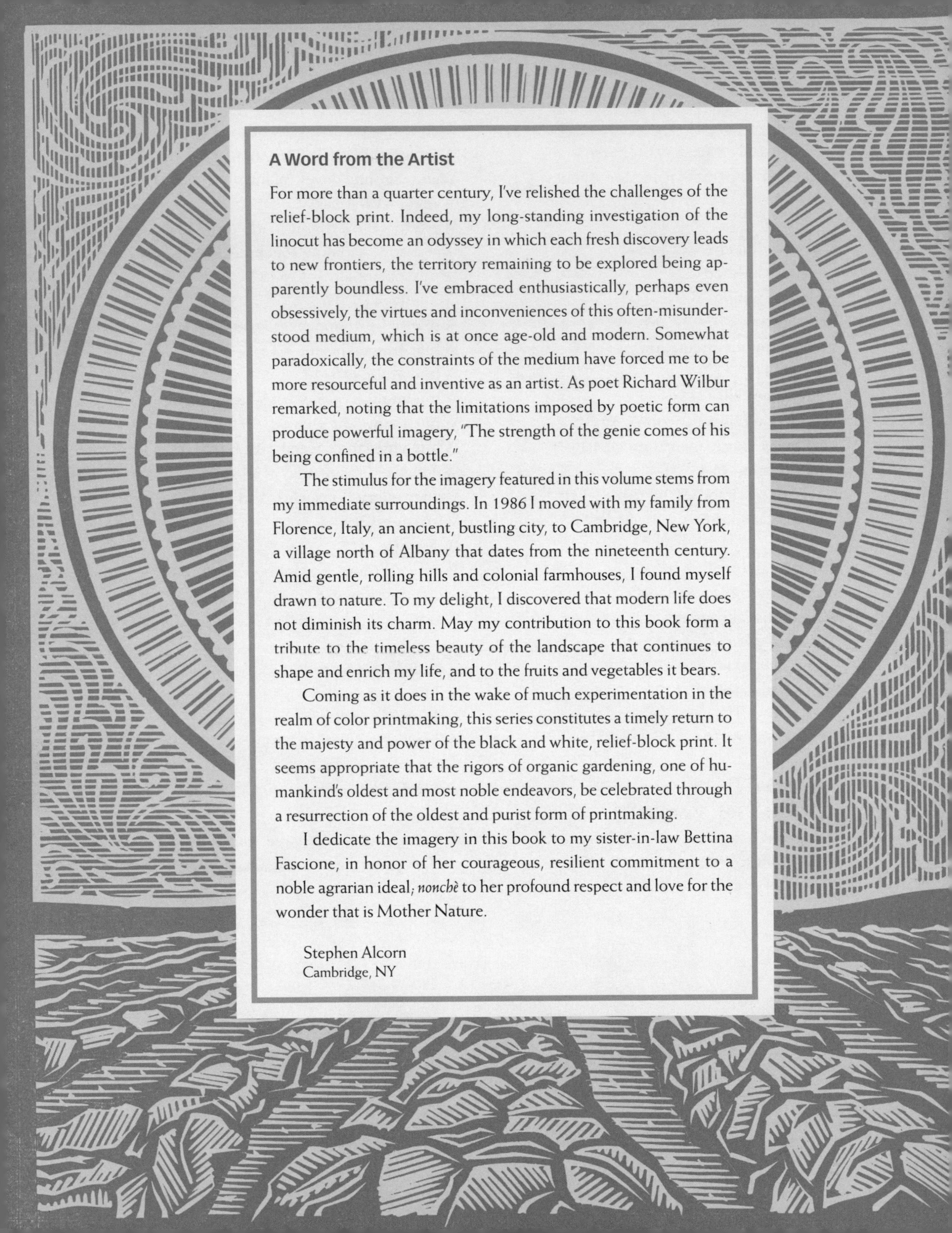

A Word from the Artist

For more than a quarter century, I've relished the challenges of the relief-block print. Indeed, my long-standing investigation of the linocut has become an odyssey in which each fresh discovery leads to new frontiers, the territory remaining to be explored being apparently boundless. I've embraced enthusiastically, perhaps even obsessively, the virtues and inconveniences of this often-misunderstood medium, which is at once age-old and modern. Somewhat paradoxically, the constraints of the medium have forced me to be more resourceful and inventive as an artist. As poet Richard Wilbur remarked, noting that the limitations imposed by poetic form can produce powerful imagery, "The strength of the genie comes of his being confined in a bottle."

The stimulus for the imagery featured in this volume stems from my immediate surroundings. In 1986 I moved with my family from Florence, Italy, an ancient, bustling city, to Cambridge, New York, a village north of Albany that dates from the nineteenth century. Amid gentle, rolling hills and colonial farmhouses, I found myself drawn to nature. To my delight, I discovered that modern life does not diminish its charm. May my contribution to this book form a tribute to the timeless beauty of the landscape that continues to shape and enrich my life, and to the fruits and vegetables it bears.

Coming as it does in the wake of much experimentation in the realm of color printmaking, this series constitutes a timely return to the majesty and power of the black and white, relief-block print. It seems appropriate that the rigors of organic gardening, one of humankind's oldest and most noble endeavors, be celebrated through a resurrection of the oldest and purist form of printmaking.

I dedicate the imagery in this book to my sister-in-law Bettina Fascione, in honor of her courageous, resilient commitment to a noble agrarian ideal; *nonchè* to her profound respect and love for the wonder that is Mother Nature.

Stephen Alcorn
Cambridge, NY